I0063621

Nanohybrids

Future Materials for Biomedical Applications

Edited by

Dr. Gaurav Sharma   **Dr. Alberto García-Peñas**

Shoolini University   University Carlos III of Madrid

India   Spain

Copyright © 2021 by the authors

Published by **Materials Research Forum LLC**
Millersville, PA 17551, USA

All rights reserved. No part of the contents of this book may be reproduced or transmitted in any form or by any means without the written permission of the publisher.

Published as part of the book series
Materials Research Foundations
Volume 87 (2021)
ISSN 2471-8890 (Print)
ISSN 2471-8904 (Online)

Print ISBN 978-1-64490-106-9
eBook ISBN 978-1-64490-107-6

This book contains information obtained from authentic and highly regarded sources. Reasonable efforts have been made to publish reliable data and information, but the author and publisher cannot assume responsibility for the validity of all materials or the consequences of their use. The authors and publishers have attempted to trace the copyright holders of all material reproduced in this publication and apologize to copyright holders if permission to publish in this form has not been obtained. If any copyright material has not been acknowledged please write and let us know so we may rectify this in any future reprints.

Distributed worldwide by

Materials Research Forum LLC
105 Springdale Lane
Millersville, PA 17551
USA
https://www.mrforum.com

Manufactured in the United States of America
10 9 8 7 6 5 4 3 2 1

Table of Contents

Preface

The discovery of (nano) hybrid materials opened new interesting possibilities for all the fields of applications, and especially in biomedicine due to strict and specific characteristics that can be obtained. The key of this success can be understood from the facile modulation of resulting materials, where the composition, the ratio between compounds, the use of covers or the designed structure play an important role in order to get a final application and responding to inalienable requirements as biocompatibility.

Recent advances can be continuously observed in the hospitals and clinics where an number of surgeries have been reduced and the effect of some harmful treatments have been mitigated. An important part of this success is clearly associated with the (nano) hybrid materials, which can improve the effectiveness of some drugs, promote a high cell growth in new scaffolds, and lead to biodegradable surgical suture. Probably, pretty soon, these implementations will change the concept of the current medicine.

This book collects an important part of the recent advances on (nano) hybrids in the field of biomedical applications. The use of hybrid magneto-plasmonic nanoparticles, including their bases, their synthetic procedures and their applications is detailed along this book. The relevance associated with these technologies could lead to non-invasive therapies through the use of magnetic nanohybrid materials.

There are different classifications of hybrid materials for biomedical applications depending on the type of base material. In this case, the book takes in account the most promising materials as reference, and consequently established some chapters in function of the type of base material: silica nanostructures, polymers, bioresorbable metals, liposomes, biopolymeric electrospun nanofibers, graphene, and gelatine. These chapters show some of the most important materials, and their biomedical applications. In addition, the development of biomaterials for cell regeneration is widely described in this book, as well as the (nano) hybrids for wound healing applications.

In general, this book focuses on current progress associated with the preparation, designing and utilization of (nano) hybrids materials for biomedical application and future perspectives. Obviously, the great number of materials and applications makes it impossible to collect all the materials and applications, and consequently this book can be a good beginning for non-specialized people, which are looking for general knowledge. On the other hand, it could be an interesting handbook for connecting researchers and industry looking for new materials according requirements from governments and society.

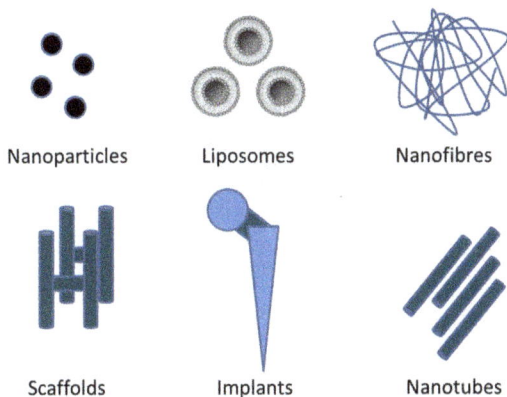

Figure1. Some biomedical applications based on (nano)nybrid materials.

We thank all the contributors for their efforts. Specifically, we want to thank the help obtained from Prof. Pilar Herrasti, and Dr. Amit Kumar

This work is dedicated to our wonderful parents who are taking care of us from the beginning. Our life is always changing, but our hearts will continue working on the most important thing of this life: Love. If we have love, we will avoid frontiers, and we will build a world of peace.

Editors:

Dr. Gaurav Sharma

Shoolini University

India

Dr. Alberto García-Peñas

University Carlos III of Madrid

Spain

Nanohybrids
Materials Research Foundations **87** (2021) 1-54

Materials Research Forum LLC
https://doi.org/10.21741/9781644901076-1

Chapter 1

Hybrid Magneto-Plasmonic Nanoparticles in Biomedicine: Fundamentals, Synthesis and Applications

J.G. Ovejero[1*], P. Herrasti[2]

[1]King's College London, Randall Centre of Cell and Mollecular Biophysics, New Hunt's House, SE1 1UL, London, UK

[2]Universidad Autónoma de Madrid, Facultad de Ciencias, Departamento de Química Física Aplicada, 28049 Madrid, Spain

jesusgovejero@gmail.com*

Abstract

Magneto-plasmonic nanoparticles have attracted increasing attention from the scientific community due to their promising properties and high applicability. Thanks to the development of new synthesis routes, it has been possible to use this kind of nanostructures in different biomedical applications such as dual imaging, combined treatments or biodetection. However, there is still a lack of biocompatible materials with suitable magneto-plasmonic features to translate all these advances to clinical studies.

Keywords

Magneto-Plasmonic Nanoparticles, Au/Fe_3O_4, Biosensing, Therapy, Theragnosis, Bioimaging

Contents

1.   Introduction

In recent years, the advancements in the synthesis processes of inorganic nanoparticles (NPs) has boosted the development of multifunctional NPs in which different complementary properties are combined. The magneto-plasmonic NPs, that combines magnetic and plasmonic responses simultaneously, are one of the most attractive families of multifunctional NPs for material science. Although the study of this kind of NPs is recent, the number of publications referring to the magneto-plasmonic NPs has grown significantly in the last decade, as well as the number of citations. Despite the enormous interest that these nanostructures arouse, the lack of materials with suitable magnetic and plasmonic (metallic) properties in a single phase represents a substantial handicap for their use in biomedicine. The transition metals of the 3d group (Fe, Co or Ni) are occasionally considered in this kind of studies, as well as their alloys with noble metals such as CoPt or FeAu. However, the chemical instability of these elements and alloys - especially at the nanoscale- may lead to the oxidization and/or segregation of the phases, degrading their physicochemical properties.

To address these problems, most of the magneto-plasmonic studies in biomedicine are carried out using composite NPs in which a noble metal phase and a ferromagnetic (or ferrimagnetic) phase are hybridized. Among them, Au / FeO_x and Au / FeO_x/SiO_2 hybrid magneto-plasmonic nanoparticles (HMPNPs) present suitable properties in terms of biocompatibility and functionalization, the present chapter will focus on this type of nanostructures. Special emphasis will be given to the plasmonic properties of Au nanostructures and the superparamagnetic properties of iron oxide NPs, as well as the possible hybridization effects. Finally, the chapter summarizes the most relevant biomedical works in which this kind of nanocomposites have been applied.

2. Fundamentals in magneto-plasmonic properties of nanoparticles

It is common to find in the specialized literature terms such us magnetic or plasmonic NPs. However, every material in nature presents a certain response to static magnetic field and electromagnetic waves. Therefore, it is necessary to define figures of merit that define the suitability of a certain material to the specific application in which these NPs are planned to be used.

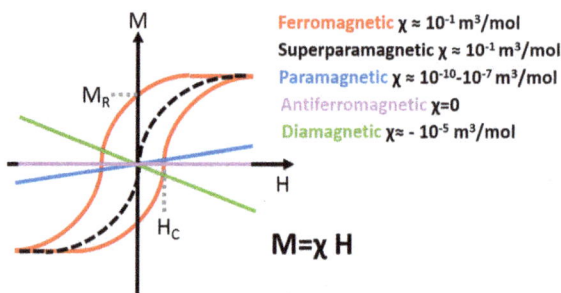

Figure 1. Magnetization curves for different magnetic responses: Ferromagnetic, superparamagnetic, paramagnetic, antiferromagnetic and diamagnetic. The legend shows the typical maximum values of magnetic susceptibility (χ) in m^3/mol for each magnetic response.

2.1 Nanomagnetism

In terms of magnetic response, materials are usually classified in four general magnetic families: Ferromagnetic, antiferromagnetic, paramagnetic and diamagnetic. Figure 1 shows the magnetic moment (M) induced in each kind of material by an applied magnetic field (H). The ratio between these two magnitudes is known as susceptibility (χ) and its maximum value in the curve can vary from $\chi_{max} \approx - 10^{-10}$ m^3/mol for diamagnetic materials to $\chi_{max} \approx 10^{-1}$ m^3/mol in the case of ferromagnetic materials. The magnetization curve shows that the magnetic moment of the materials with ferromagnetic response does not disappear when the magnetic field is removed. It stead of that, a certain magnetization

Materials Research Forum LLC
https://doi.org/10.21741/9781644901076-1

remanence (M_R) is preserved in the material at H=0. The field required to cancel this magnetization is called coercivity field (H_C). The area inside the loop (A_{loop}) represents the amount of energy dissipated during the magnetization reversal in ferromagnetic materials[a] (Equation 1).

$$E_{dis} = \oint \mu_0 H dM = A_{loop} \tag{1}$$

When the dimensions of the materials are reduced to the nanoscale, new magnetic behaviors emerge associated to effects size confinement, magnetic correlation length, surface anisotropy, etc. [1] One of the most relevant phenomena for biomedical applications is the superparamagnetism. This phenomenon is a consequence of the magnetization disorder induced by thermal fluctuations. The superparamagnetic response, plotted in the Figure 1a with a dash black line, presents certain similarities with paramagnetic response ($\chi > 0$ and $H_C = 0$), but much higher susceptibility. The maximum χ values for superparamagnetic response results comparable to the one observed in ferromagnetic materials.

The superparamagnetic response depends on the ratio between the thermal relaxation time and time scale in which the material is studied. We use the term superparamagnetic response instead of superparamagnetic material because most ferromagnetic material would present null remanence and coercivity if they are analyzed in a secular time scale. Therefore, to define properly a NP as superparamagnetic it is crucial to define time scale in which it is going to be studied.

To clarify this point, we will consider as a first simple approximation a collection of small NPs in which all the spins rotate coherently (monodomain) and with only one easy magnetization axis (uniaxial anisotropy). Under these conditions, we can assume that in each NP the magnetization reversal take place by a coherent rotation of all the spins (macrospin model), discarding other processes such as domain wall motion or incoherent rotation.

Equation 2 describes the potential magnetic energy for a single NP as a function of the angle between the magnetic moment and the easy axis (θ) when a magnetic field ($\mu_0 H$) is applied along its easy axis. K_A is the magnetic anisotropy constant (dependent on the

---

[a] There is a special kind of ferromagnetic material, like some iron oxides, that present two uncompensated opposite lattices of spins (ferrimagnetic), but they can be macroscopically treated as ferromagnetic materials with smaller magnetization.

material and shape of the NP), V_{NP} is the volume of the NP and M_S is the saturation magnetization of the system.

$$E(\theta) = K_A V_{NP} sen^2(\theta) + \mu_0 H M_S \cos(\theta) \tag{2}$$

Figure 2 shows the distribution of the magnetic moment orientation at a fixed temperature for a collection of NPs like the one described. The left column shows magnetization reversal of a superparamagnetic system (non-dissipative) in which the field is inverted slowly and thus the magnetic relaxation time (τ_r) is smaller than the time of inversion of the field (τ_m). On the contrary, the right column shows the magnetization reversal in a "blocked" system in which τ_r is larger than τ_m and part of the magnetic energy is dissipated during reversal process.

Both systems begin with all the magnetic moments in $\theta=0$, parallel to the direction of the applied field (Figure 2a and 2a′). When the field is removed, the superparamagnetic system ($\tau_r < \tau_m$) has time to reach the thermodynamic equilibrium populating both minimums of energy with the same amount of magnetic moments, what cancels the global magnetization (Figure 2b). In this way, the system can invert the magnetization without any remanence or coercivity, generating a closed magnetization curve (Figure 2d). On the contrary, we find that in the blocked system the thermal fluctuations are not able to equalize the population of both minima at the time the field is cancelled (Figure 2b′). In this case, the system presents a remanent magnetization when the field is cancelled, and it is necessary to apply an additional field opposite to the magnetization to complete the reversal process (Figure 2c′). As a result of this forced reversal, the magnetization curve results an open loop and the system dissipates energy according to Equation 1.

In an analogue way, fixing the time of field inversion (τ_m) it is possible to define a temperature at which the thermal fluctuations are strong enough to overcome the energy barrier and perform a spontaneous relaxation of the magnetic moments faster than the field inversion. This temperature, denominated "blocking temperature", is commonly used to characterize the superparamagnetic response of NPs, and it is defined as in Equation 3.

$$T_B = \frac{K_A V_{NP}}{K_B \ln\left(\frac{\tau_m}{\tau_0}\right)} \tag{3}$$

Where K_B is the Boltzman constant and τ_0 is a parameter that collects intrinsic properties of the material (spin gyromagnetic ratio, attenuation, etc.) and certain experimental conditions (temperature, field strength). τ_0 is in the order of the nanosecond [2,3].

It is worth to mention that there are other magnetic phenomena that come up at the nanoscale such as spin canting, exchange bias, etc. that are out of the scope of this chapter but can be found in specialized reviews [4,5]. These effects are especially relevant in ultrasmall magnetic NPs but play a secondary role in the size range of the magnetic NPs usually employed in biomedical applications.

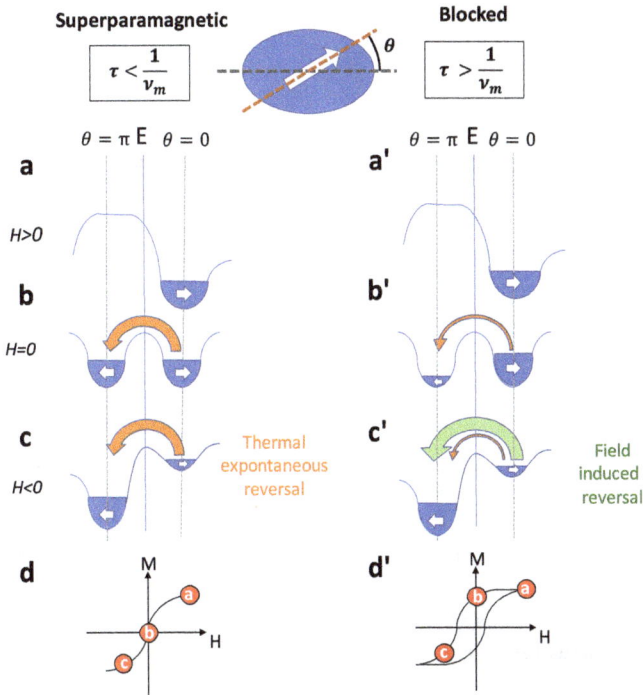

Figure 2. Schematic response of a ferromagnetic NP in superparamagnetic regime (left) and blocked regime (right). The magnetic field is applied parallel to the easy axis indicated in grey. a, a', b, b', c and c' represent the potential magnetic energy (Equation 2) as a function of θ for maximum field (a and a'), zero field (b and b') and negative intermediate field (c and c'). The blue shapes indicate the population of the minima. Figures d and d' display the magnetization curves for superparamagnetic and blocked regimes respectively.

Nanohybrids   Materials Research Forum LLC
Materials Research Foundations **87** (2021) 1-54   https://doi.org/10.21741/9781644901076-1

2.2   Localized surface plasmon resonance

The metallic materials also develop unique optical properties if their dimensions are reduced to the nanoscale. When an electro-magnetic wave penetrates a metallic material, the free electrons are accelerated in the opposite direction to the electric field. In massive materials, the electrons are linearly displaced until they are scattered by interactions with the crystal lattice (usually with phonons) or other electrons. The distance that traveled before being scattered is known as "average free path" and it is in the order of tens of nanometers in good conductors such as gold.

However, when the dimensions of the material are reduced to a size comparable to the average free path the movement of the electrons is restricted by the limits of the NP, what produces a load decompensation in the network. As shown in Figure 3a, the displacement of electrons towards one of the limits of the NP leaves an unbalanced positive charge on the opposite side generated by the ions of the net. As a consequence of this polarization, an electric recovery field appears, and the electrons are attracted towards the positive region of the material. In first approximation, it can be considered that the sum of the excitation field and the recovery field produces a harmonic oscillation of the electrons around the surfaces of the NP. This collective oscillation of electrons is known as localized surface plasmon resonance (LSPR). The frequency at which the incident wave and the recovery field resonates generating a maximum oscillation of the charge, is known as plasmon frequency (ω_p). Alternatively, we can define a corresponding wavelength called plasmon wavelength (λ_p).

According to electrodynamics laws, accelerated or decelerated electrons dissipate energy in form of radiation. In the case of the LSPR, the oscillating electrons radiate an electric field that modifies with the field of the incident light. As a result of the interference between them, the global electric field results enhanced in the vicinity of the NP. Figure 3b shows the electric field distribution around a single Au nanorod when a LSPR is excited in its longitudinal direction. It can be observed that the local field is more intense around the sharpest regions of the nanorod as a consequence of the accumulation of charge in these areas [6]. This effect results specially dramatic in sharp geometries such as nanostars or nanotriangles [7].

The term nanolenses or nanoantennas are commonly applied to metallic NPs with plasmonic response to describe the effect of such electric field concentration. The use of nanolenses permits to overcome the classical far-field resolution limits and concentrate the incident radiation to subnanometer scales [8]. A second consequence of the field enhancement is the enlargement of the cross-section of the NP (its optical size) by several orders of magnitude respect to the geometric volume. Thanks to this effect, even small

NPs present an intense absorption and scattering of the incident light. The extraordinary properties of metallic NPs with plasmonic response make them excellent candidates for applications such as solar panels [9,10], non-linear optical systems [11] or biosensors [12,13].

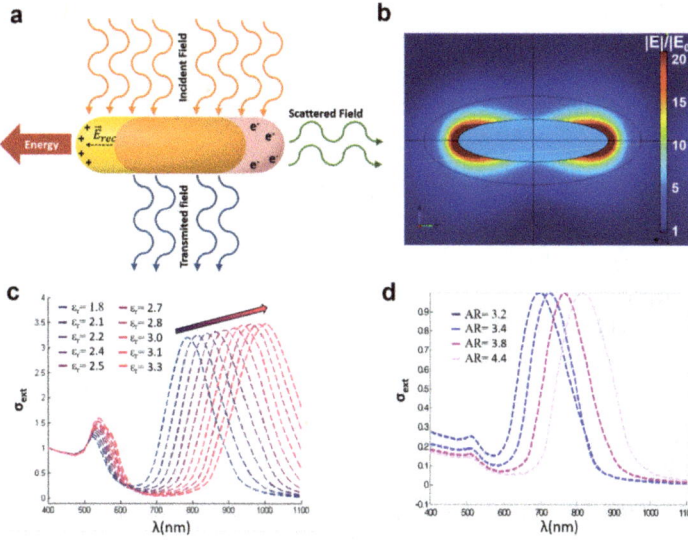

Figure 3. (a)Schematic representation of the localized surface plasmon resonance (LSPR) of a metallic NP. (b)Field enhancement (Respect to incident field E_0) in the proximities of a metallic spheroid. (c) Extinction cross-section for AuNR with an aspect ratio (AR) of 3.8 surrounded by a dielectric media of increasing permittivity. (d) Normalized extinction cross-section for Au NR in water (ε_r) with different AR.

The first theoretical approach to describe the optical response of LSPR and the cross-section enhancement generated in metallic NPs was carried out by Mie in 1908 [14]. He solved the Maxwell equations on a metallic sphere of volume V and complex permittivity ($\varepsilon(\omega) = \varepsilon_1(\omega) + i\varepsilon_2(\omega)$) surrounded by an infinite dielectric media of permittivity $\varepsilon_m(\omega)$ for an incident radiation of frequency ω.

The main approximation assumed by Mie [14] to solve this problem was to consider that the size of the sphere was much smaller than the wavelength of the incident radiation (quasistatic approximation). This hypothesis is generally valid in the case of the NPs smaller than 50 nm in the visible and IR range of the spectrum. Based on this approximation he obtained that the polarizability of the sphere is defined by the Equation 4.

$$\alpha(\omega) = 3\varepsilon_m(\omega)\frac{\varepsilon(\omega)-\varepsilon_m(\omega)}{\varepsilon(\omega)+2\varepsilon_m(\omega)}\ V_{NP} \tag{4}$$

Substituting the real and imaginary components of the permittivity, it is not complicated to calculate the extinction cross section (σ_{ext}) of the sphere, expressed in Equation 5.

$$\sigma_{ext} = 18\pi\frac{\omega}{c}\varepsilon_m^{3/2}(\omega)\frac{\varepsilon_2(\omega)}{[\varepsilon_1(\omega)+2\varepsilon_m(\omega)]^2+\varepsilon_2(\omega)^2}\ V_{NP} \tag{5}$$

This equation relates the optical size (σ_{ext}) and the volume of the NP (V_{NP}) by a frequency dependent factor that includes the permittivity of the NP and the surrounding media. The frequency at which prefactor is maximum (ω_p) is defined by Fröhlich condition (Equation 6).

$$\varepsilon_1(\omega_p) \approx 2\varepsilon_m(\omega_p) \tag{6}$$

According to Frölich condition, the frequency at which the resonance occurs will depend on the real permittivity (ε_1) of the NP and the permittivity of the dielectric media in which the NP is embedded (ε_m). Figure 3c shows the theoretical cross-section spectrum of a single Au nanorod embedded in a dielectric media with different permittivity values. It can be observed that the absorption peak grows in intensity and λ_p shifts to larger wavelengths when ε_m increases.

On the other hand, the complex component of the permittivity (ε_2) is related to the energy loses of the electric currents in the metallic NP. Therefore, it can be used to define how quickly the intensity of the LSPR decays. The suitability of a spherical metallic NP to sustain a LSPR is generally defined with the quality factor (Q_{LSPR}) displayed in Equation 7.

$$Q_{LSPR} = -\frac{\varepsilon_1(\omega)}{\varepsilon_2(\omega)} \tag{7}$$

This expression should be slightly modified for non-spherical geometries [15] but results useful for the analysis of energy dissipation in LSPR. It shows clearly that only materials

Materials Research Forum LLC
https://doi.org/10.21741/9781644901076-1

with low resistivity, i.e. low ε_2, such as Au or Ag present a strong plasmonic response. Considering Equation 6 and 7 it is possible to conclude that the optimum plasmonic response takes place when ε_2 is minimum at the resonance frequency of LSPR (ω_p).

Another feature of the LSPR especially relevant for biomedical imaging was included in Figure 2a: the scattering of incident light. The amount of incident radiation that results scattered can be also determined by Mie´s theory using a parameter called scattering cross section (σ_{sca}). Equation 8 shows that in this case σ_{sca} depends on the square of the volume of the NP.

$$\sigma_{sca} = \frac{3}{2\pi}\left(\frac{\omega^4}{c}\right)\varepsilon_m^2(\omega)\frac{\left(\varepsilon_1(\omega)-\varepsilon_m(\omega)\right)^2+\varepsilon_2(\omega)^2}{\left(\varepsilon_1(\omega)+2\,\varepsilon_m(\omega)\right)^2+\varepsilon_2(\omega)^2}\,V_{NP}^2 \qquad (8)$$

Therefore, the scattering cross section can also overcome the geometrical size of the NP. Thanks to this property, the light scattered by a plasmonic NP can be used to overcome the resolution limitations of the traditional optical microscopy and achieve the detection of single NP [16].

In 1912 Gans extended the analytical approach of Mie to the case spheroids with asymmetric geometries, imposing specific contour conditions [17]. He obtained a correction factor to the Mie equation in which the different axes of the NP present different intensities and ω_p. In the case of metal spheroids, these parameters depend on the relative size of the axes (aspect ratios).

Gans calculations showed that LSPR can be also modified by tuning the geometry of the metallic NP. Maybe the best examples are the Au nanorods [16,18]. Figure 3d shows the absorption spectrum of a Au nanorod when the aspect ratio between its long and short axes is increased from 3 to 6. The possibility of tuning the position of the absorption peak using geometrical factors has attracted great attention and the synthesis routes of metallic NPs has been developed to modify their size and shape. Due to the intensive investigation in Au colloidal chemistry, a plethora of geometries can be synthesized nowadays in a controlled and reproducible way [19–22].

Geometries of higher complexity such as, nanostars or nanocages, does not present enough symmetries to achieve an analytical solution to Maxwell equations. In those cases, it is necessary to apply numerical methods (discrete dipole approximation, finite-difference time domain methods, finite elements methods, etc.) to reproduce the plasmonic response of the NPs [23].

2.3 Magneto-plasmonic response

The term "magneto-plasmonic NPs" is usually applied to materials that combines plasmonic and ferromagnetic responses. However, there are intrinsic magneto-plasmonic effects in which the excitation of the LSPR induces a magnetic moment in the NPs, or alternatively, phenomena in which magnetic fields can modify the LSPR of metallic NP. In this section we will review the physical fundamentals of these phenomena.

The electric fields used to excite the LSPR sustain always an orthogonal magnetic field. However, this magnetic field is orders of magnitude weaker due to the relative value of the fine-structure constants [24]. Therefore, the magnetic response induce by an electro-magnetic wave in isolated NPs is generally weak and results hidden by the LSPR associated to the electric field [25]. Besides, the ω_p is usually outside the range of optical frequencies [15]. It is necessary to create assemblies of metallic NPs spaced by small gaps to couple their electronic oscillations and observe an effective magneto-plasmonic response. In this kind excitations, known as magnetic plasmon resonances (MPR) [26], the magnetic moment is sustained by the coupling of the LSPRs present in the assembly.

Figure 4 shows an example of a magnetic dipole excitation induced in an assembly of four Au NPs with a square arrangement. In this work, Shafiei et al. studied the magnetic modes of the Fano resonances induced in the assembly by the coupling of the four LSPR [25]. Figure 4a shows the charge oscillations that sustain the electric (left) and magnetic (right) dipole resonances. In the former case, the charges are distributed like the classical LSPR presented in Figure 3a. In the MPR, the charge oscillates in a ring shape trajectory that induces a magnetic moment perpendicular to the plane.

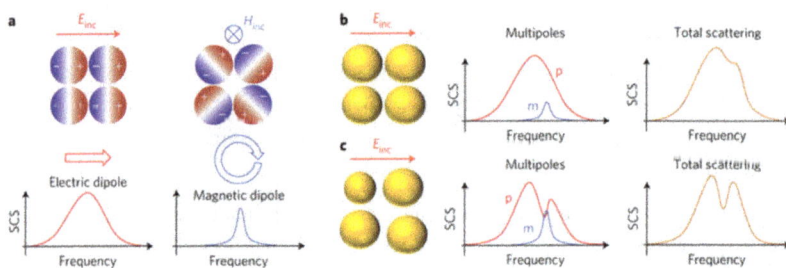

Figure 4. (a) Plasmon resonances of a square assembly of Au NPs in response to an in-plane electric field (red) and orthogonal magnetic field (blue). The graphs bellow each mode show the corresponding scattering cross sections(SCS). Electric (red), magnetic (blue) and total (orange) SCS for: (b) symmetric and (c) asymmetric assembly. Reprinted with permission from Shafiei et al.[25] Copyright© 2013 Springer Nature.

The intensity of the MPR is smaller than the electric LSPR for perfectly symmetric assemblies and it can only be appreciated as a bump in the global scattering spectrum (Figure 4b). But this kind of resonances are extremely sensitive to small structural asymmetries. Variations in the shape of the NPs [27] or in the size of the gaps [28] modify dramatically the spectrum of the assembly. Figure 4c shows how the relative intensity of the MPR grows significantly when the assembly is formed with NPs of different sizes and gaps between NPs. For this geometry LSPR and MPR can be clearly distinguished in the global spectra. The possibility of inducing LSPR and MPR in metallic NP assemblies results particularly interesting to create biosensors but also metamaterials with negative diffractive index in the visible range [29].

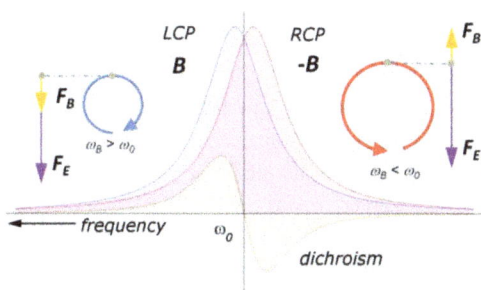

Figure 5. Circular polarized dichroism in a metallic NP induced by a static magnetic field B: Scheme of the process leading to the diamagnetic line shape in the magnetic circular dichroism (orange) resulting from the difference between the left-handed (blue) and right-handed (red) circular plasmonic modes. The insets show the Lorentz force generated by B (F_B) on the circular currents induce by LCP and RCP polarized light. Reprinted with permission from Pineider et al. [31] Copyright© 2013 American Chemical Society.

Alternatively, a magnetic field can be used to modified the LSPR associated with circular currents induced in metallic NPs. According to Faraday´s Law, circular electronic currents induce dipolar magnetic fields perpendicular to the plane in which they are contained. This principle is also present at the nanoscale and can be used to excite magnetic dipolar moment in metallic NPs using circularly polarized light. In fact, such circulating currents has been postulated as a plausible explanation of the giant diamagnetism observed in Au nanorods [30].

Pineider et al. studied in 2013 the different propagation of left and right circularly polarized (LCP and RCP) lights through metallic NPs when a static magnetic field is applied in parallel to the direction of the incident light [31]. They explain this effect, known as magnetic circular dichroism (MCD), as a consequence of the Lorentz force (yellow arrows in Figure 5) generated by the static field on the electrons of the circular currents. This force confines or expands the circular trajectories depending on their helicity and shifts the LSPR to lower and higher frequencies of the spectrum, respectively. This effect can be enhanced using composite nanomaterials [32].

2.4   Magneto-plasmonic effects in hybrid materials

A second category of magneto-plasmonic effects appears when materials with magnetic and plasmonic response form a hybrid magneto-plasmonic nanostructure (HMPNP) or nanocomposite. The interaction in the interface of those materials modifies the magnetic and plasmonic response of the individual phases. These sort of effects are also called "magnetic-optic" effects to differentiate them from intrinsic magneto-plasmonic effect.

As mentioned before, plasmonic oscillations are enormously sensitive to the nature of their dielectric environment, since the enhancement of near electric field extend several nanometers beyond the boundaries of the material [33]. For this reason, it is common to observe changes in the plasmonic resonance of the metallic phase as a consequence of the change in the permittivity of the dielectric environment [34]. This effect is observed in metallic NPs when are partially or fully coated with a magnetic material [35,36] but also in metallic shells when the core filled with a magnetic material of different dielectric constants [37,38].

Other effects such as the change in electron density in the metallic surface can also alter the plasmonic behavior of this phase. If, during hybridization, the bonding of the two materials induces a charge displacement in the interface, the number of electrons in the metal surface is reduced and the collective oscillations are compromised [39]. This fact is especially relevant in metal junctions with semiconductors [40,41] but there are evidences of similar effects in metal juntcions with magnetic oxides [42–44]. A third phenomenon observed in this type of interfaces is the charge delocalization of the metallic phase [45]. This effect is especially important in anisotropic nanostructures that accumulate a large number of defects near the interface between the two phases [46].

On the other hand, the magnetic properties can also be altered because of their hybridization with metallic materials, either by direct modification of their surface or by longer-range effects. Studies on the magnetism of NPMPH show that the presence of a metallic phase can induce changes in the H_C and M_S of the magnetic material [47,48]; the processes of magnetization reversal; and even generate new exchange-mediated anisotropies (Exchange bias) [34,40,49–51]. The causes behind these phenomena are diverse and some of them are still a matter of debate [52]. Some interface phenomena that are postulated as the possible origin of these interactions are: changes in lattice parameters induced by the mismatch of the lattices parameters in the interphase [49], surface spinning canting [53] or a charge transfer from the metallic phase [44]. These effects result enhanced when the two components of the HMPNP present a large interface/volume ratio. Two examples of this kind of nanostructures are ferromagnetic core-metallic shell NPs (e.g. Fe_3O_4@Au) and metallic core-ferromagnetic satellite NPs (Au @ Fe_3O_4) [54].

Due to the mutual interference of magnetic and plasmonic phases it is necessary to perform a careful magneto-plasmonic and structural characterization of the HMPNPs after hybridization. Different ratios of phases or arrangements can lead to a completely new magneto-plasmonic response. For this reason, we will review the possible arrangement of HMPNPs based on $Fe_3O_4/Au/SiO_2$ explored in the literature.

3. Synthesis of magneto-plasmonic nanoparticles

3.1 The shortage of magneto-plasmonic elements and alloys

The main limitation for the research and development of magneto-plasmonic NPs is the lack of materials that combine a strong ferromagnetic response with proper plasmonic features. A quality factor Q_{LSPR} was defined in **Section 2.2**, to evaluate the suitability of metallic elements as LSPR holders according to its permittivity, excluding the effects of NP size, shape, etc. In a recent review [55], Blaber et al. redefine this parameter to include the effect of the number of electrons involved in interband transition using the equation 8:

$$Q_{LSPR} = \frac{2(\omega_p^2 - 2\gamma^2[1+2\sqrt{\pi}\alpha\sigma/\mu])^{3/2}}{3\gamma\omega_p^2\sqrt{3(1+2\sqrt{\pi}\alpha\sigma/\mu)}} \tag{9}$$

Where μ, σ and α are parameters related to the frequency, distribution and number of electrons associated to interband transitions respectively; and γ is a damping factor associated to scattering of the free electrons. The authors summarizes the implications of Equation 9 with the following criteria: "the number of electrons involved in interband transitions must be low, and at the highest possible frequency" [55].

The periodic table shown in Figure 6, reflects the Q_{LSPR} values obtained by Balber et al. for the different metallic elements together with the magnetic susceptibility of all the elements at room temperature. The classification of the plasmonic performance was simplified to gold background for $10 > Q_{SPLR} > 3.5$ (excellent), silver background for $Q_{SPLR} > 10$ (good) and bronze background for $Q_{SPLR} < 3.5$ (poor). Although these limits are subjective, they offer a clear picture of the elements generally employed in plasmonic studies. The number displayed in the center of the case indicates the value of Q_{SPLR} at their optimum plasmon frequencies.

The magnetic response at room temperature is indicated in the outline of each element following the color code of the magnetization curves presented in Figure 1. The paramagnetic elements -mainly on the left-hand side of the table- were identified with a blue outline and the edges of diamagnetic elements were colored in green. It can observe that there are only three ferromagnetic elements highlighted in red (Fe, Co, Ni) and one antiferromagnetic with purple (Cr) outline. The numbers in the bottom of the case, indicate the susceptibility of para- y diamagnetic elements (10^{-9} m^3 mol^{-1}), and the magnetization in Bohr magnetons in the case of ferro- and antiferromagnetic elements.

This table offers a clear outlook of the magnetic and plasmonic properties of metallic elements and shows the scarcity of pure metallic phases with an appropriate magneto-plasmonic response. On the one hand, elements with good plasmonic features such as Au, Ag or Al, presents either a diamagnetic or a paramagnetic response that results technologically unsuitable. On the other hand, ferromagnetic elements present Q_{LSPR} values smaller than 3 and the plasmon frequencies at the mid infrared region of the spectrum.

The best compromise might be found in Ni, which presents a QLSPR of 2.71 and a magnetic moment of 0.61 Bohr magnetons. For that reason, the majority of magneto-plasmonic studies with single phase magneto-plasmonic NPs employ this element [56,57], albeit other alternatives has been explored [58]. For example, the magneto-plasmonic response of Ni was recently studied using a meta-surface of Ni nanoantennas in different magneto-plasmonic Kerr Effect geometries [59,60]. However, in the case of biomedical applications, the high reactivity of Ni and its low biocompatibility, demands a careful coating of the NP to prevent the surface oxidation or generation of reactive oxygen species [61,62].

Metal alloys of magnetic and plasmonic metals (e.g. Au-Fe or Au-Co) has been explored as an alternative for the synthesis of magneto-plasmonic NPs. However, it is generally accepted that alloying noble metals with non-noble metals depress their plasmonic properties. Theoretical studies predicts that the absorption of AuNP can be enhanced with small doping of Fe [63] but, to the best of our knowledge this prediction has not yet been experimentally reported [64–66].

The magnetic properties of ferromagnetic materials result also negatively affected by the alloying, since the directed exchange of the magnetic atoms is generally substituted by other indirect exchange mechanisms (e.g. RKKY) [67] or even disappears at very low concentrations. FePt NPs is one of the exceptions to this general rule. The layer by layer distribution of Fe and Pt atoms in the $L1_0$ tetragonal phase of FePt generates NP with high magnetization and one strongest magnetic anisotropy [68,69]. An important limitation of noble and 3d metal alloys is the low thermodynamic stability. that usually degenerates to a phase segregation [70,71].

Therefore, the synthesis HMPNPs or magneto-plasmonic nanocomposites has been found as the most successful approach to create magneto-plasmonic NPs. The majority of HMPNPs are nanocomposites of Au and iron oxides due to their good plasmonic and magnetic properties of this materials. Besides, both compounds are chemically stable, highly biocompatible [72–75] and easily functionalizable [76], what makes them excellent candidates for use in biomedicine.

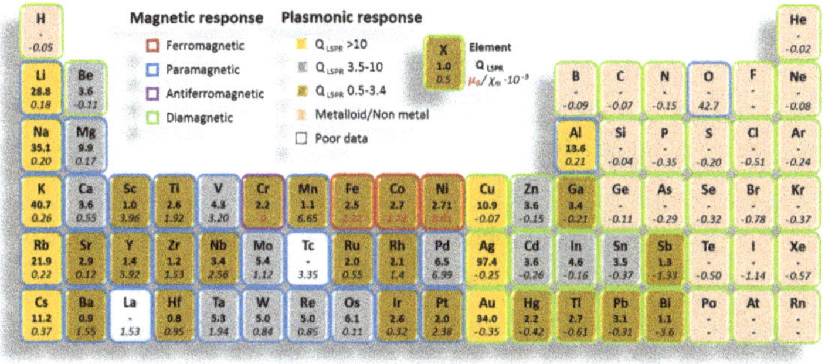

Figure 6. Magneto-plasmonic periodic table. The filling color of the cages indicate the quality of the plasmonic properties (metalloid/nonmetal (pink), poor (bronze), good (silver), excellent (gold)) and the outline indicates the magnetic response (ferromagnetic (red), paramagnetic (blue), antiferromagnetic (purple) and diamagnetic (green)). In the middle of the case appears the maximum values of Q_{LSPR}. In the bottom of the case appears the magnetic susceptibility (10^{-9} m^3mol^{-1}) in the case of para- and diamagnetic materials (black), and the magnetization in Bohr magnetons for ferromagnetic and antiferromagnetic materials (red).

Iron oxide NPs of a maghemite (γ-Fe_2O_3) or magnetite (Fe_3O_4) have demonstrated a strong potential for diagnostic and therapeutic applications due to their unique magnetic properties [77–79]. They present a relatively high susceptibility and magnetization at room temperature [80], as well as an appropriate magnetocrystalline anisotropy to create superparamagnetic NPs in a range between 10-50 nm. Indeed, magnetite and maghemite are the only magnetic NPs approved by the American Food and Drug Administration (FDA) for human use [81].These iron oxide NPs have been successfully used as contrast enhancement agents for imaging techniques such as magnetic resonance imaging (MRI) or magnetic particle imaging (MPI) and NP-based therapies such as magnetic hyperthermia [82–84].

Au NPs, in comparison to other noble metal, present a low chemical reactivity and oxidative effects in cellular environments [85], although the literature shows some controversy about the pharmacokinetics of Au NPs [73]. It has been shown that Au NPs of sizes smaller than 3 nm are rapidly expelled by the excretory system, but some studies show that larger sizes can obstruct nephrons and accumulate in the kidneys [86] and the routes of assimilation by the organism are still unclear for bigger Au NPs [75]. Recently, Rosa et al. has published a comparative review in which they conclude that the cytotoxic effect of AuNPs in cellular systems is small for concentrations lower than 75 μg / mL [87].

A third biocompatible material widely used in the synthesis of HMPNP is the silicon oxide or silica (SiO_2). Despite it does not present neither plasmonic nor ferromagnetic properties, it is commonly used to couple magnetic and plasmonic phases preserving their individual properties. Some advantages of using silica in the synthesis of HMPNP are:

- The surface chemistry of SiO_2 is versatile and well-known. Silica coated NPs can be easily functionalized to increase the colloidal stability or introduce a specific interaction with other biomolecules.

- The finely tunable porosity of this material make SiO_2 NP and SiO_2 coatings an ideal cargo to be loaded with drugs or fluorescent dyes [88,89].

- The well-developed sol-gel chemistry for the coatings of inorganic NPs with SiO_2, which allows improving the hydrophobicity of the coated NPs, reducing their toxicity and preserving the physicochemical properties, among many other functionalities.

Regarding its discussed cytotoxicity, Lu et al. [90] determined by serological, hepathological and histopathological studies in mouse models that the amount of silica required to observe harmful effects in these systems must be greater than 1 mg per mouse (50 mg Kg^{-1}), an amount that is above of silica commonly used in treatments with this type of material in HMPNP.

3.2   Synthesis of hybrid magneto plasmonic nanoparticles

Many different chemical routes have been developed so far for the synthesis of HMPNPs and new methods are continuously created to exploit the possibilities of this kind of nanoagents. One of the first and most successful approach was published by the group of Sun et al. in 2005. They performed a controlled decomposition of Fe and Au precursors

Materials Research Forum LLC
https://doi.org/10.21741/9781644901076-1

at high temperature [42] to create dimers of Fe_3O_4 /Au (Figure 7a) but also more complex structures like $Au@Fe_3O_4$ core-satellites HMPNPs (Figure 7c).

This paper, published in 2005, represents a seminal work for the later development of HMPNPs and the study of their magnetic and plasmonic properties [49,91]. Soon after the release of Sun's paper many alternative routes were proposed for the synthesis of HNPMPs such as direct and inverse microemulsions [92,93], layer-to-layer depositions [94], redox reactions [95], laser ablation [96,97], reverse micelles [98] or sonochemical pathways [99].

At present, the control of the synthesis parameters permits to design HMPNPs with different geometry and arrangement between the phases [40,42,47,102–113,116]. Figure 7 aims to offer a general scheme of the multiple geometries developed for compact HMPNPs "geometries developed for compact HMPNPs based on gold and iron oxide (hollow structures were excluded) based on Au and iron oxide. Geometrical features such as the shape, the relative arrangement or the proportion between the phases can result critical for the magneto-plasmonic properties of the nanostructure due to the interaction effects mentioned before [34,36,117].

Among all the synthesis routes shown in Figure 7, the chemical deposition of metal precursors is certainly the most commonly employed [100]. In this protocol, the hybridization starts with the deposition of seeds on previously formed cores and continues with a controlled growth of the seeds to form a more or less continuous coating around them in a second step [101]. This method presents a large versatility in terms of core and shell geometries.

More recently, organic surfactants has been used as mediators to promote the deposition of one phase over the other creating either an homogeneous layer [47,108,110] , small satellites around the core [104,111,113] or even more complicated geometries [106]. A less explored method is the electrostatic assembling of individual NP into an hybrid nanostructure [114]. The main limitations of the electrostatic assembly are the lack of stability in biological media and the low specificity in the assembly [115].

Incorporating a third phase in the synthesis of Au/FeO_x HMPNP introduces a new design variable that further expands the range of nanocomposites available today. This third phase can be simply used as a matrix to connect both phases avoiding some of the interfacial interactions mentioned in Section 2.3 [118], but it can also play a functional role in the nanostructure as in the case of photosensible polymers [119].

The materials chosen for this third phase goes from organic polymers (polyethienglycol, polyethyleneimine, etc.) to inorganic materials such as SiO_2 or TiO_2 [120]. However, silica is one of the most common choice due to its versatility as well as the functional reasons mentioned before. Figure 8 classifies this type of hybrid nanostructures as core-shell structures, core-satellite structures, and nonspherical nanostructures. It shows how the possibilities of design increases significantly in the case of three-component HMPNPs and suggests the innumerable possibilities that remain still unexplored.

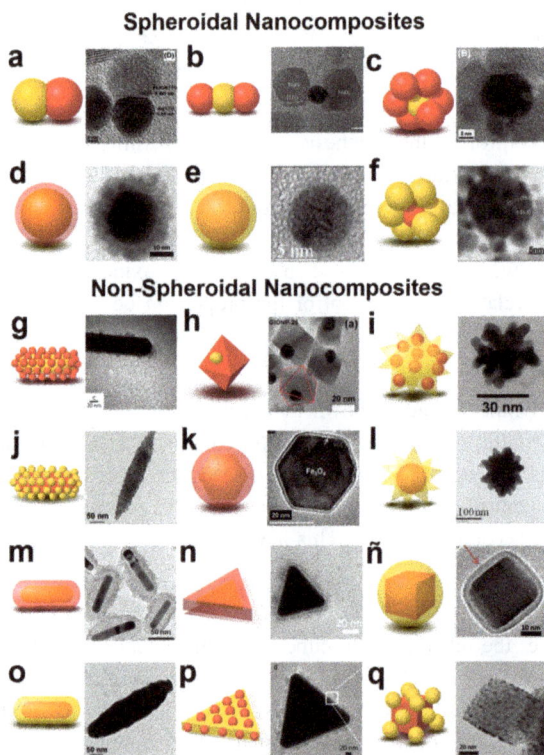

Figure 7. Hybrid magneto-plasmonic nanoparticles based on gold (yellow) and iron oxide (red): (a) Dimer Au/FeO$_x$ [42], (b) trimer FeO$_x$/Au/FeO$_x$ [40], (c) core-satellites Au@FeO$_x$ [42],(d) core-shell Au@FeO$_x$ [102],(e) core-shell FeO$_x$@Au [47],(f) core-satellites FeO$_x$@Au [103], (g) core-satellites Au nanorod(NR)@FeO$_x$ [104],(h) Dimer Au/FeO$_x$ nanoctahedron [105], (i) multicore-shell FeO$_x$@AuNS [106],(j) core-satellites FeO$_x$NR@Au [107],(k) core-shell FeO$_x$ nanohexagon@Au [108], (l) core-shell FeO$_x$@AuNS [109], (m) core-shell AuNR@FeO$_x$ [110],(n) core-shell Au nanotriangle(NT)@ FeO$_x$ [110], (ñ) core-shell FeO$_x$ nanocube(NC)@Au [108],(o) core-shell FeO$_x$NR@Au [107],(p) core-satellites AuNT@FeO$_x$ [111], (q) core-satellites FeO$_x$NC@Au [112]. TEM images adapted with permission from the references indicated.

Core-shell spheroidal nanocomposites

Core-satelite spheroidal nanocomposites

Non-spheroidal nanocomposites

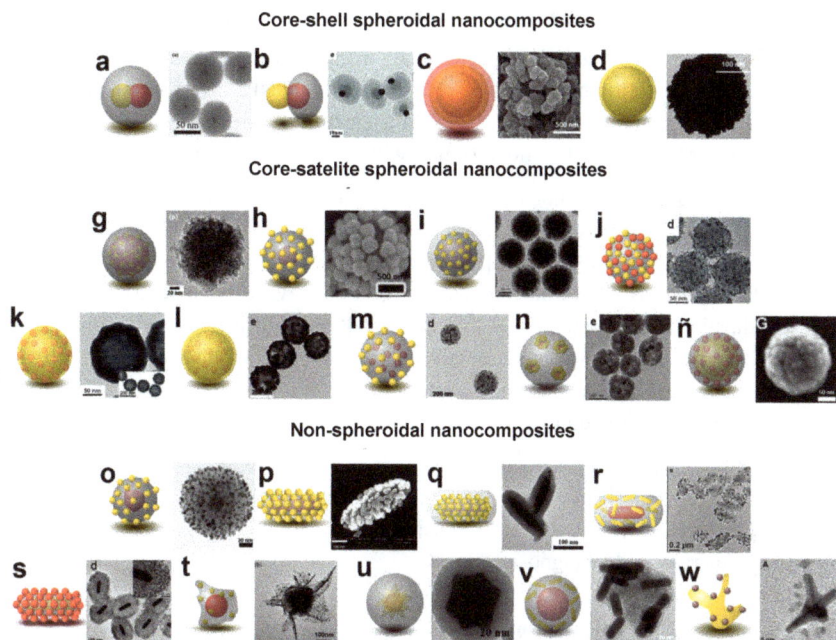

Figure 8. Magneto-plasmonic nanocomposites based on Au (yellow), FeO$_x$ (red) and SiO$_2$ (grey): (a) Dimer@shell (Au/FeO$_x$)@SiO$_2$ [121], (b) Dimer/core@shell Au/(FeO$_x$@SiO$_2$) [122], (c) core-shell-shell SiO$_2$@Au@FeO$_x$ [123], (d) core-shell-shell FeO$_x$ @SiO$_2$@Au [124], (e) core-satellites-shell FeO$_x$@SiO$_2$@Au [125], (f) core-shell-satellites FeO$_x$@SiO$_2$@Au [126], (g) core-shell-satellite-shell FeO$_x$@SiO$_2$@Au@SiO$_2$ [127], (h) core-satellites SiO$_2$@FeO$_x$/Au [128], (i) core-shell-satellites SiO$_2$@Au@ FeO$_x$ [129], (j) core-satellites-shell SiO$_2$@FeO$_x$@Au [128], (k) multicore-shell-shell m-FeO$_x$@SiO$_2$@Au [130], (l) multicore-shell-satellite m-FeO$_x$@SiO$_2$@Au [130], (m) multicore(core-satellites)-shell m-(FeO$_x$@Au)@SiO$_2$ [131], (n) core-shell-satellites-shell SiO$_2$@Au@FeO$_x$@SiO$_2$ [129], (o) core-shell-satellites FeO$_x$NR@SiO$_2$@Au [132], (p) core-shell-satellites-shell FeO$_x$NR@SiO$_2$@Au@SiO$_2$ [133], (q) core-shell-satellites FeO$_x$NR@SiO$_2$@AuNR [134], (r) core-satellites-shell AuNR@FeO$_x$@SiO$_2$ [135] ,(s) core-shell-satellites AuNR@ SiO$_2$@FeO$_x$ [136],(t) core-shell-satellites FeO$_x$NR@SiO$_2$NS@Au [137], (u) core-shell-shell FeO$_x$NR@AuNS@SiO$_2$ [138], (v) core-satellites-shell FeO$_x$@AuNR@SiO$_2$ [139],(w) core-satellites(core-shell) AuNS@(FeO$_x$@SiO$_2$)[140], (x) core-shell-satellites FeO$_x$NC@SiO$_2$@Au [141]. TEM images adapted with permission from the references indicated.

4.   Applications of hybrid magneto-plamonic nanoparticles in biomedicine

NPs and HMPNPs have gained significant momentum in the recent years as mediator agents in the field of the biomedicine. The small size of NPs offers unique advantages such as the possibility of intravenous injection, the interaction with individual biomolecules or the surface functionalization with active biomolecules.

The biocompatibility is a major issue for the use of HMPNP in biological systems. But mentioned before, there are few biocompatible materials that presents suitable magnetic and plasmonic properties [33,80,142]. Therefore, the usual approach to design HMPNP is to hybridize materials whose biocompatibility has been widely demonstrated. However, the hybridization of multiple phases in a single NP tends to increase the volume of the nanostructure and may complicate the stabilization and biofunctionalization of HMPNP if different phases are exposed in their surface (Janus-like NPs). Thus, the synthesis of compact HMPNPs with suitable structural and magneto-plasmonic features results still challenging.

In this part of the chapter, we will review some applications of Au/Fe_3O_4 and $Au/Fe_3O_4/SiO_2$ HMPNP in the biomedical fields displayed in Figure 9: Bioimaging, biosensing, therapy and theragnosis.

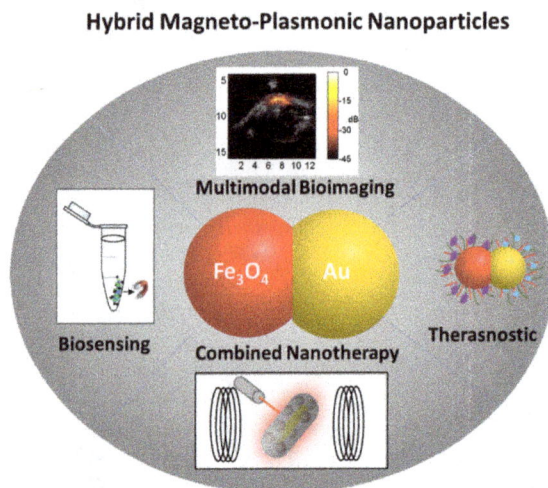

Figure 9. Scheme of the most important biomedical applications of hybrid magneto-plasmonic nanoparticles based on $Au/Fe_3O_4/SiO_2$.

Materials Research Forum LLC

https://doi.org/10.21741/9781644901076-1

4.1 Bioimaging

Magnetic resonance imaging (MRI) is one of the most common imaging techniques used in clinical medicine. To summarize the fundaments of this technique, the signal collected during the MRI measurement is produced by the magnetic relaxation of the nucleus spins of the water protons. Under the action of an external field (B_0), the nuclear spin of the protons is aligned in the direction of the field, giving rise to a net magnetic moment. After the application of a radiofrequency transverse pulse, perpendicular to B_0, the nuclear spins are excited and begin to precess in the plane perpendicular to B_0. After the subtraction of the RF pulse, the nuclear spins gradually recover their equilibrium state parallel to B_0 through a relaxation process. The contrast agents are characterized by modifying the relaxation time of the surrounding media, there are two types of relaxation mechanisms: the longitudinal relaxation process (T_1), i.e. the recovery of the magnetic moment along the direction B_0, and the process of transverse relaxation (T_2), i.e. the loss of signal in the perpendicular plane. The presence of superparamagnetic NPs in an organ or tissue generates an additional magnetic field that modifies the speed of transverse proton relaxation, resulting in a negative (obscuration) or positive (brighten) contrast of the tissue [143].

On the other hand, metallic NPs with plasmonic response have been also widely employed as contrast agents in optical image techniques such as computed tomography (CT), dark field microscopy (DFM), optical coherence tomography (OCT) or photoacoustic (PA) imaging. As it was explained in section 2.2, the plasmonic NPs (such as Au NPs) present extraordinary properties for the absorption and scattering of light in the region of the visible and near infrared (NIR). The plasmonic response of these NPs can be used to overcome the resolution limits of traditional imaging techniques. Besides, the strong X-ray absorption of metallic NPs, like Au NPs, can be used to create contrast agent for computed CT [144].

An interesting application of the HMPNP, is the combination of different image techniques in a single protocol. This is known as multimodal image [145,146]. By incorporating several components into a single nanostructure, it is possible to generate images of the patient that provides complementary information for diagnosis [147]. One of the first *in vivo* applications of HMPNP was the use of Au/Fe_3O_4 as dual contrast agents for MRI and CT [148]. The magnetic response of magnetite and the high atomic weight of the Au were combined in dimer NPs to generate a dual contrast agent. In this example, the Au phase was used as for the absorption of X-ray, but the plasmonic properties of the metallic phase can also be exploited to combine the MRI with other optical techniques such as MRI-OCT [149], MRI-Raman spectroscopy [150], MRI-photoluminescence [151] or MRI-PA image [152]. HMPNPs has been also used to

Materials Research Forum LLC
https://doi.org/10.21741/9781644901076-1

combine CT with other magnetic-based imaging techniques such as magnetic particle imaging (MPI) [153]. Those are some example of the multitude of imaging techniques in which HMPNP can be used as dual contrast agents.

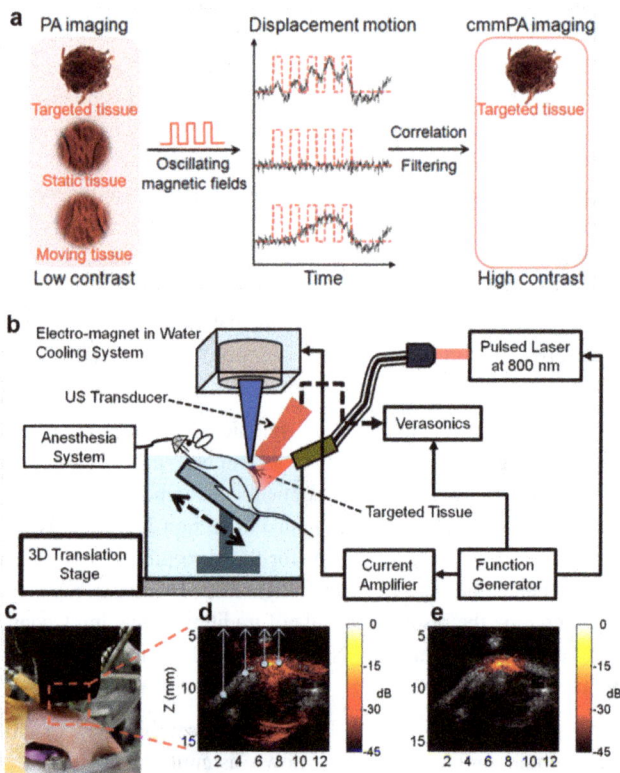

Figure 10. (a) Background subtraction principle for cyclic magneto-motive photoacoustic imaging (cmmPA). Studying the correlation between the cyclic magnetic field and the movement of the biomarkers it is possible to distinguish the targeted tissue from static tissues and tissues moving at frequencies different than the field applied. (b) Image of the cmmPA in-vivo setup for mouse models. (c) PA image of mouse back (indicated with red square) superimposed over the US image of the tissue. (d) cmmPA image of the targeted tissue after background subtraction. Adapted with permission from Li et al.[154] Copyright©2015 American Chemical Society.

Materials Research Forum LLC

https://doi.org/10.21741/9781644901076-1

Besides multimodal imaging, HMPNPs offer functionalities that cannot be achieved with two individual contrast agents. Taking advantage of the concomitance of magnetic and plasmonic response in the HMPNPs, it is possible to improve the resolution of imaging techniques based on the plasmonic response such as PA imaging [154,155] or DFM using a magneto-motive contrast [109,156]. In this approach a magnetic pulse is applied in the region marked with magnetic NPs to induce the displacement of the NPs toward the regions of stronger field (magneto-motive). Filtering the movement of the pixels (or voxels) in the image by the frequency of the applied pulse, it is possible to differentiate the tissue labeled with magneto-motive contrast agents from the background produce by the unlabeled environment, as displayed in Figure 10a.

Although this technique can be applied to imaging techniques such as magneto-motive ultrasound (mmUS) using just magnetic NPs [157,158], the background subtraction results specially efficient in the case of dual contrast agents. The magneto-motive PA (mmPA) imaging is a good example of it. To obtain a PA image, the analyzed region is irradiated with short laser pulses that are transformed into pressure waves by the photosensitive elements of the tissue Figure 10b. These waves are registered with an ultrasound transducer and used to reconstructed a 3D image of irradiated region. Plasmonic NPs outstands as PA contrast agents for their strong light absorption and good thermal conductivity, especially in the near IR [159]. Comparing Figure 10d and Figure 10e it is possible to observe how mmPAI removes the PA background and helps to define the limits of a tissue targeted with magneto-optic contrast agents with higher resolution [160].

4.2 . Therapy

The use of multifunctional NPs in biomedicine is not restricted medical imaging. There is a wide field of research in the development of nanotherapies. Among them, one of the areas that focuses most efforts is cancer treatment. Organic and inorganic NPs have been widely used as transport platform for targeted drug delivery due to its size [161–163] and easy biofunctionalization [164–167]. HMPNPs result particularly interesting because, thanks to their magneto-plasmonic response, they can be externally activated once they have reached the target using light, magnetic fields or both simultaneously.

One of the most common approaches to the use of HMPNPs is the so-called hyperthermia treatment of cancer cells. It consists in raising the local temperature of the tumor tissues to temperatures over 42 °C, temperature at which the cancer cells suffer an apoptotic death. The local nanoheating can be induced by applying high frequency magnetic field to superparamagnetic NPs (magnetic hyperthermia) as well as by exciting the LSPR of metallic NPs with laser radiation (photothermal therapy). Below we will review briefly

Materials Research Forum LLC
https://doi.org/10.21741/9781644901076-1

the mechanisms by which the heat is produced in each kind of hyperthermia. We will present also some examples of the combined use of HMPNP in the treatment of cancer.

The heat dissipated in superparamagnetic NPs is governed by the ratio between the frequency of the applied field and the magnetization relaxation time of the NP. As it was exposed in the section 2.1, the magnetic NPs present a superparamagnetic response (no remanence) for field inversion time longer than it τ_R, and a magnetically blocked response -in which they dissipate energy- for shorter field inversion time, i.e. high frequencies. Thanks to the lack of remanence in the superparamagnetic regime, the NPs can be injected in the blood stream avoiding the problems of magnetic aggregation and vessel obstruction, but they can also be used as nanoheaters once they reach the tumor region by applying high frequency magnetic fields.

The amount of heat released by the superparamagnetic NP is the product of the area of the hysteresis cycle (energy dissipated) times the frequency of the applied field. This area depends on factors such as the anisotropy of the material, the magnetic and hydrodynamic volumes, and the frequency and amplitude of the applied field. The amount of heat released is usually estimated empirically using a parameter called specific absorption rate (SAR) or specific loss power (SLP), displayed in Equation 10.

$$SAR = \frac{C}{m_{NP}} \frac{\Delta T}{\Delta t} \tag{10}$$

Where C calorific capacity of the system, m_{NP} is the concentration of NPs used and $\Delta T/\Delta t$ is the rise of the temperature in K s^{-1}. The intrinsic loss power (ILP) is an alternative normalized parameter to determine the performance of NPs as nanoheaters independently of the magnetic field conditions of intensity (H) and frequency (f) according to Equation 11 [168].

$$ILP = \frac{SAR}{f \cdot H^2} \tag{11}$$

Since the pioneering work of Gilchrist et al. in 1957 [169], important advances have been made in clinical therapy. The development of chemical routes for the synthesis of Fe_3O_4 NPs as well as the optimization of magnetic field induction systems [170,171] have promoted the progress of this kind of clinical treatments. Multicore Fe_3O_4 NPs [172] or linear arrangements of Fe_3O_4 NP in silica shells [173] are some examples of the NP engineering developed for this purpose. The recent approval of hyperthermia treatments at Charite Hospital in Berlin (Nanotherm teraphy)[174], demonstrates the interest that this type of treatment arises.

On the other hand, the photothermia therapy use the energy dissipated by metallic NPs when the LSPR is optically excited to rise the temperature of the surrounding medium. The energy absorbed by the electronic cloud of the NPs is dissipated in three steps[175]. Firstly, the electric field of the radiation accelerate the electrons creating electron-electron collisions that rise their temperature up to values over 1000K. Using ultrafast spectroscopy, it has been determined that this process take place in few femtoseconds. In a second step, the energy of the electrons is relaxed by the interaction with the ion lattice creating phonons in the NP. The time scale for this step is in the order of picoseconds. Finally, the thermal energy of the lattice is transmitted to the medium. The last step depends on the shape of the NP, the composition of the NPs and coatings, and the nature of the medium. When the transmission of these phonons is not fast enough to the medium, the energy accumulated can generate the partial fusion or reshaping of the metallic NPs [176]. Silica coatings of these metallic NPs can avoid the formation of a gas phase in the surface and promote the effective transfer of thermal energy to the medium depending on the porosity [177]. Since the approval by the FDA of the clinical trial AuroLase® based on core-shell SiO_2@Au NPs of 150 nm (NCT00848042, NTC01679470), different modalities of phototherapy have been developed. An example of this is the NANOM FIM study, which successfully completed preclinical stages I and II (NTC01270139), showing a clear regression of coronary atherosclerosis in human patients [178,179].

Although a priori both methodologies, magnetic hyperthermia and photothermia are promising approaches to the treatment of cancer in shallow tissues, the intensity of high frequency magnetic fields and visible-NIR radiation decays significantly in deeper regions of the body. Under this perspective, in recent years the combination of these two types of structures has led to the creation of magneto-plasmonic hyperthermia. In this sense, it is worth highlighting the work of Espinosa et al. on magneto-plasmonic liposomes and HMPNP [166,180–182] which meant an enormous advance in the field. The combination of magnetic properties and optical response produces SAR values above 5000 W g^{-1}, which means a considerable increment in heating power compared to the values obtained in individual magnetic and plasmonic hyperthermia. This impressive value of SAR may compensate the loss of radiation intensity in deep organs and provide an effective treatment even at low doses of NPs. HMPNP are ideal nanoagents to conduct a combined magneto-palsmonic hyperthermia [135]. However, the use of these magneto-plasmonic nanostructures still requires further development for their full application in medical trials.

Hyperthermia is probably the most studied approach for the treatment of cancer mediated by NPs. But there are complementary alternatives like photodynamic therapy [181] or radio sensation [183] in the case of plasmonic NPs and magnetic concentration [184] or mechanical damage in the case of superparamagnetic NPs [185–188]. A suitable combination of these approaches can be design to overcome limitations like the thermos-resistance of tumor tissues using HMPNP [189,190].

4.3   Biosensing

Biosensing is one the most advanced application of plasmonic NPs in biomedicine. Some commercial sensors like pregnancy test are currently based in the specific binding of functionalized Au NPs [191]. The strong sensitivity of the LSPR to the dielectric environment of the NP and the easy functionalization of Au surfaces offer ideal conditions for trapping and detecting analytes even at low concentrations. Besides, the LSPR of individual NPs result severely modified when they are assembled because of a cross-linking aggregation [192]. Using this methodologies, it has been possible to create nanosensors based on protein-DNA interactions [193], antibody antigen [194] among many others.

The concentration of the electric field in the proximity of the NP surface mentioned in section 2.2 is other interesting property of LSPR for biosensing. This phenomenon results specially interesting for detection systems based on Raman scattering. The Raman spectroscopy is a technique that analyzes the inelastic scattering of photons in vibrating systems like biomolecules. It is an efficient technique for the detection and identification of organic and inorganic compounds in aqueous environments, even at very low concentrations.

The main limitation of this technique is the low probability of inelastic photon scattering. To observe a detectable Raman scattering it is convenient to multiply the probability of this events by enhancing the local field. If the analyte is close to the metallic surface, it is possible to use the field concentration produce by LSPR to increase the Raman signal by several order of magnitude [195]. This technique is known as surface enhanced Raman spectroscopy (SERS) [196,197].

One important application of SERS is the detection and identification of pathogens. The detection at low concentration is a crucial aim for the diagnosis and prevention of certain diseases. In this sense, L. Zhan et al. [198] studied the use of HMPNPs for the detection and identification of bacteria using magnetic concentration. The concept of concentration of bacterial can be observed schematically in the Figure 11. On the silicon surface a thin layer of NPs and bacterial cells is deposited. The NPs are attracted to the center with a permanent magnet. In this way the bacterial cells are dragged to the detection region.

Materials Research Forum LLC

https://doi.org/10.21741/9781644901076-1

This methodology takes advantage of the magnetic response of Fe_3O_4 phase to concentrate the bacteria and the plasmonic properties Au phase to create a detectable SERS signal. A detection limit better than 0.1 ppb was achieved thanks to the magnetic concentration.

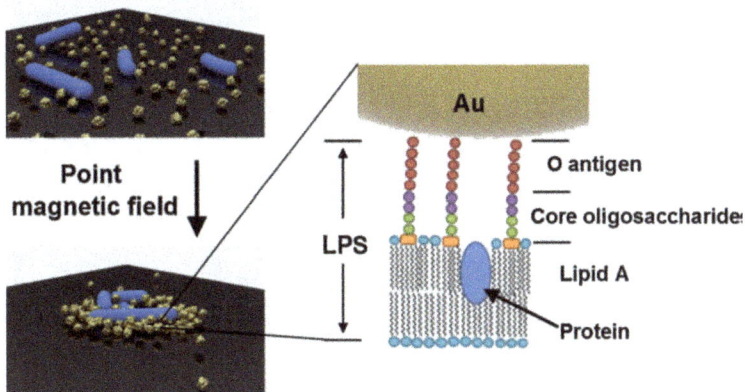

Figure 11. Schematics of the magnetic concentration process of Au-Fe_3O_4 NPs and bacteria (left) and the biomolecular characteristics of the bacterial cell wall that make possible the detection by SERS (right). Reprinted from Zhang et al. [198], copyright © 2012 with permission from Elsevier.

An additional functionality of HMPNPs is the possibility of eliminating pathogens with photothermal treatment of bacteria. C. Wang et al. [199] optimized the magneto-plasmonic properties of AuNR-Fe_3O_4 HMPNPs by tuning the size and aspect ratio of AuNRs. In this work, they concluded that the magnetic separation and the photokilling of multiple pathogens can be achieved in a single step. A scheme of the process carried out is shown in the Figure 12.

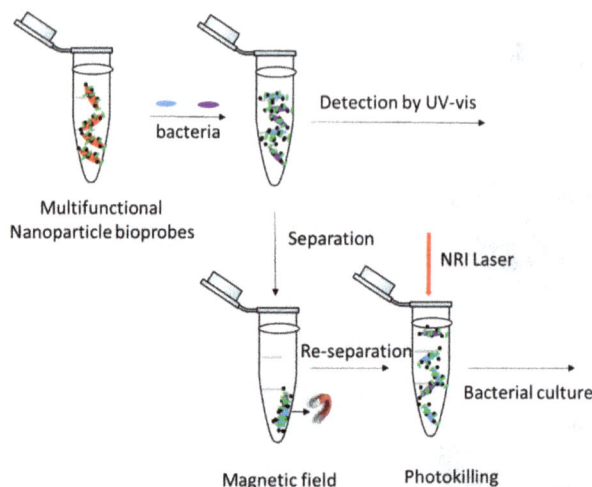

Figure 12. Detection, separation, and thermal ablation of multiple bacterial strains using HMPNPs. EDC: 1-ethyl-3-(3-dimethylaminopropyl)carbodiimide hydrochloride, NHS: N-hydroxysuccinimide. Adapted from Wang et al. [199] *Copyright © 2010 with permission from John Wiley and Sons.*

4.4   Theragnosis

The word theragnosis refers to the combination of diagnosis and treatment using multifunctional platforms and represents one of the main goals for the development of personalized medical therapies. The multifunctional NPs are excellent theragnostic platforms since they can be injected and targeted to a specific tissue. Nonetheless, before using NPs in theragnosis, it is necessary to understand pharmacokinetics parameters such as the mean circulation time in the blood stream, the accumulation in targeted areas or the biodegradation of the nanomaterials.

HMPNPs are attracting great attention as theragnostic agents since they combine magnetic and plasmonic properties that can be applied multiple imaging and therapy techniques using a single multipurpose HMPNP [200–202]. Fe_3O_4 and Au are two of the most common materials for therapeutic and diagnostic dual use [203–205]. In the last decades, multiple examples of fluorescent markers, vector molecules or drugs has been attached to the surface of Au NPs to create multifunctional agents for theragnosis [206]. Fe_3O_4 NPs also present a rich surface chemistry for the functionalization with bioactive molecules for studies of MRI guided nanotherapy [207]. A general strategy to design an all-in-one theragnostic agents consist on: (I) Optimize fundamental properties, i.e. improve the crystallinity, shape, size and bulk-like magnetic properties, and combine with efficient stabilization of NPs in water by functionalization with different compounds. (II) Study the properties for biomedical research, such as biocompatibility and internalization in tumour cells in vitro and in vivo. (III) Combine the magneto-plasmonic properties of the HMPNP with other tracers like fluorescent dyes and test their viability in vitro and in vivo. (IV) Functionalization of Fe_3O_4 and Au surfaces with antitumor drugs. However, there are several challenges that needs to be solved for the use of HMPNP in theragnosis.

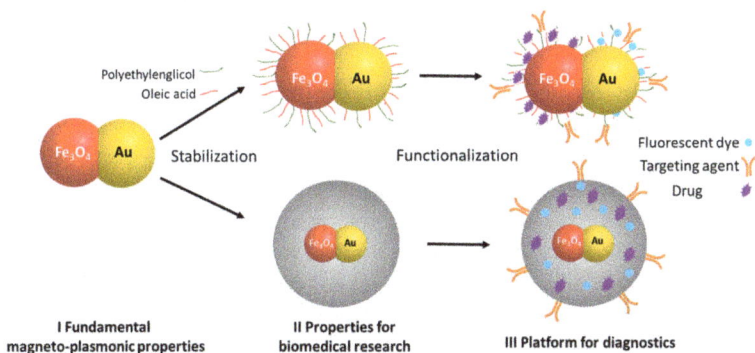

Figure 13. Fe_3O_4-Au HMPNPs represent a unique theragnostic platform. The silica coating of can be employed as a strategy for the stabilization of HMPNPs, but it can also be charged with drugs or fluorescent probes. Inspired in Efremova et al. 2018. [208]

One of them is the complexity of the surface in HMPNP. It generally requires a careful process stabilization and functionalization. Figure 13 shows the general design protocol of Fe_3O_4-Au HMPNPs from the first steps of synthesis and characterization, stabilization, and a final step of *in vivo* application [208]. This "journey" can be simplified including a third biocompatible material like silica that homogenize the surface of the HMPNP [209,210] and offers a well understood surface chemistry [211]. Besides, porous silica can be loaded with fluorescent dyes [212,213] or drugs [88,214] in a controlled and simple manner.

One of the most common approaches found in the literature for *in vivo* use of theragnostic HMPNPs is the MRI guided photothermal treatments [215]. Geometries such as core-shell $Fe_3O_4@Au$ NP [216], core-satellite-shell $SiO_2@Fe_3O_4@Au$ [149], core-shell-shell $Fe_3O_4@SiO_2@Au$ [217] or even yolk-shell $Fe_3O_4@Au$ [152] have already demonstrate a great potential for this field.

There is still a multitude of alternatives yet to be explored such as PT therapy with PL and MRI image [212], dosing of drugs with multiphotonic image and MRI [218] or photothermal therapy guided with MRI and PA imaging [136].

Conclusions

This chapter provides an overview of the application of hybrid magneto-plasmonic nanomaterials, based on iron oxide, gold and silica, in different fields of biomedicine. The fundaments of superparamagnetic and plasmonic response have been expose in detail to highlight the potential of magnetic and plasmonic properties of the NP, but also the new functionalities of HMPNPs. It was also shown there are phenomena in which magnetic and plasmonic responses are interconected. In the case of HMPNPs, the interference between the magnetic and plasmonic phases can modify the properties of the system. Therefore, it is necessary to understand the interaction effect between the phases to optimize the performance of this kind of nanostructures. Physical and chemical properties can be tuned by modifying their morphology and composition. We showed that, including a third component like silica in the HMPNP, the design variability and the functionalities of the HMPNP result widely increased.

Colloidal stability and biocompatibility are crucial parameters for the use of HMPNPs in biomedical applications. This was the main reason to focus the chapter in iron oxide, gold and silica HMPNPs. The synthesis routes of these kinds of hybrid nanomaterials offer a plethora of morphologies whose properties can be adjusted to the final application. The combination optical and magnetic properties of gold and iron oxide make thie HMPNPs a powerful tool for magnetic and optical imaging, hyperthermia and photothermia therapy; biosensing and all-in-one platforms for theranostics. In conclusion, although significant successes have been already achieved, to continue and advance in the knowledge of these promising hybrid materials and expand their range of applicability, it is necessary the interdisciplinary cooperation of different branches of science, chemistry, materials science, physics and biology.

Acknowledgements

The authors want to acknowledge the financial support of: the Spanish Ministry of Economy and Competitiveness under projects MAT2015-67557-C2-2-P, MAT2015 - 67557-C1-2-P and PGC2018-09-5642-B-100.

Nanohybrids
Materials Research Foundations **87** (2021) 1-54

Materials Research Forum LLC
https://doi.org/10.21741/9781644901076-1

References

[1] C. Binns, Nanomagnetism: Fundamentals and Applications, in: Frontiers of Nanoscience, Elsevier, Oxford, 2014, pp 1-29. https://doi.org/10.1016/B978-0-08-098353-0.00001-4

[2] W.T. Coffey, D.S.F. Crothers, J.L. Dormann, Y.P. Kalmykov, E.C. Kennedy, W. Wernsdorfer, Thermally Activated Relaxation Time of a Single Domain Ferromagnetic Particle Subjected to a Uniform Field at an Oblique Angle to the Easy Axis: Comparison with Experimental Observations, Phys. Rev. Lett. 80 (1998) 5655–5658. https://doi.org/10.1103/PhysRevLett.80.5655

[3] Y.P. Kalmykov, The relaxation time of the magnetization of uniaxial single-domain ferromagnetic particles in the presence of a uniform magnetic field, J. Appl. Phys. 96 (2004) 1138–1145. https://doi.org/10.1063/1.1760839

[4] G. Kandasamy, D. Maity, Recent advances in superparamagnetic iron oxide nanoparticles (SPIONs) for in vitro and in vivo cancer nanotheranostics, Int. J. Pharm. 496 (2015) 191–218. https://doi.org/10.1016/j.ijpharm.2015.10.058

[5] M.-H. Phan, J. Alonso, H. Khurshid, P. Lampen-Kelley, S. Chandra, K. Stojak Repa, Z. Nemati, R. Das, Ó. Iglesias, H. Srikanth, Exchange Bias Effects in Iron Oxide-Based Nanoparticle Systems, Nanomaterials 6 (2016) 221. https://doi.org/10.3390/nano6110221

[6] R. Marty, G. Baffou, A. Arbouet, C. Girard, R. Quidant, Charge distribution induced inside complex plasmonic nanoparticles, Opt. Express 18 (2010) 3035. https://doi.org/10.1364/OE.18.003035

[7] E. Hao, G.C. Schatz, J.T. Hupp, Synthesis and optical properties of anisotropic metal nanoparticles, J. Fluoresc. 14 (2004) 331–341. https://doi.org/10.1023/B:JOFL.0000031815.71450.74

[8] J.A. Lloyd, S.H. Ng, A.C.Y. Liu, Y. Zhu, W. Chao, T. Coenen, J. Etheridge, D.E. Gómez, U. Bach, Plasmonic Nanolenses: Electrostatic Self-Assembly of Hierarchical Nanoparticle Trimers and Their Response to Optical and Electron Beam Stimuli, ACS Nano 11 (2017) 1604–1612. https://doi.org/10.1021/acsnano.6b07336

[9] R.A. Pala, J. White, E. Barnard, J. Liu, M.L. Brongersma, Design of plasmonic thin-film solar cells with broadband absorption enhancements, Adv. Mater. 21 (2009) 3504–3509. https://doi.org/10.1002/adma.200900331

[10] H.A. Atwater, A. Polman, Plasmonics for improved photovoltaic devices, Nat.

Mater. 9 (2010) 865–865. https://doi.org/10.1038/nmat2866

[11] J. Butet, T. V Raziman, K.-Y. Yang, G.D. Bernasconi, O.J.F. Martin, Controlling the nonlinear optical properties of plasmonic nanoparticles with the phase of their linear response, Opt. Express 24 (2016) 17138–48. https://doi.org/10.1364/OE.24.017138

[12] J.N. Anker, W.P. Hall, O. Lyandres, N.C. Shah, J. Zhao, R.P. Van Duyne, Biosensing with plasmonic nanosensors, Nat. Mater. 7 (2008) 442–453. https://doi.org/10.1038/nmat2162

[13] T. Sannomiya, J. Vörös, Single plasmonic nanoparticles for biosensing, Trends Biotechnol. 29 (2011) 343–351. https://doi.org/10.1016/j.tibtech.2011.03.003

[14] G. Mie, Beiträge zur Optik trüber Medien, speziell kolloidaler Metallösungen, Ann. Phys. 330 (1908) 377–445. https://doi.org/10.1002/andp.19083300302

[15] V. Amendola, R. Pilot, M. Frasconi, O.M. Maragò, M.A. Iatì, Surface plasmon resonance in gold nanoparticles: a review, J. Phys. Condens. Matter 29 (2017) 203002. https://doi.org/10.1088/1361-648X/aa60f3

[16] M.A. Garcia, Surface plasmons in metallic nanoparticles: fundamentals and applications, J. Phys. D 44 (2011) 283001. https://doi.org/10.1088/0022-3727/44/28/283001

[17] R. Gans, Über die Form ultramikroskopischer Goldteilchen, R. Ann. Phys. 342 (1912) 881. https://doi.org/10.1002/andp.19123420503

[18] H. Chen, L. Shao, Q. Li, J. Wang, Gold nanorods and their plasmonic properties, Chem. Soc. Rev. 42 (2013) 2679–2724. https://doi.org/10.1039/C2CS35367A

[19] V. Grazú, J.M. De La Fuente, Nanobiotechnology - Inorganic Nanoparticles vs Organic Nanoparticles, in: Front. Nanosci. 4 Elsevier, Oxford ,2012, pp 443–485.

[20] M. Grzelczak, J. Perez-Juste, P. Mulvaney, L.M. Liz-Marzan, Shape control in gold nanoparticle synthesis, Chem Soc Rev 37 (2008) 1783–1791. https://doi.org/10.1039/b711490g

[21] N. Khlebtsov, V. Bogatyrev, L. Dykman, B. Khlebtsov, S. Staroverov, A. Shirokov, L. Matora, V. Khanadeev, T. Pylaev, N. Tsyganova, G. Ter-entyuk, Analytical and Theranostic Applications of Gold Na- noparticles and Multifunctional Nanocomposites, Theranostics 3 (2013) 167–180. https://doi.org/10.7150/thno.5716

[22] J. a Webb, R. Bardhan, Emerging advances in nanomedicine with engineered gold nanostructures, Nanoscale (2014) 2502–2530. https://doi.org/10.1039/c3nr05112a

[23] Y. Davletshin, Modeling the optical properties of a single gold nanorod for use in biomedical applications, Theses and dissertations, Ryeson University, 2010, pp 1035.

[24] L.D. Landau, L. P. Pitaevskii, E.M. Lifshitz, Scattering of electromagnetic waves, in: Electrodynamics of Continuous Media, Pergamon Press, Oxford, 1960,pp 299-310

[25] F. Shafiei, F. Monticone, K.Q. Le, X.-X. Liu, T. Hartsfield, A. Alù, X. Li, A subwavelength plasmonic metamolecule exhibiting magnetic-based optical Fano resonance, Nat. Nanotechnol. 8 (2013) 95–99. https://doi.org/10.1038/nnano.2012.249

[26] A.K. Sarychev, G. Shvets, V.M. Shalaev, Magnetic plasmon resonance, Phys. Rev. E 73 (2006) 036609. https://doi.org/10.1103/PhysRevE.73.036609

[27] N.J. Greybush, V. Pacheco-Peña, N. Engheta, C.B. Murray, C.R. Kagan, Plasmonic Optical and Chiroptical Response of Self-Assembled Au Nanorod Equilateral Trimers, ACS Nano (2019). https://doi.org/10.1021/acsnano.8b07619

[28] S.N. Sheikholeslami, A. García-Etxarri, J.A. Dionne, Controlling the interplay of electric and magnetic modes via Fano-like plasmon resonances, Nano Lett. 11 (2011) 3927–3934. https://doi.org/10.1021/nl202143j

[29] H. Liu, D.A. Genov, D.M. Wu, Y.M. Liu, Z.W. Liu, C. Sun, S.N. Zhu, X. Zhang, Magnetic plasmon hybridization and optical activity at optical frequencies in metallic nanostructures, Phys. Rev. B 76 (2007) 073101. https://doi.org/10.1103/PhysRevB.76.073101

[30] A. Hernando, A. Ayuela, P. Crespo, P.M. Echenique, Giant diamagnetism of gold nanorods, New J. Phys. 16 (2014) 073043. https://doi.org/10.1088/1367-2630/16/7/073043

[31] F. Pineider, G. Campo, V. Bonanni, C.D.J. Fernández, G. Mattei, A. Caneschi, D. Gatteschi, C. Sangregorio, Circular magnetoplasmonic modes in gold nanoparticles, Nano Lett. 13 (2013) 4785–4789. https://doi.org/10.1021/nl402394p

[32] P. Varytis, N. Stefanou, A. Christofi, N. Papanikolaou, Strong circular dichroism of core-shell magnetoplasmonic nanoparticles, J. Opt. Soc. Am. B 32 (2015) 1063. https://doi.org/10.1364/JOSAB.32.001063

[33] M.B. Cortie, A.M. Mcdonagh, Synthesis and Optical Properties of Hybrid and Alloy Plasmonic Nanoparticles, Chem. Rev. 111 (2011) 3713–3735. https://doi.org/10.1021/cr1002529

[34] H. Zeng, S. Sun, Syntheses, properties, and potential applications of multicomponent

magnetic nanoparticles, Adv. Funct. Mater. 18 (2008) 391–400.
https://doi.org/10.1002/adfm.200701211

[35] K. Korobchevskaya, C. George, A. Diaspro, L. Manna, R. Cingolani, A. Comin, Ultrafast carrier dynamics in gold/iron-oxide nanocrystal heterodimers, Appl. Phys. Lett. 99 (2011) 011907. https://doi.org/10.1063/1.3609324

[36] V. Velasco, L. Muñoz, E. Mazarío, N. Menéndez, P. Herrasti, A. Hernando, P. Crespo, Chemically synthesized Au–Fe 3 O 4 nanostructures with controlled optical and magnetic properties, J. Phys. D. Appl. Phys. 48 (2015) 035502. https://doi.org/10.1088/0022-3727/48/3/035502

[37] E.A. Chaffin, S. Bhana, R.T. O'Connor, X. Huang, Y. Wang, Impact of Core Dielectric Properties on the Localized Surface Plasmonic Spectra of Gold-Coated Magnetic Core–Shell Nanoparticles, J. Phys. Chem. B 118 (2014) 14076–14084. https://doi.org/10.1021/jp505202k

[38] E.A. Kwizera, E. Chaffin, X. Shen, J. Chen, Q. Zou, Z. Wu, Z. Gai, S. Bhana, R. Oconnor, L. Wang, H. Adhikari, S.R. Mishra, Y. Wang, X. Huang, Size- and Shape-Controlled Synthesis and Properties of Magnetic-Plasmonic Core-Shell Nanoparticles, J. Phys. Chem. C 120 (2016) 10530–10546. https://doi.org/10.1021/acs.jpcc.6b00875

[39] M.C. Daniel, D. Astruc, Gold Nanoparticles: Assembly, Supramolecular Chemistry, Quantum-Size-Related Properties, and Applications Toward Biology, Catalysis, and Nanotechnology, Chem. Rev. 104 (2004) 293–346. https://doi.org/10.1021/cr030698+

[40] W. Shi, H. Zeng, Y. Sahoo, T.Y. Ohulchanskyy, Y. Ding, Z.L. Wang, M. Swihart, P.N. Prasad, A general approach to binary and ternary hybrid nanocrystals, Nano Lett. 6 (2006) 875–881. https://doi.org/10.1021/nl0600833

[41] T. Mokari, E. Rothenberg, I. Popov, R. Costi, U. Banin, Selective Growth of Metal Tips onto Semiconductor Quantum Rods and Tetrapods, Science (80-.). 304 (2004) 1787–1790. https://doi.org/10.1126/science.1097830

[42] H. Yu, M. Chen, P.M. Rice, S.X. Wang, R.L. White, S. Sun, Dumbbell-like Bifunctional Au−Fe3O4 Nanoparticles, Nano Lett. 5 (2005) 379–382. https://doi.org/10.1021/nl047955q

[43] Y. Lee, M.A. Garcia, N.A. Frey Huls, S. Sun, Synthetic Tuning of the Catalytic Properties of Au-Fe3O4 Nanoparticles, Angew. Chemie Int. Ed. 49 (2010) 1271–1274. https://doi.org/10.1002/anie.200906130

[44] F. Pineider, C. De Julián Fernández, V. Videtta, E. Carlino, A. Al Hourani, F. Wilhelm, A. Rogalev, P.D. Cozzoli, P. Ghigna, C. Sangregorio, Spin-polarization transfer in colloidal magnetic-plasmonic au/iron oxide hetero-nanocrystals, ACS Nano 7 (2013) 857–866. https://doi.org/10.1021/nn305459m

[45] A. Comin, K. Korobchevskaya, C. George, A. Diaspro, L. Manna, Plasmon bleaching dynamics in colloidal gold-iron oxide nanocrystal heterodimers, Nano Lett. 12 (2012) 921–926. https://doi.org/10.1021/nl2039875

[46] C. George, A. Genovese, F. Qiao, K. Korobchevskaya, A. Comin, A. Falqui, S. Marras, A. Roig, Y. Zhang, R. Krahne, L. Manna, Optical and electrical properties of colloidal (spherical Au)-(spinel ferrite nanorod) heterostructures, Nanoscale 3 (2011) 4647. https://doi.org/10.1039/c1nr10768b

[47] L. Wang, H.-Y. Park, S.I.-I. Lim, M.J. Schadt, D. Mott, J. Luo, X. Wang, C.-J. Zhong, Core@shell nanomaterials: gold-coated magnetic oxide nanoparticles, J. Mater. Chem. 18 (2008) 2629. https://doi.org/10.1039/b719096d

[48] I.C. Chiang, D.H. Chen, Synthesis of monodisperse FeAu nanoparticles with tunable magnetic and optical properties, Adv. Funct. Mater. 17 (2007) 1311–1316. https://doi.org/10.1002/adfm.200600525

[49] S. Chandra, N.A. Frey Huls, M.H. Phan, S. Srinath, M.A. Garcia, Y. Lee, C. Wang, S. Sun, O. Iglesias, H. Srikanth, Exchange bias effect in Au-Fe $_3$ O $_4$ nanocomposites, Nanotechnology 25 (2014) 055702. https://doi.org/10.1088/0957-4484/25/5/055702

[50] J.P. Pierce, M.A. Torija, Z. Gai, J. Shi, T.C. Schulthess, G.A. Farnan, J.F. Wendelken, E.W. Plummer, J. Shen, Ferromagnetic stability in Fe nanodot assemblies on cu(111) induced by indirect coupling through the substrate, Phys. Rev. Lett. 92 (2004) 237201–1. https://doi.org/10.1103/PhysRevLett.92.237201

[51] M. Feygenson, J.C. Bauer, Z. Gai, C. Marques, M.C. Aronson, X. Teng, D. Su, V. Stanic, V.S. Urban, K.A. Beyer, S. Dai, Exchange bias effect in Au-Fe3O4 dumbbell nanoparticles induced by the charge transfer from gold, Phys. Rev. B 92 (2015) 054416. https://doi.org/10.1103/PhysRevB.92.054416

[52] G. Cheng, A.R. Hight Walker, Synthesis and characterization of cobalt/gold bimetallic nanoparticles, J. Magn. Magn. Mater. 311 (2007) 31–35. https://doi.org/10.1016/j.jmmm.2006.11.164

[53] N.A. Frey, S. Srinath, H. Srikanth, C. Wang, S. Sun, Static and Dynamic Magnetic Properties of Composite Au-Fe3O4 Nanoparticles, IEEE Trans. Magn. 43 (2007)

3094–3096. https://doi.org/10.1109/TMAG.2007.893846

[54] N. a. Frey, M.H. Phan, H. Srikanth, S. Srinath, C. Wang, S. Sun, Interparticle interactions in coupled Au–Fe3O4 nanoparticles, J. Appl. Phys. 105 (2009) 07B502. https://doi.org/10.1063/1.3056582

[55] M.G. Blaber, M.D. Arnold, M.J. Ford, A review of the optical properties of alloys and intermetallics for plasmonics, J. Phys. Condens. Matter 22 (2010) 143201. https://doi.org/10.1088/0953-8984/22/14/143201

[56] K. Nouneh, M. Oyama, R. Diaz, M. Abd-Lefdil, I.V. Kityk, M. Bousmina, Nanoscale synthesis and optical features of metallic nickel nanoparticles by wet chemical approaches, J. Alloys Compd. 509 (2011) 5882–5886. https://doi.org/10.1016/j.jallcom.2011.02.164

[57] H. Amekura, Y. Takeda, N. Kishimoto, Criteria for surface plasmon resonance energy of metal nanoparticles in silica glass, Nucl. Instruments Methods Phys. Res. Sect. B Beam Interact. with Mater. Atoms 222 (2004) 96–104. https://doi.org/10.1016/j.nimb.2004.01.003

[58] J. Zhang, C.Q. Lan, Nickel and cobalt nanoparticles produced by laser ablation of solids in organic solution, Mater. Lett. 62 (2008) 1521–1524. https://doi.org/10.1016/j.matlet.2007.09.038

[59] K. Lodewijks, N. Maccaferri, T. Pakizeh, R.K. Dumas, I. Zubritskaya, J. Åkerman, P. Vavassori, A. Dmitriev, Magnetoplasmonic design rules for active magneto-optics, Nano Lett. 14 (2014) 7207–7214. https://doi.org/10.1021/nl504166n

[60] V. Bonanni, S. Bonetti, T. Pakizeh, Z. Pirzadeh, J. Chen, J. Nogués, P. Vavassori, R. Hillenbrand, J. Åkerman, A. Dmitriev, Designer magnetoplasmonics with nickel nanoferromagnets, Nano Lett. 11 (2011) 5333–8. https://doi.org/10.1021/nl2028443

[61] J. Perez, M.F. Contreras, E. Vilanova, T. Ravasi, J. Kosel, Cytotoxicity and Effects on Cell Viability of Nickel Nanowires, 2013 Int. Conf. Biol. Med. Chem. Eng. (2013) 178–184.

[62] R. Karmhag, G.A. Niklasson, M. Nygren, Oxidation kinetics of nickel nanoparticles, J. Appl. Phys. 89 (2001) 3012–3017. https://doi.org/10.1063/1.1325002

[63] V. Amendola, R. Saija, O.M. Maragò, M.A. Iatì, Superior plasmon absorption in iron-doped gold nanoparticles †, Nanoscale 7 (2015) 8782. https://doi.org/10.1039/C5NR00823A

[64] V. Amendola, S. Scaramuzza, S. Agnoli, S. Polizzi, M. Meneghetti, Strong

dependence of surface plasmon resonance and surface enhanced Raman scattering on the composition of Au–Fe nanoalloys, Nanoscale 6 (2014) 1423–1433. https://doi.org/10.1039/C3NR04995G

[65] P. Mohan, M. Takahashi, K. Higashimine, D. Mott, S. Maenosono, AuFePt Ternary Homogeneous Alloy Nanoparticles with Magnetic and Plasmonic Properties, Langmuir 33 (2017) 1687–1694. https://doi.org/10.1021/acs.langmuir.6b04363

[66] S. Scaramuzza, F. Carraro, E. Cattaruzza, Formation of alloy nanoparticles by laser ablation of Au/Fe multilayer films in liquid environment, J. Colloid Interface Sci. 489 (2017) 18–27. https://doi.org/10.1016/j.jcis.2016.10.023

[67] V. Amendola, M. Meneghetti, O.M. Bakr, P. Riello, S. Polizzi, D.H. Anjum, S. Fiameni, P. Arosio, T. Orlando, C. De Julian Fernandez, F. Pineider, C. Sangregorio, A. Lascialfari, Coexistence of plasmonic and magnetic properties in Au89Fe 11 nanoalloys, Nanoscale 5 (2013) 5611–5619. https://doi.org/10.1039/c3nr01119d

[68] S. Sun, C.B. Murray, D. Weller, L. Folks, A. Moser, Monodisperse FePt nanoparticles and ferromagnetic FePt nanocrystal superlattices, Science (80-.). 287 (2000) 1989–1992. https://doi.org/10.1126/science.287.5460.1989

[69] Y. Ding, S. Yamamuro, D. Farrell, S.A. Majetich, Phase transformation and magnetic moment in FePt nanoparticles, J. Appl. Phys. 93 (2003) 7411. https://doi.org/10.1063/1.1544495

[70] V. Velasco, D. Pohl, A. Surrey, A. Bonatto-Minella, A. Hernando, P. Crespo, B. Rellinghaus, On the stability of AuFe alloy nanoparticles, Nanotechnology 25 (2014) 215703. https://doi.org/10.1088/0957-4484/25/21/215703

[71] M. Cui, H. Lu, H. Jiang, Z. Cao, X. Meng, Phase Diagram of Continuous Binary Nanoalloys: Size, Shape and Segregation Effects, Sci. Rep. 7 (2017) 41990. https://doi.org/10.1038/srep41990

[72] A.G. Roca, R. Costo, A.F. Rebolledo, S. Veintemillas-Verdaguer, P. Tartaj, T. González-Carreño, M.P. Morales, C.J. Serna, Progress in the preparation of magnetic nanoparticles for applications in biomedicine, J. Phys. D. Appl. Phys. 42 (2009) 224002. https://doi.org/10.1088/0022-3727/42/22/224002

[73] Y.F. Li, C. Chen, Fate and toxicity of metallic and metal-containing nanoparticles for biomedical applications, Small 7 (2011) 2965–2980. https://doi.org/10.1002/smll.201101059

[74] N. Khlebtsov, L. Dykman, Biodistribution and toxicity of engineered gold

nanoparticles: a review of in vitro and in vivo studies, Chem. Soc. Rev 40 (2011) 1647–1671. https://doi.org/10.1039/C0CS00018C

[75] M. Zhu, G. Nie, H. Meng, T. Xia, A. Nel, Y. Zhao, Physicochemical properties determine nanomaterial cellular uptake, transport, and fate, Acc. Chem. Res. 46 (2013) 622–31. https://doi.org/10.1021/ar300031y

[76] C.E. Hoyle, C.N. Bowman, Thiol-ene click chemistry, Angew. Chemie - Int. Ed. 49 (2010) 1540–1573. https://doi.org/10.1002/anie.200903924

[77] S. Kuan Yen, P. Padmanabhan, S. Tamil Selvan, Multifunctional Iron Oxide Nanoparticles for Diagnostics, Therapy and Macromolecule Delivery, Theranostics 3 (2013) 986–1003. https://doi.org/10.7150/thno.4827

[78] N. Lee, D. Yoo, D. Ling, M.H. Cho, T. Hyeon, J. Cheon, Iron Oxide Based Nanoparticles for Multimodal Imaging and Magnetoresponsive Therapy, Chem. Rev. 115 (2015) 10637–10689. https://doi.org/10.1021/acs.chemrev.5b00112

[79] R.A. Revia, M. Zhang, Magnetite nanoparticles for cancer diagnosis, treatment, and treatment monitoring: Recent advances, Mater. Today 19 (2015) 157–168. https://doi.org/10.1016/j.mattod.2015.08.022

[80] P. Tartaj, M.P. Morales, T. Gonzalez-Carreño, S. Veintemillas-Verdaguer, C.J. Serna, The iron oxides strike back: From biomedical applications to energy storage devices and photoelectrochemical water splitting, Adv. Mater. 23 (2011) 5243–5249. https://doi.org/10.1002/adma.201101368

[81] A.P. Khandhar, R.M. Ferguson, J.A. Simon, K.M. Krishnan, Tailored magnetic nanoparticles for optimizing magnetic fluid hyperthermia, J. Biomed. Mater. Res. Part A 100A (2012) 728–737. https://doi.org/10.1002/jbm.a.34011

[82] R. Ghosh, L. Pradhan, Y.P. Devi, S.S. Meena, R. Tewari, A. Kumar, S. Sharma, N.S. Gajbhiye, R.K. Vatsa, B.N. Pandey, R.S. Ningthoujam, Induction heating studies of Fe3O4 magnetic nanoparticles capped with oleic acid and polyethylene glycol for hyperthermia, J. Mater. Chem. 21 (2011) 13388. https://doi.org/10.1039/c1jm10092k

[83] R.P. Araújo-Neto, E.L. Silva-Freitas, J.F. Carvalho, T.R.F. Pontes, K.L. Silva, I.H.M. Damasceno, E.S.T. Egito, A.L. Dantas, M.A. Morales, A.S. Carriço, Monodisperse sodium oleate coated magnetite high susceptibility nanoparticles for hyperthermia applications, J. Magn. Magn. Mater. 364 (2014) 72–79. https://doi.org/10.1016/j.jmmm.2014.04.001

[84] E.M. Múzquiz-Ramos, V. Guerrero-Chávez, B.I. Macías-Martínez, C.M. López-

Materials Research Forum LLC
https://doi.org/10.21741/9781644901076-1

Badillo, L.A. García-Cerda, Synthesis and characterization of maghemite nanoparticles for hyperthermia applications, Ceram. Int. 41 (2015) 397–402. https://doi.org/10.1016/j.ceramint.2014.08.083

[85] Nagender Reddy Panyala, Eladia María Peña-Méndez, Josef Havel, Gold and nano-gold in medicine: overview, toxicology and perspectives, J Appl Biomed 7 (2009) 75–91. https://doi.org/10.32725/jab.2009.008

[86] J. Liu, M. Yu, C. Zhou, J. Zheng, Renal clearable inorganic nanoparticles: a new frontier of bionanotechnology, Mater. Today 16 (2013) 477–486. https://doi.org/10.1016/j.mattod.2013.11.003

[87] S. Rosa, C. Connolly, G. Schettino, K.T. Butterworth, K.M. Prise, Biological mechanisms of gold nanoparticle radiosensitization, Cancer Nanotechnol. 8 (2017) 2. https://doi.org/10.1186/s12645-017-0026-0

[88] J. Liu, C. Detrembleur, M.-C.C. De Pauw-Gillet, S.S. Mornet, C. Jérôme, E. Duguet, C. Jérôme, E. Duguet, C. Jerome, E. Duguet, Gold Nanorods Coated with Mesoporous Silica Shell as Drug Delivery System for Remote Near Infrared Light-Activated Release and Potential Phototherapy, Small 11 (2015) 2323–2332. https://doi.org/10.1002/smll.201402145

[89] B. Sahoo, K.S.P. Devi, S. Dutta, T.K. Maiti, P. Pramanik, D. Dhara, Biocompatible mesoporous silica-coated superparamagnetic manganese ferrite nanoparticles for targeted drug delivery and MR imaging applications, J. Colloid Interface Sci. 431 (2014) 31–41. https://doi.org/10.1016/j.jcis.2014.06.003

[90] J. Lu, M. Liong, Z. Li, J.I. Zink, F. Tamanoi, Biocompatibility, Biodistribution, and Drug-Delivery Efficiency of Mesoporous Silica Nanoparticles for Cancer Therapy in Animals, Small 6 (2010) 1794–1805. https://doi.org/10.1002/smll.201000538

[91] P.K. Jain, Y. Xiao, R. Walsworth, A.E. Cohen, Surface plasmon resonance enhanced magneto-optics (SuPREMO): Faraday rotation enhancement in gold-coated iron oxide nanocrystals, Nano Lett. 9 (2009) 1644–1650. https://doi.org/10.1021/nl900007k

[92] M. Mikhaylova, D. Kyung Kim, N. Bobrysheva, M. Osmolowsky, V. Semenov, T. Tsakalakos, M. Muhammed, Superparamagnetism of Magnetite Nanoparticles: Dependence on Surface Modification, Lagmuir 20 (2004) 2472–2477. https://doi.org/10.1021/la035648e

[93] J.L. Lyon, D.A. Fleming, M.B. Stone, P. Schiffer, M.E. Williams, Synthesis of Fe oxide Core/Au shell nanoparticles by iterative hydroxylamine seeding, Nano Lett. 4

(2004) 719–723. https://doi.org/10.1021/nl035253f

[94] M. Spasova, V. Salgueiriño-Maceira, A. Schlachter, M. Hilgendorff, M. Giersig, L.M. Liz-Marzán, M. Farle, Magnetic and optical tunable microspheres with a magnetite/gold nanoparticle shell, J. Mater. Chem. 15 (2005) 2095. https://doi.org/10.1039/b502065d

[95] W.R. Lee, M.G. Kim, J.R. Choi, J. Il Park, S.J. Ko, S.J. Oh, J. Cheon, Redox-transmetalation process as a generalized synthetic strategy for core-shell magnetic nanoparticles, J. Am. Chem. Soc. 127 (2005) 16090–16097. https://doi.org/10.1021/ja053659j

[96] J. Zhang, M. Post, T. Veres, Z.J. Jakubek, J. Guan, D. Wang, F. Normandin, Y. Deslandes, B. Simard, Laser-assisted synthesis of superparamagnetic Fe@Au core-shell nanoparticles, J. Phys. Chem. B 110 (2006) 7122–7128. https://doi.org/10.1021/jp0560967

[97] K. Kawaguchi, J. Jaworski, Y. Ishikawa, T. Sasaki, N. Koshizaki, Preparation of gold/iron-oxide composite nanoparticles by a unique laser process in water, J. Magn. Magn. Mater. 310 (2007) 2369–2371. https://doi.org/10.1016/j.jmmm.2006.11.109

[98] S.-J. Cho, B.R. Jarrett, A.Y. Louie, S.M. Kauzlarich, Gold-coated iron nanoparticles: a novel magnetic resonance agent for T 1 and T 2 weighted imaging, Nanotechnology 17 (2006) 640–644. https://doi.org/10.1088/0957-4484/17/3/004

[99] W. Wu, Q. He, H. Chen, J. Tang, L. Nie, Sonochemical synthesis, structure and magnetic properties of air-stable Fe 3 O 4 / Au nanoparticles, Nanotechnology 18 (2007) 145609. https://doi.org/10.1088/0957-4484/18/14/145609

[100] E.A. Kwizera, E. Chaffin, Y. Wang, X. Huang, Synthesis and properties of magnetic-optical core-shell nanoparticles, RSC Adv. 7 (2017) 17137–17153. https://doi.org/10.1039/C7RA01224A

[101] S. Sabale, P. Kandesar, V. Jadhav, R. Komorek, R.K. Motkuri, X.Y. Yu, Recent developments in the synthesis, properties, and biomedical applications of core/shell superparamagnetic iron oxide nanoparticles with gold, Biomater. Sci. 24 (2017) 2212–2225. https://doi.org/10.1039/C7BM00723J

[102] H. Sun, J. He, J. Wang, S.Y. Zhang, C. Liu, T. Sritharan, S. Mhaisalkar, M.Y. Han, D. Wang, H. Chen, Investigating the multiple roles of polyvinylpyrrolidone for a general methodology of oxide encapsulation, J. Am. Chem. Soc. 135 (2013) 9099–9110. https://doi.org/10.1021/ja4035335

[103] D. Caruntu, B.L. Cushing, G. Caruntu, C.J. Connor, Attachment of gold nanograins onto colloidal magnetite nanocrystals, Chem. Mater. 17 (2005) 3398–3402. https://doi.org/10.1021/cm050280n

[104] A. Gole, J.W. Stone, W.R. Gemmill, H.-C. zur Loye, C.J. Murphy, Iron oxide coated gold nanorods: synthesis, characterization, and magnetic manipulation, Langmuir 24 (2008) 6232–6237. https://doi.org/10.1021/la703975y

[105] L. Li, Y.M. Du, K.Y. Mak, C.W. Leung, P.W.T. Pong, Novel hybrid Au/Fe3O4 magnetic octahedron-like nanoparticles with tunable size, IEEE Trans. Magn. 50 (2014) 23–27. https://doi.org/10.1109/TMAG.2014.2299393

[106] L.L. Ma, M.D. Feldman, J.M. Tam, A.S. Paranjape, K.K. Cheruku, T.A. Larson, J.O. Tam, D.R. Ingram, V. Paramita, J.W. Villard, J.T. Jenkins, T. Wang, G.D. Clarke, R. Asmis, K. Sokolov, B. Chandrasekar, T.E. Milner, K.P. Johnston, Small multifunctional nanoclusters (Nanoroses) for targeted cellular imaging and therapy, ACS Nano 3 (2009) 2686–2696. https://doi.org/10.1021/nn900440e

[107] H. Wang, D.W. Brandl, F. Le, P. Nordlander, N.J. Halas, Nanorice: A hybrid plasmonic nanostructure, Nano Lett. 6 (2006) 827–832. https://doi.org/10.1021/nl060209w

[108] M. Abdulla-Al-Mamun, Y. Kusumoto, T. Zannat, Y. Horie, H. Manaka, Au-ultrathin functionalized core–shell (Fe3O4@Au) monodispersed nanocubes for a combination of magnetic/plasmonic photothermal cancer cell killing, RSC Adv. 3 (2013) 7816. https://doi.org/10.1039/c3ra21479f

[109] H.M. Song, Q. Wei, Q.K. Ong, A. Wei, Plasmon-resonant nanoparticles and nanostars with magnetic cores: Synthesis and magnetomotive imaging, ACS Nano 4 (2010) 5163–5173. https://doi.org/10.1021/nn101202h

[110] Z. Yang, X. Ding, J. Jiang, Facile synthesis of magnetic–plasmonic nanocomposites as T1 MRI contrast enhancing and photothermal therapeutic agents, Nano Res. 9 (2016) 787–799. https://doi.org/10.1007/s12274-015-0958-9

[111] S. Yu, J.A. Hachtel, M.F. Chisholm, S.T. Pantelides, A. Laromaine, A. Roig, Magnetic gold nanotriangles by microwave-assisted polyol synthesis, Nanoscale 7 (2015) 14039–46. https://doi.org/10.1039/C5NR03113C

[112] M. Abbas, S. RamuluTorati, C. Kim, Multifunctional Fe $_3$ O $_4$ /Au core/satellite nanocubes: an efficient chemical synthesis, characterization and functionalization of streptavidin protein, Dalt. Trans. 46 (2017) 2303–2309. https://doi.org/10.1039/C6DT04486G

[113] E.N. Esenturk, A.R. Hight Walker, Gold nanostar @ iron oxide core-shell nanostructures: Synthesis, characterization, and demonstrated surface-enhanced Raman scattering properties, J. Nanoparticle Res. 15 (2013) 1364. https://doi.org/10.1007/s11051-012-1364-9

[114] R.L. Truby, S.Y. Emelianov, K.A. Homan, Ligand-mediated self-assembly of hybrid plasmonic and superparamagnetic nanostructures, Langmuir 29 (2013) 2465–2470. https://doi.org/10.1021/la3037549

[115] F. Bertorelle, M. Ceccarello, M. Pinto, G. Fracasso, D. Badocco, V. Amendola, P. Pastore, M. Colombatti, M. Meneghetti, Efficient AuFeOx nanoclusters of laser-ablated nanoparticles in water for cells guiding and surface-enhanced resonance Raman scattering imaging, J. Phys. Chem. C 118 (2014) 14534–14541. https://doi.org/10.1021/jp503725w

[116] U. Tamer, I.H. Boyacİ, E. Temur, A. Zengin, I. Dincer, Y. Elerman, Fabrication of magnetic gold nanorod particles for immunomagnetic separation and SERS application, J. Nanoparticle Res. 13 (2011) 3167–3176. https://doi.org/10.1007/s11051-010-0213-y

[117] D. Yang, X. Pang, Y. He, Y. Wang, G. Chen, W. Wang, Z. Lin, Precisely Size-Tunable Magnetic/Plasmonic Core/Shell Nanoparticles with Controlled Optical Properties, Angew. Chemie 127 (2015) 12259–12264. https://doi.org/10.1002/ange.201504676

[118] E.D. Smolensky, M.C. Neary, Y. Zhou, T.S. Berquo, V.C. Pierre, Fe3O4@organic@Au: Core-shell nanocomposites with high saturation magnetisation as magnetoplasmonic MRI contrast agents, Chem. Commun. 47 (2011) 2149–2151. https://doi.org/10.1039/C0CC03746J

[119] C.-W. Chen, W.-J. Syu, T.-C. Huang, Y.-C. Lee, J.-K. Hsiao, K.-Y. Huang, H.-P. Yu, M.-Y. Liao, P.-S. Lai, Encapsulation of Au/Fe$_3$O$_4$ nanoparticles into a polymer nanoarchitecture with combined near infrared-triggered chemo-photothermal therapy based on intracellular secondary protein understanding, J. Mater. Chem. B 5 (2017) 5774–5782. https://doi.org/10.1039/C7TB00944E

[120] S. V. Salihov, Y.A. Ivanenkov, S.P. Krechetov, M.S. Veselov, N. V. Sviridenkova, A.G. Savchenko, N.L. Klyachko, Y.I. Golovin, N. V. Chufarova, E.K. Beloglazkina, A.G. Majouga, Recent advances in the synthesis of Fe3O4@AU core/shell nanoparticles, J. Magn. Magn. Mater. 394 (2015) 173–178. https://doi.org/10.1016/j.jmmm.2015.06.012

[121] X.F. Zhang, L. Clime, H.Q. Ly, M. Trudeau, T. Veres, Multifunctional Fe 3 O 4 −Au/Porous Silica@Fluorescein Core/Shell Nanoparticles with Enhanced Fluorescence Quantum Yield, J. Phys. Chem. C 114 (2010) 18313–18317. https://doi.org/10.1021/jp1051112

[122] B. Wu, S. Tang, M. Chen, N. Zheng, Amphiphilic modification and asymmetric silica encapsulation of hydrophobic Au–Fe 3 O 4 dumbbell nanoparticles, Chem. Commun. 50 (2014) 174–176. https://doi.org/10.1039/C3CC47634K

[123] S.N.A. Keivani, M. Naderi, G. Amoabediny, Superparamagnetic plasmonic nanocomposites: Synthesis and characterization studies, Chem. Eng. J. 264 (2015) 66–76. https://doi.org/10.1016/j.cej.2014.11.059

[124] V. Salgueiriño-Maceira, M.A. Correa-Duarte, F. Michael, A. López-Quintela, K. Sieradzki, R. Diaz, Bifunctional gold-coated magnetic silica spheres, Chem. Mater. 18 (2006) 2701–2706. https://doi.org/10.1021/cm0603001

[125] X. He, L. Tan, D. Chen, X. Wu, X. Ren, Y. Zhang, X. Meng, F. Tang, Fe3O4– Au@mesoporous SiO2 microspheres: an ideal artificial enzymatic cascade system, Chem. Commun. 49 (2013) 4643. https://doi.org/10.1039/c3cc40622a

[126] Y. Hu, Y. Sun, Stable magnetic hot spots for simultaneous concentration and ultrasensitive surface-enhanced Raman scattering detection of solution analytes, J. Phys. Chem. C 116 (2012) 13329–13335. https://doi.org/10.1021/jp303775m

[127] J. Ge, Q. Zhang, T. Zhang, Y. Yin, Core-Satellite Nanocomposite Catalysts Protected by a Porous Silica Shell: Controllable Reactivity, High Stability, and Magnetic Recyclability, Angew. Chemie Int. Ed. 47 (2008) 8924–8928. https://doi.org/10.1002/anie.200803968

[128] S.I. Stoeva, F. Huo, J.S. Lee, C.A. Mirkin, Three-layer composite magnetic nanoparticle probes for DNA, J. Am. Chem. Soc. 127 (2005) 15362–15363. https://doi.org/10.1021/ja055056d

[129] R. Bardhan, W. Chen, C. Perez-Torres, M. Bartels, R.M. Huschka, L.L. Zhao, E. Morosan, R.G. Pautler, A. Joshi, N.J. Halas, Nanoshells with Targeted Simultaneous Enhancement of Magnetic and Optical Imaging and Photothermal Therapeutic Response, Adv. Funct. Mater. 19 (2009) 3901–3909. https://doi.org/10.1002/adfm.200901235

[130] X. Wang, H. Liu, D. Chen, X. Meng, T. Liu, C. Fu, N. Hao, Y. Zhang, X. Wu, J. Ren, F. Tang, Multifunctional Fe 3 O 4 @P(St/MAA)@Chitosan@Au Core/Shell Nanoparticles for Dual Imaging and Photothermal Therapy, ACS Appl. Mater.

Interfaces 5 (2013) 4966–4971. https://doi.org/10.1021/am400721s

[131] F. Chen, Q. Chen, S. Fang, Y. Sun, Z. Chen, G. Xie, Y. Du, Multifunctional nanocomposites constructed from Fe3O4–Au nanoparticle cores and a porous silica shell in the solution phase, Dalt. Trans. 40 (2011) 10857. https://doi.org/10.1039/c1dt10374a

[132] X. Xu, M.B. Cortie, Precious metal core-shell spindles, J. Phys. Chem. C 111 (2007) 18135–18142. https://doi.org/10.1021/jp076425q

[133] Z. Chen, Y. Liang, J. Hao, Z.-M. Cui, Noncontact Synergistic Effect between Au Nanoparticles and the Fe2O3 Spindle Inside a Mesoporous Silica Shell as Studied by the Fenton-like Reaction, Langmuir 32 (2016) 12774–12780. https://doi.org/10.1021/acs.langmuir.6b03235

[134] M. Ma, H. Chen, Y. Chen, X. Wang, F. Chen, X. Cui, J. Shi, Au capped magnetic core/mesoporous silica shell nanoparticles for combined photothermo-/chemo-therapy and multimodal imaging, Biomaterials 33 (2012) 989–998. https://doi.org/10.1016/j.biomaterials.2011.10.017

[135] J.G. Ovejero, I. Morales, P. de la Presa, N. Mille, J. Carrey, M.A. Garcia, A. Hernando, P. Herrasti, Hybrid nanoparticles for magnetic and plasmonic hyperthermia, Phys. Chem. Chem. Phys. 20 (2018) 24065–24073. https://doi.org/10.1039/C8CP02513D

[136] L. Huang, L. Ao, D. Hu, W. Wang, Z. Sheng, W. Su, Magneto-Plasmonic Nanocapsules for Multimodal-Imaging and Magnetically Guided Combination Cancer Therapy, Chem. Mater. 28 (2016) 5896–5904. https://doi.org/10.1021/acs.chemmater.6b02413

[137] S. Xuan, F. Wang, X. Gong, S.-K. Kong, J.C. Yu, K.C.-F. Leung, Hierarchical core/shell Fe3O4@SiO2@γ-AlOOH@Au micro/nanoflowers for protein immobilization, Chem. Commun. 47 (2011) 2514. https://doi.org/10.1039/c0cc05390b

[138] W.P. Li, P.Y. Liao, C.H. Su, C.S. Yeh, Formation of oligonucleotide-gated silica shell-coated Fe3O4</inf>-Au core-shell nanotrisoctahedra for magnetically targeted and near-infrared light-responsive theranostic platform, J. Am. Chem. Soc. 136 (2014) 10062–10075. https://doi.org/10.1021/ja504118q

[139] D.-W. Wang, X.-M. Zhu, S.-F. Lee, H.-M. Chan, H.-W. Li, S.K. Kong, J.C. Yu, C.H.K. Cheng, Y.-X.J. Wang, K.C.-F. Leung, Folate-conjugated Fe3O4@SiO2@gold nanorods@mesoporous SiO2 hybrid nanomaterial: a

theranostic agent for magnetic resonance imaging and photothermal therapy, J. Mater. Chem. B 1 (2013) 2934. https://doi.org/10.1039/c3tb20090f

[140] P. Pallavicini, E. Cabrini, A. Casu, G. Dacarro, Y. Antonio Diaz-Fernandez, A. Falqui, C. Milanese, F. Vita, P. Lecante, A. Mosset, J. Osuna, T.O. Ely, C. Amiens, B. Chaudret, Silane-coated magnetic nanoparticles with surface thiol functions for conjugation with gold nanostars, Dalt. Trans. 44 (2015) 21088–21098. https://doi.org/10.1039/C5DT02812D

[141] C.-L. Fang, K. Qian, J. Zhu, S. Wang, X. Lv, S.-H. Yu, Monodisperse α-Fe(2)O(3)@SiO(2)@Au core/shell nanocomposite spheres: synthesis, characterization and properties, Nanotechnology 19 (2008) 125601. https://doi.org/10.1088/0957-4484/19/12/125601

[142] S. Laurent, D. Forge, M. Port, A. Roch, C. Robic, L. Vander Elst, R.N. Muller, Magnetic iron oxide nanoparticles: Synthesis, stabilization, vectorization, physicochemical characterizations and biological applications, Chem. Rev. 108 (2008) 2064–2110. https://doi.org/10.1021/cr068445e

[143] Pankhurst, Connolly, Jones, Dobson, Applications of magnetic nanoparticles in biomedicine, J. Physics-London-D Appl. Phys. 36 (2003) 167–181. https://doi.org/10.1088/0022-3727/36/13/201

[144] D. Xi, S. Dong, X. Meng, Q. Lu, L. Meng, J. Ye, Gold nanoparticles as computerized tomography (CT) contrast agents, RSC Adv. 2 (2012) 12515. https://doi.org/10.1039/c2ra21263c

[145] A. Espinosa, R. Di Corato, J. Kolosnjaj-Tabi, P. Flaud, T. Pellegrino, C. Wilhelm, Duality of Iron Oxide Nanoparticles in Cancer Therapy: Amplification of Heating Efficiency by Magnetic Hyperthermia and Photothermal Bimodal Treatment, ACS Nano 10 (2016) 2436–2446. https://doi.org/10.1021/acsnano.5b07249

[146] B.R. Smith, S.S. Gambhir, Nanomaterials for in Vivo Imaging, Chem. Rev. 117 (2017) 901–986. https://doi.org/10.1021/acs.chemrev.6b00073

[147] L. Martí-Bonmatí, R. Sopena, P. Bartumeus, P. Sopena, Multimodality imaging techniques, Contrast Media Mol. Imaging 5 (2010) 180–189. https://doi.org/10.1002/cmmi.393

[148] J. Zhu, Y. Lu, Y. Li, J. Jiang, L. Cheng, Z. Liu, L. Guo, Y. Pan, H. Gu, Synthesis of Au-Fe3O4 heterostructured nanoparticles for in vivo computed tomography and magnetic resonance dual model imaging, Nanoscale 6 (2014) 199–202. https://doi.org/10.1039/C3NR04730J

[149] J. Kim, S. Park, J.E. Lee, S.M. Jin, J.H. Lee, I.S. Lee, I. Yang, J.S. Kim, S.K. Kim, M.H. Cho, T. Hyeon, Designed fabrication of multifunctional magnetic gold nanoshells and their application to magnetic resonance imaging and photothermal therapy, Angew. Chemie - Int. Ed. 45 (2006) 7754–7758. https://doi.org/10.1002/anie.200602471

[150] V. Amendola, S. Scaramuzza, L. Litti, M. Meneghetti, G. Zuccolotto, A. Rosato, E. Nicolato, P. Marzola, G. Fracasso, C. Anselmi, M. Pinto, M. Colombatti, Magneto-plasmonic Au-Fe alloy nanoparticles designed for multimodal SERS-MRI-CT imaging, Small 10 (2014) 2476–2486. https://doi.org/10.1002/smll.201303372

[151] X. Ji, R. Shao, A.M. Elliott, R.J. Stafford, E. Esparza-Coss, J.A. Bankson, G. Liang, Z.-P. Luo, K. Park, J.T. Markert, C. Li, Bifunctional Gold Nanoshells with a Superparamagnetic Iron Oxide–Silica Core Suitable for Both MR Imaging and Photothermal Therapy, J. Phys. Chem. C 111 (2007) 6245–6251. https://doi.org/10.1021/jp0702245

[152] Y. Huang, T. Wei, J. Yu, Y. Hou, K. Cai, X.J. Liang, Multifunctional metal rattle-type nanocarriers for MRI-guided photothermal cancer therapy, Mol. Pharm. 11 (2014) 3386–3394. https://doi.org/10.1021/mp500006z

[153] A. Tomitaka, H. Arami, Z. Huang, A. Raymond, E. Rodriguez, Y. Cai, M. Febo, Y. Takemura, M. Nair, Hybrid magneto-plasmonic liposomes for multimodal image-guided and brain-targeted HIV treatment, Nanoscale 10 (2017) 184. https://doi.org/10.1039/C7NR07255D

[154] J. Li, B. Arnal, C.W. Wei, J. Shang, T.M. Nguyen, M. O'Donnell, X. Gao, Magneto-optical nanoparticles for cyclic magnetomotive photoacoustic imaging, ACS Nano 9 (2015) 1964–1976. https://doi.org/10.1021/nn5069258

[155] C. Wei, J. Xia, I. Pelivanov, C. Jia, S. Huang, X. Hu, X. Gao, M.O. Donnell, Magnetomotive photoacoustic imaging : in vitro studies of magnetic trapping with simultaneous photoacoustic detection of rare circulating tumor cells, 522 (2013) 513–522. https://doi.org/10.1002/jbio.201200221

[156] Q. Wei, H.M. Song, A.P. Leonov, J. a Hale, D. Oh, Q.K. Ong, K. Ritchie, A. Wei, Gyromagnetic imaging: Dynamic optical contrast using gold nanostars with magnetic cores, J. Am. Chem. Soc. 131 (2009) 9728–9734. https://doi.org/10.1021/ja901562j

[157] J. Oh, M.D. Feldman, J. Kim, C. Condit, S. Emelianov, T.E. Milner, Detection of magnetic nanoparticles in tissue using magneto-motive ultrasound, Nanotechnology 17 (2006) 4183–4190. https://doi.org/10.1088/0957-4484/17/16/031

[158] M. Mehrmohammadi, J. Oh, L. Ma, E. Yantsen, T. Larson, S. Mallidi, S. Park, K.P. Johnston, K. Sokolov, T. Milner, S. Emelianov, Imaging of iron oxide nanoparticles using magneto-motive ultrasound, in: Proc. - IEEE Ultrason. Symp., 2007: pp. 652–655. https://doi.org/10.1109/ULTSYM.2007.169

[159] L.-C. Chen, C.-W. Wei, J.S. Souris, S.-H. Cheng, C.-T. Chen, C.-S. Yang, P.-C. Li, L.-W. Lo, Y.-S. Chen, W. Frey, S. Kim, K. Homan, P. Kruizinga, K. Sokolov, S. Emelianov, Enhanced photoacoustic stability of gold nanorods by silica matrix confinement, J. Biomed. Opt. 18 (2012) 8867. https://doi.org/10.1364/OE.18.008867

[160] M. Qu, M. Mehrmohammadi, R. Truby, I. Graf, K. Homan, S. Emelianov, Contrast-enhanced magneto-photo-acoustic imaging in vivo using dual-contrast nanoparticles, Photoacoustics 2 (2014) 55–62. https://doi.org/10.1016/j.pacs.2013.12.003

[161] H. Maeda, J. Wu, T. Sawa, Y. Matsumura, K. Hori, Tumor vascular permeability and the EPR effect in macromolecular therapeutics: A review, J. Control. Release 65 (2000) 271–284. https://doi.org/10.1016/S0168-3659(99)00248-5

[162] H. Yin, L. Liao, J. Fang, Enhanced Permeability and Retention (EPR) Effect Based Tumor Targeting : The Concept , Application and Prospect, JSM Clin Oncol Res 2 (2014) 1–5.

[163] A. Wicki, D. Witzigmann, V. Balasubramanian, J. Huwyler, Nanomedicine in cancer therapy: Challenges, opportunities, and clinical applications, J. Control. Release 200 (2015) 138–157. https://doi.org/10.1016/j.jconrel.2014.12.030

[164] I.H. El-Sayed, X.H. Huang, M.A. El-Sayed, Surface plasmon resonance scattering and absorption of anti-EGFR antibody conjugated gold nanoparticles in cancer diagnostics: Applications in oral cancer, Nano Lett. 5 (2005) 829–834. https://doi.org/10.1021/nl050074e

[165] A. Kumar, H. Ma, X. Zhang, K. Huang, S. Jin, J. Liu, T. Wei, W. Cao, G. Zou, X.-J. Liang, Gold nanoparticles functionalized with therapeutic and targeted peptides for cancer treatment, Biomaterials 33 (2012) 1180–1189. https://doi.org/10.1016/j.biomaterials.2011.10.058

[166] R. a Sperling, W.J. Parak, Surface modification, functionalization and bioconjugation of colloidal inorganic nanoparticles, Philos. Trans. A. Math. Phys. Eng. Sci. 368 (2010) 1333–1383. https://doi.org/10.1098/rsta.2009.0273

[167] A.J. Cole, V.C. Yang, A.E. David, Cancer theranostics: the rise of targeted magnetic nanoparticles, Trends Biotechnol. 29 (2011) 323–332.

https://doi.org/10.1016/j.tibtech.2011.03.001

[168] R.R. Wildeboer, P. Southern, Q.A. Pankhurst, On the reliable measurement of specific absorption rates and intrinsic loss parameters in magnetic hyperthermia materials, J. Phys. D. Appl. Phys. 47 (2014) 495003. https://doi.org/10.1088/0022-3727/47/49/495003

[169] R.K. Gilchrist, R. Medal, W.D. Shorey, R.C. Hanselman, J.C. Parrott, C.B. Taylor, Selective Inductive Heating of Lymph Nodes *, Ann. Surg. 146 (1957) 596–606. https://doi.org/10.1097/00000658-195710000-00007

[170] I. Obaidat, B. Issa, Y. Haik, Magnetic Properties of Magnetic Nanoparticles for Efficient Hyperthermia, Nanomaterials 5 (2015) 63–89. https://doi.org/10.3390/nano5010063

[171] E.A. Périgo, G. Hemery, O. Sandre, D. Ortega, E. Garaio, F. Plazaola, F.J. Teran, Fundamentals and advances in magnetic hyperthermia, Appl. Phys. Rev. 2 (2015) 041302. https://doi.org/10.1063/1.4935688

[172] L. Gutiérrez, R. Costo, C. Grüttner, F. Westphal, N. Gehrke, D. Heinke, A. Fornara, Q. a Pankhurst, C. Johansson, M.P. Morales, Synthesis methods to prepare single- and multi-core iron oxide nanoparticles for biomedical applications, Dalt. Trans. (2015) 2943–2952. https://doi.org/10.1039/C4DT03013C

[173] I. Andreu, E. Natividad, L. Solozábal, O. Roubeau, Nano-objects for addressing the control of nanoparticle arrangement and performance in magnetic hyperthermia, ACS Nano 9 (2015) 1408–1419. https://doi.org/10.1021/nn505781f

[174] K. Maier-Hauff, F. Ulrich, D. Nestler, H. Niehoff, P. Wust, B. Thiesen, H. Orawa, V. Budach, A. Jordan, Efficacy and safety of intratumoral thermotherapy using magnetic iron-oxide nanoparticles combined with external beam radiotherapy on patients with recurrent glioblastoma multiforme, J. Neurooncol. 103 (2011) 317–324. https://doi.org/10.1007/s11060-010-0389-0

[175] S. Link, M.A. El-sayed, Shape and size dependence of radiative, non-radiative and photothermal properties of gold nanocrystals, Int. Rev. Phys. Chem. 19 (2000) 409–453. https://doi.org/10.1080/01442350050034180

[176] S. Link, C. Burda, B. Nikoobakht, M.A. El-Sayed, Laser-Induced Shape Changes of Colloidal Gold Nanorods Using Femtosecond and Nanosecond Laser Pulses, J. Phys. Chem. B 104 (2000) 6152–6163. https://doi.org/10.1021/jp000679t

[177] S.C. Nguyen, Q. Zhang, K. Manthiram, X. Ye, J.P. Lomont, C.B. Harris, H.

Materials Research Forum LLC
https://doi.org/10.21741/9781644901076-1

Weller, A.P. Alivisatos, Study of Heat Transfer Dynamics from Gold Nanorods to the Environment via Time-Resolved Infrared Spectroscopy, ACS Nano 10 (2016) 2144–2151. https://doi.org/10.1021/acsnano.5b06623

[178] P. Pedrosa, R. Vinhas, A. Fernandes, P. Baptista, Gold Nanotheranostics: Proof-of-Concept or Clinical Tool?, Nanomaterials 5 (2015) 1853–1879. https://doi.org/10.3390/nano5041853

[179] R.S. Riley, E.S. Day, Gold nanoparticle-mediated photothermal therapy: applications and opportunities for multimodal cancer treatment, Wiley Interdiscip. Rev. Nanomedicine Nanobiotechnology 9 (2017) e1449. https://doi.org/10.1002/wnan.1449

[180] J.L. Markman, A. Rekechenetskiy, E. Holler, J.Y. Ljubimova, Nanomedicine therapeutic approaches to overcome cancer drug resistance, Adv. Drug Deliv. Rev. 65 (2013) 1866–1879. https://doi.org/10.1016/j.addr.2013.09.019

[181] R. Di Corato, G. Béalle, J. Kolosnjaj-Tabi, A. Espinosa, O. Clément, A.K.A. Silva, C. Ménager, C. Wilhelm, Combining magnetic hyperthermia and photodynamic therapy for tumor ablation with photoresponsive magnetic liposomes, ACS Nano 9 (2015) 2904–2916. https://doi.org/10.1021/nn506949t

[182] A. Espinosa, M. Bugnet, G. Radtke, S. Neveu, G.A. Botton, C. Wilhelm, A. Abou-Hassan, V. Budach, A. Jordan, P. Wust, R. Bazzi, E. Pereira, Can magneto-plasmonic nanohybrids efficiently combine photothermia with magnetic hyperthermia?, Nanoscale 7 (2015) 18872–18877. https://doi.org/10.1039/C5NR06168G

[183] S. Jain, D.G. Hirst, J.M. O'Sullivan, Gold nanoparticles as novel agents for cancer therapy, Br. J. Radiol. 85 (2012) 101–13. https://doi.org/10.1259/bjr/59448833

[184] S. Bhana, G. Lin, L. Wang, H. Starring, S.R. Mishra, G. Liu, X. Huang, Near-infrared-absorbing gold nanopopcorns with iron oxide cluster core for magnetically amplified photothermal and photodynamic cancer therapy, ACS Appl. Mater. Interfaces 7 (2015) 11637–11647. https://doi.org/10.1021/acsami.5b02741

[185] D.-H. Kim, E.A. Rozhkova, I. V Ulasov, S.D. Bader, T. Rajh, M.S. Lesniak, V. Novosad, Biofunctionalized magnetic-vortex microdiscs for targeted cancer-cell destruction, Nat. Mater. 9 (2010) 165–71. https://doi.org/10.1038/nmat2591

[186] E. Zhang, M.F. Kircher, M. Koch, L. Eliasson, S.N. Goldberg, E. Renström, Dynamic magnetic fields remote-control apoptosis via nanoparticle rotation, ACS Nano 8 (2014) 3192–3201. https://doi.org/10.1021/nn406302j

Materials Research Forum LLC
https://doi.org/10.21741/9781644901076-1

[187] M. Domenech, I. Marrero-Berrios, M. Torres-Lugo, C. Rinaldi, Lysosomal membrane permeabilization by targeted magnetic nanoparticles in alternating magnetic fields, ACS Nano 7 (2013) 5091–5101. https://doi.org/10.1021/nn4007048

[188] A. Vegerhof, E. Barnoy, M. Motiei, D. Malka, Y. Danan, Z. Zalevsky, R. Popovtzer, Targeted Magnetic Nanoparticles for Mechanical Lysis of Tumor Cells by Low-Amplitude Alternating Magnetic Field, Materials (Basel). 9 (2016) 943. https://doi.org/10.3390/ma9110943

[189] J.R. Lepock, H.E. Frey, A.M. Rodahl, J. Kruuv, Thermal analysis of CHL V79 cells using differential scanning calorimetry: Implications for hyperthermic cell killing and the heat shock response, J. Cell. Physiol. 137 (1988) 14–24. https://doi.org/10.1002/jcp.1041370103

[190] N. Iovino, A.C. Bohorquez, C. Rinaldi, Magnetic nanoparticle targeting of lysosomes: a viable method of overcoming tumor resistance?, Nanomedicine 9 (2014) 937–939. https://doi.org/10.2217/nnm.14.52

[191] J. Su, Z. Zhou, H. Li, S. Liu, Quantitative detection of human chorionic gonadotropin antigen via immunogold chromatographic test strips, Anal. Methods 6 (2014) 450–455. https://doi.org/10.1039/C3AY41708E

[192] H. Aldewachi, T. Chalati, M.N. Woodroofe, N. Bricklebank, B. Sharrack, P. Gardiner, Gold nanoparticle-based colorimetric biosensors, Nanoscale Rev. Cite This Nanoscale 10 (2018) 18. https://doi.org/10.1039/C7NR06367A

[193] V. Hegde, M. Wang, W.A. Deutsch, Characterization of human ribosomal protein S3 binding to 7,8-dihydro-8-oxoguanine and abasic sites by surface plasmon resonance, DNA Repair (Amst). 3 (2004) 121–126. https://doi.org/10.1016/j.dnarep.2003.10.004

[194] S.F. Chou, W.L. Hsu, J.M. Hwang, C.Y. Chen, Development of an immunosensor for human ferritin, a nonspecific tumor marker, based on surface plasmon resonance, Biosens. Bioelectron. 19 (2004) 999–1005. https://doi.org/10.1016/j.bios.2003.09.004

[195] L. Lu, A. Eychmüller, Ordered macroporous bimetallic nanostructures: Design, characterization, and applications, Acc. Chem. Res. 41 (2008) 244–253. https://doi.org/10.1021/ar700143w

[196] M. Fleischmann, P.J. Hendra, A.J. McQuillan, Raman spectra of pyridine adsorbed at a silver electrode, Chem. Phys. Lett. 26 (1974) 163–166. https://doi.org/10.1016/0009-2614(74)85388-1

[197] D.L. Jeanmaire, R.P. Van Duyne, Surface raman spectroelectrochemistry Part I Heterocyclic, aromatic, and aliphatic amines adsorbed on the anodized silver electrode, J. Electroanal. Chem. 84 (1977) 1–20. https://doi.org/10.1016/S0022-0728(77)80224-6

[198] L. Zhang, J. Xu, L. Mi, H. Gong, S. Jiang, Q. Yu, Multifunctional magnetic-plasmonic nanoparticles for fast concentration and sensitive detection of bacteria using SERS, Biosens. Bioelectron. 31 (2012) 130–136. https://doi.org/10.1016/j.bios.2011.10.006

[199] C. Wang, J. Irudayaraj, Multifunctional magnetic-optical nanoparticle probes for simultaneous detection, separation, and thermal ablation of multiple pathogens, Small 6 (2010) 283–289. https://doi.org/10.1002/smll.200901596

[200] B. Pelaz, C. Alexiou, R.A. Alvarez-Puebla, F. Alves, A.M. Andrews, S. Ashraf, L.P. Balogh, L. Ballerini, A. Bestetti, C. Brendel, S. Bosi, M. Carril, W.C.W. Chan, C. Chen, X. Chen, X. Chen, Z. Cheng, D. Cui, J. Du, C. Dullin, et al., Diverse Applications of Nanomedicine, ACS Nano 11 (2017) 2313–2381. https://doi.org/10.1021/acsnano.6b06040

[201] X. Nan, X. Zhang, Y. Liu, M. Zhou, X. Chen, X. Zhang, Dual-Targeted Multifunctional Nanoparticles for Magnetic Resonance Imaging Guided Cancer Diagnosis and Therapy, ACS Appl. Mater. Interfaces 9 (2017) 9986–9995. https://doi.org/10.1021/acsami.6b16486

[202] R.X. Zhang, T. Ahmed, L.Y. Li, J. Li, A.Z. Abbasi, X.Y. Wu, Design of nanocarriers for nanoscale drug delivery to enhance cancer treatment using hybrid polymer and lipid building blocks, Nanoscale 9 (2017) 1334–1355. https://doi.org/10.1039/C6NR08486A

[203] L. Li, S. Fu, C. Chen, X. Wang, C. Fu, S. Wang, W. Guo, X. Yu, X. Zhang, Z. Liu, J. Qiu, H. Liu, Microenvironment-Driven Bioelimination of Magnetoplasmonic Nanoassemblies and Their Multimodal Imaging-Guided Tumor Photothermal Therapy, ACS Nano 10 (2016) 7094–7105. https://doi.org/10.1021/acsnano.6b03238

[204] H. Wang, G. Cao, Z. Gai, K. Hong, P. Banerjee, S. Zhou, Magnetic/NIR-responsive drug carrier, multicolor cell imaging, and enhanced photothermal therapy of gold capped magnetite-fluorescent carbon hybrid nanoparticles, Nanoscale 7 (2015) 7885–7895. https://doi.org/10.1039/C4NR07335E

[205] B. Sanavio, F. Stellacci, Recent Advances in the Synthesis and Applications of Multimodal Gold-Iron Nanoparticles, Curr. Med. Chem. 24 (2017) 497–511.

https://doi.org/10.2174/092986732366616082911531

[206] D.A. Giljohann, D.S. Seferos, W.L. Daniel, M.D. Massich, P.C. Patel, C.A. Mirkin, Gold nanoparticles for biology and medicine, Angew. Chemie - Int. Ed. 49 (2010) 3280–3294. https://doi.org/10.1002/anie.200904359

[207] R.R. Qiao, C.H. Yang, M.Y. Gao, Superparamagnetic iron oxide nanoparticles: from preparations to in vivo MRI applications, J. Mater. Chem. 19 (2009) 6274–6293. https://doi.org/10.1039/b902394a

[208] M. V. Efremova, V.A. Naumenko, M. Spasova, A.S. Garanina, M.A. Abakumov, A.D. Blokhina, P.A. Melnikov, A.O. Prelovskaya, M. Heidelmann, Z.A. Li, Z. Ma, I. V. Shchetinin, Y.I. Golovin, I.I. Kireev, A.G. Savchenko, V.P. Chekhonin, N.L. Klyachko, M. Farle, A.G. Majouga, U. Wiedwald, Magnetite-Gold nanohybrids as ideal all-in-one platforms for theranostics, Sci. Rep. 8 (2018) 11295. https://doi.org/10.1038/s41598-018-29618-w

[209] G.A. Sotiriou, A.M. Hirt, P.Y. Lozach, A. Teleki, F. Krumeich, S.E. Pratsinis, Hybrid, silica-coated, Janus-like plasmonic-magnetic nanoparticles, Chem. Mater. 23 (2011) 1985–1992. https://doi.org/10.1021/cm200399t

[210] G. a. Sotiriou, F. Starsich, A. Dasargyri, M.C. Wurnig, F. Krumeich, A. Boss, J.C. Leroux, S.E. Pratsinis, Photothermal killing of cancer cells by the controlled plasmonic coupling of silica-coated Au/Fe2O3 nanoaggregates, Adv. Funct. Mater. 24 (2014) 2818–2827. https://doi.org/10.1002/adfm.201303416

[211] L.T. Zhuravlev, The surface chemistry of amorphous silica Zhuravlev model, Colloids Surfaces A Physicochem. Eng. Asp. 173 (2000) 1–38. https://doi.org/10.1016/S0927-7757(00)00556-2

[212] R. Bardhan, S. Lal, A. Joshi, N.J. Halas, Theranostic Nanoshells: From Probe Design to Imaging and Treatment of Cancer, Acc Chem Res. 44 (2011) 936–946. https://doi.org/10.1021/ar200023x

[213] R. Bardhan, N.K. Grady, J.R. Cole, A. Joshi, N.J. Halas, Fluorescence Enhancement by Au Nanostructures:Nanoshells and Nanorods, ACS Nano 3 (2009) 744–752. https://doi.org/10.1021/nn900001q

[214] B. Sahoo, K.S.P. Devi, S. Dutta, T.K. Maiti, P. Pramanik, D. Dhara, Biocompatible mesoporous silica-coated superparamagnetic manganese ferrite nanoparticles for targeted drug delivery and MR imaging applications, J. Colloid Interface Sci. 431 (2014) 31–41. https://doi.org/10.1016/j.jcis.2014.06.003

[215] Y. Bayazitoglu, S. Kheradmand, T.K. Tullius, An overview of nanoparticle
       assisted laser therapy, Int. J. Heat Mass Transf. 67 (2013) 469–486.
       https://doi.org/10.1016/j.ijheatmasstransfer.2013.08.018

[216] T. a Larson, J. Bankson, J. Aaron, K. Sokolov, Hybrid plasmonic magnetic
       nanoparticles as molecular specific agents for MRI/optical imaging and photothermal
       therapy of cancer cells, Nanotechnology 18 (2007) 325101.
       https://doi.org/10.1088/0957-4484/18/32/325101

[217] X. Ji, R. Shao, A.M. Elliott, R.J. Stafford, E. Esparza-Coss, J.A. Bankson, G.
       Liang, Z.-P. Luo, K. Park, J.T. Markert, C. Li, Bifunctional Gold Nanoshells with a
       Superparamagnetic Iron Oxide−Silica Core Suitable for Both MR Imaging and
       Photothermal Therapy, J. Phys. Chem. C 111 (2007) 6245–6251.
       https://doi.org/10.1021/jp0702245

[218] Y.M. Zhou, H.B. Wang, M. Gong, Z.Y. Sun, K. Cheng, X. kai Kong, Z. Guo,
       Q.W. Chen, M. Zhang, M.H. Cho, T. Hyeon, Yolk-type Au@Fe3O4@C nanospheres
       for drug delivery, MRI and two-photon fluorescence imaging, Dalt. Trans. 42 (2013)
       9906. https://doi.org/10.1039/c3dt50789k

Nanohybrids
Materials Research Foundations **87** (2021) 55-68

Materials Research Forum LLC
https://doi.org/10.21741/9781644901076-2

Chapter 2

Biomaterials for Cell Regeneration

A. Saravanan[1], P. Senthil Kumar[2,3*], R.V. Hemavathy[1], S. Jeevanantham[1]

[1]Department of Biotechnology, Rajalakshmi Engineering College, Chennai 602105, India

[2]Department of Chemical Engineering, SSN College of Engineering, Chennai 603110, India

[3]SSN-Centre for Radiation, Environmental Science and Technology (SSN-CREST), SSN College of Engineering, Chennai 603110, India

senthilkumarp@ssn.edu.in*

Abstract

Biomaterial sciences approaches are as of now crucial systems for the improvement of regenerative cell and medication. Present day material advances take into consideration the improvement of inventive biomaterials that nearly compare to prerequisites of the current biomedical application. A few biomaterials helpful for unmistakable applications in restorative sciences, incorporating into tissue repair and organ reproduction. Natural materials for example, agarose, collagen, alginate, chitosan or fibrin completely coordinate with living tissues of the beneficiary and have low cytotoxicity. Biomaterials, for example, ceramics and metals, are now utilized as inserts to supplant or enhance the usefulness of the harmed tissue or organ. Additionally, the constant advancement of present day innovation opens new experiences of polymeric and smart material applications. Biomaterials may improve the immature microorganisms organic movement and their usage by setting up an explicit microenvironment emulating characteristic cell specialty.

Keywords

Biomaterials, Cell Regeneration, Tissue, Cytotoxicity, Bioengineering

Contents

1.   Introduction

Biomaterials can reiterate in vivo microenvironments to give a characterized specialty to refined cells, bringing about cell phenotypes that are like those in vivo. Biomaterials were initially characterized as nonviable materials intended to connect with organic frameworks when utilized in therapeutic gadgets. However, biomaterials now incorporate transplantable crossovers containing both nonviable materials and living cells; they can upgrade the regenerative procedure and advance practical cortical recuperation at times. Some biomaterials can bolster explicit cell practices, for example, homing and coordinated separation. The objective of bio-material science is to create bolsters that can give a comparative degree of motioning to Extra Cellular Matrix (ECM). ECM inferred parts are raising hotspots for the designing of biomaterials that are equipped for inciting attractive cell-explicit reactions. A wide assortment of bio-materials is utilized in the focal sensory system for the creation of a framework for tissue recovery, including both manufactured and common materials. Biomaterials required for different biomedical applications can be gotten from characteristic or engineered materials while a crossover of the two materials type is exceptionally basic inferable from the one of a kind capacity to upgrade the concoction, natural and mechanical properties. A wide range of strategies is accessible for the presentation of useful gatherings on a surface to expand the utilitarian property of the biomaterials. Specifically, the most widely recognized systems for surface adjustment in the field of biomaterials are wet-compound strategies, ozone treatment, UV-treatment and photograph uniting, self-get together, and plasma treatment. The nanoscale level gives scientists the likelihood to make biomaterials with novel physical properties. To be sure scale appears to be significant in material plan the same number of such nanomaterials have exhibited better properties thought about than their micron organized partners. Joining of nanoparticles in biomaterials themselves can prompt improved mechanical and natural material properties contrasted with simple composites without nanoparticles. In spite of the fact that the use of inorganic nanoparticles in the field of the regenerative drugs is less far-reaching, some significant commitments have been made in the previous decade. For instance, inorganic nanoparticles have been utilized as adaptable stages for improved biomolecule conveyance. The fuse of biochemical factors inside biomaterials is a well-known methodology to give inner signs to appended immature microorganisms. The utilization of a conveyance framework is frequently required in light of the fact that the biomolecules being referred to have short half-lives, corrupt quick and additionally can be destructive at high portions. This can prompt askew impacts with wild activities and even conceivable lethal symptoms. There are a few organic and inorganic materials including, cellulose, keratin, collagen, and silk to comprise a critical piece of biomaterial inquire about attributable to the good concoction, natural and mechanical properties required for biomaterial handling and creation. The accomplishment of biomolecules-inferred biomaterials is dominatingly subject to keeping up the auxiliary and utilitarian honesty of biomolecules inside the created framework or gadget followed by their powerful biomedical usage. The whole procedure of biomolecule-determined biomaterial cooperation with cell or tissue and coming about reaction establishes tissue designing set of three frameworks, wherein biomolecules act in collaboration as auxiliary structure squares of cell and practical conjugates of materials to copy the structure-work relationship of regular organic frameworks [1].

In heart tissue designing, biomaterials and additionally administrative components are joined together with the foundational microorganisms or cardiomyocytes to intently imitate the normal myocardium to improve the cell expansion, movement, and separation. Characteristic biomaterials counting collagen, fibrin, matrigel, self-assembling peptide, decellularized extracellular grid, and engineered polymers, for example, poly(lactide-co-glycolide) (PLGA), polycaprolactone (PCL), poly(glycerol-sebacate) (PGS) and polyurethane (PU) are currently generally utilized in heart tissue designing. These biomaterials ought to be biocompatible also, biodegradable. Figure 1 shows that role of biomaterials in heart regeneration.

In a perfect world, they ought to have normally happening heart tissue-like nanofibrous structures and anisotropic mechanical properties, giving an enlightening microenvironment for the cells to join, develop, relocate and separate. A portion of these biomaterials can likewise be utilized to convey protein, quality or then again RNAs together with the cells. Studies have found that by controlling the properties of the biomaterials, for example, the lattice firmness, morphology or concoction properties, heart separation of the conveyed immature microorganisms can be essentially improved

[2]. When utilizing biomaterials to develop a local like microenvironment for the seeded cells, both the concoction and physical properties of the biomaterial-based frameworks/platforms are pivotal for cell multiplication, separation and mix with the enduring myocardium tissues in harmed zones. The improvement progress of regenerative drug nanostructured items for tissue recovery in the clinic, beginning with the fundamental idea item, at that point in vitro and in vivo examinations, and finishing with clinical examinations and commercialization. Polymer nanoparticles can be utilized as transporters stacked with drugs or organic particles, which is a promising instrument for skin recovery and wound mending. Regranex, becaplermin gel, was first endorsed by the FDA for the recuperating of diabetic neuropathic foot ulcers. It is sodium carboxymethylcellulose-based gel containing recombinant platelet-determined development factor which is valuable in wound mending. In any case, a symptom has been accounted for, for instance, cancer-causing impacts which could be limited by the use of low centralizations of development factors in nanoparticles [3-5].

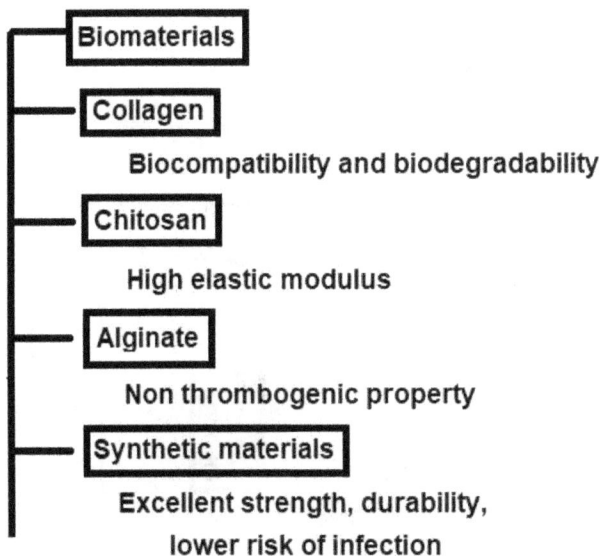

Figure 1. Role of biomaterials in heart regeneration.

This chapter describes the cell and tissue regeneration process and the role of biomaterial in the regeneration process. It also explains the nature and properties of the biomaterials, their sources and applications in regenerative medicine and other therapeutics. The current status of the biomaterials in the regenerative medicine and other applications and their future scopes and applications also included this chapter.

2.   Cell regeneration: Mechanism, methodologies

Regeneration includes the limit with respect to reestablishment or recomposition of tissues, organs or recomposition of tissues, after impressive physical damage or harm, coming about because of pathologies, tumors, intrinsic sicknesses or injuries, for instance. an outcome of tissue recovery, both the organization and the tissue properties are re-established, and the recently framed tissue is exceptionally like the first tissue. The regenerative response has been initiated once the loss of either tissue or local wound is recognised. There are several signals have been involved for the tissue regeneration process but the specific factor for the initiation of the regenerative response does not identified. Currently, the following events are considered as relevant event for the tissue regeneration. Figure 2 shows role of biomaterials in tissue regeneration.

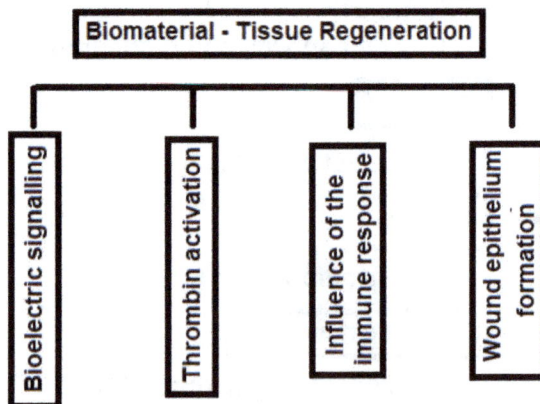

Figure 2. Role of biomaterials in tissue regeneration.

The event are; (i) Bioelectric signalling, (ii) Thrombin activation, (iii) Influence of the immune response and (iv) Wound epithelium formation [6]. Bioelectric signalling includes electric flows in injury that flows after the removal of appendages and mirrors the geometry of the altered tissue and subsequently, the electrical opposition. The removal of a lizard appendage creates a determination of low opposition toward the end of the stump, through which the ionic current streams during the principal days. Ongoing encounters in the caudal recovery of Xenopus hatchlings show the presence of significant action of the V-ATPase proton siphon. The hindrance of this siphon blocked recovery, while upkeep of the V-ATPase articulation kept up the regenerative limit [7]. Thrombin enactment is a basic controller of the response to the wound in vertebrates and has become a solid contender to start the regenerative reaction in lizards as saw in the fix of the digestive system, appendages, and heart. During recovery of the liver in vertebrates, some proof proposes that, after hepatectomy, the arrival of serotonin by the platelets is a major sign for the beginning of hepatocyte expansion. Platelet enactment is another part of the thrombin-subordinate reaction to damage [8]. Arrangement of the scar epithelium on wounds is an early reaction to the damage and comprises of the migration of epithelial cells to the removal plane or tissue injury. The injury epithelium accept a specific personality and assumes a significant job in resulting recovery occasions. Sometimes, the arrangement of epithelium on the injury doesn't happen. In crystalline focal point recovery, epithelial transdifferentiation happens at the site without including the arrangement of an injury epithelium. In heart recovery in zebrafish, there is early and summed up actuation of the epicardium, which can play out a job like that of the injury epithelium. The injury epithelium can give a distal point of confinement to institutionalize systems during recovery, despite the fact that positional character is typically viewed as a component of the mesenchymal cells. It has been proposed that the epidermal cells of various circumferential characters can relocate and frame a utilitarian injury epithelium [9].

3.   Advanced smart biomaterials

Recent advances in bioengineered tissue impersonate the qualities of characteristic tissue at the nanometre scale; have demonstrated the advantages for the cell to cell connection, expansion, separation, and network statement in vitro and acceptance in vivo. Since cells progressively communicate with their nearby condition at the nanoscale, it is important to control the properties of designed tissues at nanoscale lengths for increasingly undifferentiated from regular tissues. A few investigations have concentrated on (a) the utilization of nanostructured materials and (b) control of the mechanical properties of the frameworks. Likewise, nanostructured biomaterials can diminish provocative reaction and increment twisted mending in contrast with regular biomaterials, perhaps because of their high surface vitality influencing protein adsorption and cell attachment The bio-functionalization, for example, the covalent connection of atoms with natural action on the outside of the materials, with proteins or explicit dynamic successions of these proteins, has been the focal point of numerous investigations trying to upgrade neuron grip and development.

Currently, there are different varieties biomaterial based nanostructures such as ceramic, polymeric, metal and composite based nanostructures have been prepared via the process like fabrication for various clinical applications including cells and tissue regeneration, skin regeneration, wound healing [10]. Traditional strategies for the conveyance of nucleic acids, proteins, or medications include infections, electroporation, or nonviral transfection specialists. These techniques depend on either a coordinating infection, which is productive however may instigate insertional mutagenesis, or a nonintegrating conveyance, which regularly experiences lower proficiency and variable spatiotemporal articulation profiles. Biomaterial conveyance techniques dependent on microparticles and nanoparticles can be intended for controlled arrival of elements at wanted stoichiometry. This exactness control is cultivated through artificially controlled atomic stoichiometry and the corruption energy of particles. There are a few specialized constraints for the institutionalization of medication conveyance to cells. In any case, by institutionalizing the segments of medication conveyance frameworks, biomaterial approaches, for example, high-thickness nanoneedles are showing progressively uniform and reproducible conveyance [11-13]. The controlled control of conveyance methodologies has a few suggestions for cell therapies.

The characteristic neural undifferentiated organism's specialty gives a model to planning an incredible fake microenvironment to control the neural foundational microorganisms' destiny, which is fundamental for the focal sensory system recovery. The cells, veins, and the ECM in the neural undifferentiated organism's specialty cooperate to decide the destiny of neural foundational microorganisms. As per their various properties, biophysical and biochemical parameters can be finished up as two fundamental undifferentiated organism's administrative intimation the neural immature microorganism's specialty. The biophysical parameters contain the mechanical properties and engineering of the ECM. The biochemical parameters are made out of the synthetic and bioactive signals beginning from the solvent cytokines and development factors discharged by the adjoining cells, cell grip atoms, and extracellular framework particles. A useful led platform for focal sensory system tissue building and recovery ought to be intended to mimic the neural undifferentiated cells specialty to manage the destiny of neural foundational microorganisms.

4.   Biomaterials: Promising tools in cell regeneration

Biomaterials have been built to both restate numerous parts of in vivo conditions and fuse novel functional characteristics. Biomaterials give a well-characterized, not so much unpredictable, but rather more tuneable environmental condition than the ECM inferred substrates or feeder cells. Fruitful interpretation of cell treatments into clinical preliminaries and on to the facility has been empowered by the assembly of a few fields of research, including biomaterials, immature microorganism and formative science, immunotherapy, reconstructing, and quality altering. There are as of now a large number of dynamic clinical preliminaries for cell treatments, including treatments for malignancy, cardiovascular disease, and neurological issues. Immature microorganism and quality altering based cell treatments is a quickly developing segment inside the phone treatment industry. Undifferentiated organisms, with their capacity to self-recharge in vitro and separate into specific cell lines, are an effectively appealing cell hotspot for cell-and tissue-built treatments. The capacity to control foundational microorganism genomes makes extra open doors for adjusting cell conduct, usefulness, and clinical utility [14]. With the objective of mimicking the 3D ECM to control undifferentiated (stem cell) cells behaviour, certain characteristic biomaterials have been embraced to help stem cell proliferation and differentiation, including collagen, gelatin, hyaluronic corrosive hydrogels, fibrin, glycosaminoglycans (GAGs), alginate, matrigel, silk and hydroxyapatite (HA), and so on. These materials show explicit points of interest, including comparative mechanical and cement properties as the normal ECM. Collagen, present in all connective tissue, goes about as a fundamental segment of ECM with predominant biocompatibility, because of the way that collagen-inferred a cellular ECM would not cause genuine unfavourable invulnerable reactions. Similar to stem cell regeneration both the natural and synthetic biomaterials have been utilized for the cardiac cells regeneration and repair. The natural biomaterials including gelatin, collagen, chitosan, alginate and fibrin glue and the synthetic biomaterials such as carbon nanotubes, polyvinylchloride, poly(lactic-co-glycolic acid) and polyurethane are most commonly used biomaterials for the cardiac cell regeneration [15].

The advancement of synthetic cell-based innovations requests the sequestration and in vitro maintenance of cell organelles. As cell-based treatments become progressively many-sided and judiciously modified, biomaterials empowered exact control will assume a critical job in coordinating the procedure, from biomanufacturing to definite conveyance. Biomaterials have a few highlights that loan themselves to improving biomanufacturing forms including a huge parameter space for properties, reproducibility, and versatility. Biomaterials are especially fit to applications where physical and mechanical properties are significant, exact authority over the cell condition is required, or when complex practices might be required in vivo. As of late, built nanocarriers could hereditarily alter T cells viably in situ in mice, bringing about long haul abatement. The transition to in situ cell designing can possibly improve or even revoke the requirement for ex vivo biomanufacturing. In any case, biomaterials-based stages have their own restrictions that should be considered.

5.   Application of nanoengineered biomaterials: Cell regeneration and therapy

The engineered biomaterials have been prepared by several surface modification process including physical and chemical modification which will increases their interactions with their target and neighbouring cells. Laminin- inferred peptide is the most normally utilized synthetic change of the surface of materials utilized for nerve recovery. In creating regenerative biomaterials for hard tissue engineering, one material alone can't meet the necessity. For intently emulate the qualities of bone tissue at the nanometre scale, hydroxyapatite nanoparticles have been combined with characteristic and manufactured polymers. Wang et al. have indicated the hydroxyapatite or polyamide nanocomposite frameworks would do well to biocompatibility and broad osteoconductivity with have bone, looking at the unadulterated polyamide platforms, at the starter time frame after implantation in vivo [16]. A few investigations have indicated that cells are generally like their in vivo phenotype when they are refined in networks and platforms looking like their local physiological specialty. For instance, MSCs can be coordinated to separate toward adipogenic, myogenic, or osteogenic genealogies in vitro through refined on substrates with flexibility like fat, muscle, or bone tissue, individually. While 2D and 3D culture substrates covered with engineered biomaterials are regularly utilized for adherent cells, even suspension culture can utilize synthetic beads with tunable biomaterials to control motioning to grow specific cell types, for example, actuated T-cell subsets.

Silk biomaterials are known for biomedical and tissue engineering applications including drug conveyance and implantable gadgets attributable to their biocompatible and a wide scope of perfect physical-concoction properties. Silk is considered as a biocompatible material since it doesn't summon a long haul or constant provocative reaction, and it permits tissue ingrowth. It incites a gentle or insignificant starting provocative reaction when embedded in vivo, which dies down with time, demonstrating immunocompatible properties of the material. Being an adaptable biopolymer, silk is broadly investigated in the uses of tissue designing and regenerative prescription. Silk-based lattices grew so far have demonstrated potential in treating different sorts of wounds running from diabetic injuries, severely charred areas, benefactor site split-thickness wounds, and weight bruises in creature contemplate. Moreover, various methodologies have been applied for the functionalization of silk-based networks utilizing development variables, anti-infection agents, and other bioactive atoms throughout the previous scarcely any years. The vast majority of the present nanotechnology-based regenerative drug items are made for bone tissue recovery, fillers, and bony deformities [17]. Bioceramic, primarily nano-sized HA has been utilized in business items. As of late, some nanoparticle-based composite biomaterials additionally have been acquired from the FDA.

Polymer nanoparticles can be utilized as transporters stacked with drugs or natural particles, which is a promising instrument for skin recovery and wound mending. Silver nanoparticles have been ordinarily utilized in tissue building because of their ability to discharge silver particles which thus prompts antibacterial movement. Silver nanoparticles are the main metal nanoparticles utilized for wound mending particularly, contaminated consumes, open injuries, and constant ulcers in the clinical market because of their low fundamental lethality, antibacterial movement, biocompatibility, and minimal effort. There are some biological molecules, for example, corticosteroids and cell reinforcements have been stacked into chitosan nanoparticles in vitro and in vivo to advance wound healing process. HemCom, a chitosan-based hemostatic operator, has been generally utilized for the control of severely bleeding wounds. Be that as it may, a reaction has been accounted for including cancer-causing impact, which could be limited by the utilization of low convergences of growth factors in nanoparticles. With including nanoparticles, numerous examinations have demonstrated that a nanostructured composite can advance osteogenesis without development factors [18]. The attachment, spreading, and expansion of MC3T3-E1 mouse preosteoblast cells can be adjusted by changing the centralization of laponite nanoparticles in laponite-PEO nanocomposites [19]. Bioactive glass/biodegradable polymer composite materials have developed with the principle uses of composite coating for the implant and tissue engineering frameworks. Specifically, ongoing examinations demonstrate the use of nano-sized bioactive glass particles in composites could improve the presentation of biomedical applications in tissue designing. Couto et al. created chitosan and bioactive glass nanoparticle multilayer coatings for the application on prosthetic gadgets [20]. Achievements in cell engineering have supported the move from single quality altering to multiplexed altering and plan of engineered circuits to confer usefulness. However, to understand the exact biomanufacturing of cell treatments, a few moves should be settled. The advancement of productive, exact genome altering strategies that can be conveyed to wanted physical tissues; the deduction of great, quiet inferred iPSCs in a simple, cost effective, and effective way; and the improvement of guidelines, measures, and quality control methods that follow great producing practice are key bottlenecks [21].

6.   Potential for future applications

Profiting by the inborn material properties of inorganic nanoparticles to add good attributes to existing materials or to make totally new multifunctional materials will bring about the design of smart, controllable nano-built materials for clinical application in regenerative medication. Most commonly biomaterials have been obtained from the following biomolecules such as proteins, nucleic acids, carbohydrates and lipids. As of now, 3D and 4D bioprinting advancements are driving the improvements in tissue building and regenerative medication. Bio-inks assume a significant job in these advances to advance cell epitome, connection and separation into utilitarian tissues or organs. As opposed to standard biomaterial innovations, 3D bioprinting innovation is influenced by a few specialized provokes identified with the support of mellow cell conditions, material biocompatibility and supplement with organic constituents like development factors. Settling these issues requests the conjunction of problematic logical ideas and imaginative specialized help from the regions of materials science, building, science, science and prescription. In view of the clinical prerequisites for controllable debasement rate, delayed soundness and incredible biocompatibility, finding and executing new techniques to improve the focusing on, mechanical highlights, corruption pace, or biocompatibility, will upgrade the handiness of metal-based biomaterials for cell or tissue regeneration and treatments. 3D human tissue culture models of disease may permit testing of regenerative prescription methodologies in human science, as differentiated to the creature models at present utilized in preclinical examinations. Expanded exactness of illness models may improve the viability of regenerative prescription procedures and upgrade the interpretation to the center of promising methodologies.

7. Summary and future perspective

The fate of biomaterials and tissue designing is advancing toward finding the perfect biomaterial-cell type blend for tissue recovery. The determination of ideal cell types/mixes is indispensable for the achievement of tissue building, as the heart is a deliberately adjusted milieu of cardiomyocytes, fibroblasts, endothelial cells, and smooth muscle cells. Natural biomaterials are profoundly biocompatible, while engineered biomaterials can be soundly intended for a specific reason. The consolidation of the stem cells into organized and changed biomaterials expands the ability to re-establish and repairing dysfunction tissues. The efficient spatial properties of a biomaterial or scaffold thus can give a defensive and once in a while inducible microenvironment for the stem cell, copying the ordinary extracellular matrix. The stem cell based biomaterials have been effectively used for various tissue regeneration including heart, bone, pancreatic islet, hematopoietic system and nerve tissues. Numerous questions stay to be replied. The significant worry of biomaterial application to cardiovascular malady is the plausibility of making a substrate for arrhythmia. Limiting the immune response to forestall the exemplification of biomaterials or tissue develops is another worry, as embodiment

Materials Research Forum LLC

https://doi.org/10.21741/9781644901076-2

forestalls legitimate combination of transplanted cells with the local condition. The utilization of nanotechnology to make bioengineered scaffolds to intently mimic natural tissues in regenerative prescription has gotten expanded consideration throughout the years. It is preferably ready to plan the bioengineered framework to coordinate with the normal tissue in regards to vital structure, material piece, and surface morphology. Though biomaterials having few limitations, biomaterial and tissue-designing explore holds an extraordinary guarantee and has accomplished a lot of progress inside the previous two decades. A definitive objective is to utilize biomaterial frameworks in combination with proper cell types for incomplete or on the other hand entire organ (re)generation. This will at last decrease the requirement for an organ transplant, right presently irreversible pathologies, also, improve personal satisfaction.

References

[1]   L.P. Datta, S. Manchineella, T. Govindaraju, Biomolecules-derived Biomaterials, Biomaterials, (2019) 119633. https://doi.org/10.1016/j.biomaterials.2019.119633

[2]   H. Xing, H. Lee, L. Luo, T.R. Kyriakides, Extracellular matrix-derived biomaterials in engineering cell function, Biotechnology Advances, (2019) 107421. https://doi.org/10.1016/j.biotechadv.2019.107421

[3]   F. Klingberg, G. Chau, M. Walraven, S. Boo, A. Koehler, M.L. Chow, A.L. Olsen, M. Im, M. Lodyga, R.G. Wells, E.S. White, B. Hinz, The fibronectin ED-A domain enhances recruitment of latent TGF-β-binding protein-1 to the fibroblast matrix, Journal of Cell Science, 131 (2018) 1-12. https://doi.org/10.1242/jcs.201293

[4]   M.C. Moore, A. Van De Walle, J. Chang, C. Juran, P.S. McFetridge, Human perinatal-derived biomaterials, Advanced Healthcare Materials, 6 (2017) 130–135. https://doi.org/10.1002/adhm.201700345

[5]   F.W. Meng, P.F. Slivka, C.L. Dearth, S.F. Badylak, Solubilized extracellular matrix from brain and urinary bladder elicits distinct functional and phenotypic responses in macrophages, Biomaterials 46 (2015) 131–140. https://doi.org/10.1016/j.biomaterials.2014.12.044

[6]   A.R. Santos Jr, V.A. Nascimento, S.C. Genari, C.B. Lombello, Mechanisms of Cell Regeneration — From Differentiation to Maintenance of Cell Phenotype. Cells and Biomaterials in Regenerative Medicine, (2014). https://doi.org/10.5772/59150

[7]   D.S. Adams, A. Masi, M. Levin, H+pump-dependent changes in membrane voltage are an early mechanism necessary and sufficient to induce Xenopus tail regeneration, Development, 134 (2007) 1323-35. https://doi.org/10.1242/dev.02812

[8]   Y. Imokawa, A. Simon, J.P. Brockes, A critical role for thrombin in vertebrate lens regeneration, Philosophical Transactions of the Royal Society of London. Series B, Biological Sciences. Preface, 359 (2004) 765-776. https://doi.org/10.1098/rstb.2004.1467

[9]   L.J. Campbell, C.M. Crews, Molecular and cellular basis of regeneration and tissue repair: wound epidermis formation and function in urodele amphibian limb regeneration, Cellular and Molecular Life Sciences, 65 (2008) 73-79. https://doi.org/10.1007/s00018-007-7433-z

[10] Y. Su, I. Cockerill, Y. Wang, Y.-X. Qin, L. Chang, Y. Zheng, D. Zhu, Zinc-Based Biomaterials for Regeneration and Therapy, Trends in Biotechnology, 37 (2018) 428-441. https://doi.org/10.1016/j.tibtech.2018.10.009

[11] A.L. Facklam, L.R. Volpatti, D.G. Anderson, Biomaterials for personalized cell theraphy, Advanced Materials, (2019) 1902005. https://doi.org/10.1002/adma.201902005

[12] R. Chen, L. Li, L. Feng, Y. Luo, M. Xu, K.W. Leong, R. Yao, Biomaterial-assisted scalable cell production for cell therapy, Biomaterials, 230 (2020) 119627. https://doi.org/10.1016/j.biomaterials.2019.119627

[13] H. Kim, S.L. Kim, Y.H. Choi, Y.H. Ahn, N.S. Hwang, Biomaterials for stem cell therapy for cardiac disease, Advances in Experimental Medicine and Biology, 1064 (2018) 181-193. https://doi.org/10.1007/978-981-13-0445-3_11

[14] Y. Xu, C. Chen, P.B. Hellwarth, X. Bao, Biomaterials for stem cell engineering and biomanufacturing, Bioactive Materials, 4 (2019) 366–379. https://doi.org/10.1016/j.bioactmat.2019.11.002

[15] Z. Cui, B. Yang, R.K. Li, Application of Biomaterials in Cardiac Repair and Regeneration, Engineering, 2 (2016) 141–148. https://doi.org/10.1016/J.ENG.2016.01.028

[16] H. Wang, Y. Li, Y. Zuo, J. Li, S. Ma, L. Cheng, Biocompatibility and osteogenesis of biomimetic nano-hydroxyapatite/polyamide composite scaffolds for bone tissue engineering, Biomaterials, 28 (2007) 3338–3348. https://doi.org/10.1016/j.biomaterials.2007.04.014

[17] D. Chouhan, B.B. Mandal, Silk Biomaterials in Wound Healing and Skin Regeneration Therapeutics: from Bench to Bedside. Acta Biomaterialia. (2019). https://doi.org/10.1016/j.actbio.2019.11.050

Nanohybrids   Materials Research Forum LLC
Materials Research Foundations **87** (2021) 55-68   https://doi.org/10.21741/9781644901076-2

[18] B. Zhang, J. Huang, R. Narayan, Nanostructured biomaterials for regenerative medicine: Clinical perspectives, Nanostructured Biomaterials for Regenerative Medicine, (2020) 47–80. https://doi.org/10.1016/B978-0-08-102594-9.00003-6

[19] A.K. Gaharwar, P.J. Schexnailder, B.P. Kline, G. Schmidt, Assessment of using Laponite cross-linked poly(ethylene oxide) for controlled cell adhesion and mineralization, Acta Biomaterialia, 7 (2011) 568–577. https://doi.org/10.1016/j.actbio.2010.09.015

[20] D.S. Couto, N.M. Alves, J.F. Mano, Nanostructured multilayer coatings combining chitosan with bioactive glass nanoparticles, Journal of Nanoscience Nanotechnology. 9 (2009) 1741–1748. https://doi.org/10.1166/jnn.2009.389

[21] I. Ajioka, Biomaterial-engineering and neurobiological approaches for regenerating   the injured cerebral cortex, Regenerative Therapy, 3 (2016) 63–67. https://doi.org/10.1016/j.reth.2016.02.002

Materials Research Forum LLC
https://doi.org/10.21741/9781644901076-3

Chapter 3

Biosensors Based on Graphene

Shalini Muniandy[1], Chin Wei Lai[1*], Thiruchelvi Pulingam[1], Bey Fen Leo[2]

[1]Affiliations[1]Nanotechnology and Catalysis Research Centre, Institute for Advanced Studies, University of Malaya, 50603 Kuala Lumpur, Malaysia

[2]Central Unit of Advanced Research Imaging, Faculty of Medicine, University of Malaya, 50603 Kuala Lumpur, Malaysia

cwlai@um.edu.my*

Abstract

Today, development of rapid and sensitive methods for direct detection of foodborne pathogens appeared as crucial matter due to their impact on human health. In this manner, graphene-based nanomaterials have received much attention as reliable electrochemical biosensors due to their exceptional combination of intrinsic properties such as high conductivity, stability and biocompatibility. The scope of this chapter is to provide a brief history of the electrochemical biosensors used for the detection of microbial pathogens and recent progress of graphene used in electrochemical biosensors for foodborne pathogens detection.

Keywords

Graphene, Foodborne Pathogens, Electrochemical Biosensors, Foodborne Disease, Bacteria

Contents

Materials Research Forum LLC
https://doi.org/10.21741/9781644901076-3

1.   Introduction

The threat posed by microbial pathogens and the constraints faced by both traditional modern detection techniques to rapidly recognize them underpins the urge for continued development of novel biosensors. The substantial challenge exist in healthcare sectors is difficulty in identifying the diverse group of microbial pathogens with high specificity, selectivity and rapidity [1]. Even though, both traditional and modern detection methods such as biochemical, immunological recognition and polymerase chain reaction (PCR) serves as gold standards in the field of microbiology but these assays are still complex and requires ample time up to days [2]. These techniques leave significant room for improvement and make them incompatible for point-of-care testing which lead to the breakthrough of electrochemical biosensors as a reliable detection tool [3].

Electrochemical biosensors posed distinguished advantages such as low capital cost for equipment, miniaturization capacity and inherent sustainability which uses low amount of solvents and sample volumes [4]. Coupling electrochemical biosensors with biorecognition element and surface modification with nanomaterials greatly improves the analytical performance of the biosensors [5]. In typical biosensing process (refer with: Fig. 1), biorecognition elements binds to targeted pathogen with high affinity and specificity and produces sensitive digital signals easily upon aptamer–analyte-binding events. The interaction between biorecognition and bacterial target through electrochemical transducer relies on configurational change, conformational change or conductivity change which gives remarkable prospects for application in healthcare and food safety [6].

Materials Research Forum LLC
https://doi.org/10.21741/9781644901076-3

New emerging technology in the field of biosensing rapidly explores the adaptation of nanoparticles especially carbonaceous nanomaterials in conjunction with suitable ligands and electrochemical detection modalities to offer rapid, sensitive, and cost-effective detection techniques [7]. Graphene and graphene-based nanocomposites are most popular nanomaterials chosen to develop a functional electrochemical biosensor due to its excellent biocompatibility which improves the bio-receptor immobilization and promising conductivity that increases the signal amplification [8, 9]. Another promising characteristic of graphene is, it allows a label-free amplification of signal resulting from interaction with microbial targets at even below the sub-micromolar levels due to large surface area of the carbon nanomaterial and improved electric conductance [10, 11].

This review specifically focuses on the graphene-based electrochemical biosensors and their application in the detection of microbial pathogens. Some aspects such as, properties of graphene, synthesis methods, and surface functionalization have been discussed. Moreover, the graphene biosensor applications in detection of microbial pathogens were also focused.

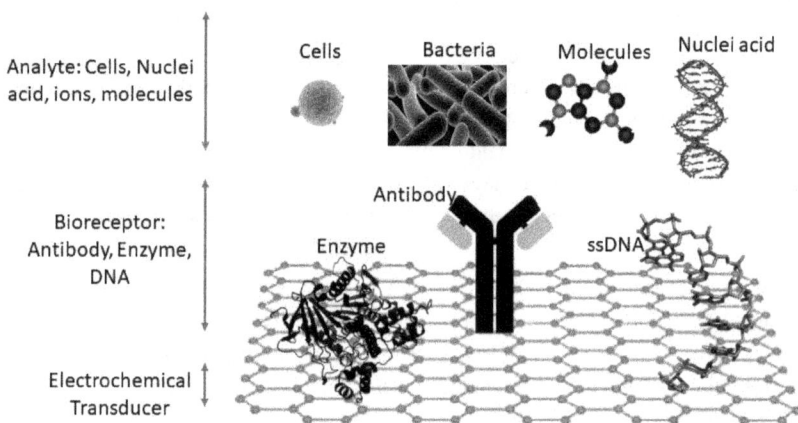

Figure 1. Biosensor components on graphene platform.

Nanohybrids   Materials Research Forum LLC
Materials Research Foundations **87** (2021) 69-102   https://doi.org/10.21741/9781644901076-3

2. Electrochemical biosensors as promising detection tool

Biosensors performs chemical or physical transduction of a biological interaction information to a digital signal. The interaction of target analytes with bio-recognition elements (antibodies, enzymes, nucleic acids, bacteria and viruses) will be measured in real-time. The bio-recognition elements can be immobilized on the surface of transducer using various approaches such as adsorption, entrapment, cross-linking, encapsulation, and covalent binding [12]. Biosensors can be categorized into direct (label-free) and indirect (labelled) detection systems. The physical change induced by target analyte's interaction were measured in real-time and product of biochemical reactions are detected by the sensor in direct and indirect detection system, respectively. Depending on its principle of signal transduction, various types of transducers such as mass-based, optical, and electrochemical sensors were used to change the biological event into a measurable signal (refer with: Table 1) [13, 14].

Among the transducers, electrochemical biosensors are more promising due to their rapid response, highly sensitive, facile, high signal-to-noise ratio, flexibility in employing recognition elements, low interference with food matrixes and instrumental simplicity [15-17]. Electrochemical biosensor is a self-contained integrated device that provides analytical information (quantitative or semi-quantitative) based on the interactions of target analytes with bioreceptors (antigen/antibody, enzymes, DNA, bacteriophages, cell structure or cells, and biomimetic bases) that in contact with electrochemical transducer [18]. The resulting electrical signal is related to the recognition process by bioreceptors and target analyte, and proportional to the analyte concentration. With the advancement of technology and miniaturization of device, electrochemical biosensor offers excellent tool for identifying and quantifying foodborne pathogens present in various food matrices with very low detection limit which is about 10^{-14} M or 10 cfu/mL [19, 20]. Electrochemical biosensors are categorized into amperometric, potentiometric, impedimetric and conductometric referring to measurable parameters such as current, potential or charge accumulation, impedence and conductivity respectively.

2.1 Amperometric biosensors

Amperometric biosensor is a well exploited electrochemical method for foodborne detection due to its superior sensitivity compared to potentiometric analysis [21]. The biological recognition element for application in amperometric food analysis varies from enzymes, antigen/antibody, bacteriophages, cellular structures/cells, nucleic acids/DNA and biomimetic bases [22, 23]. The complete system of amperometric biosensor consists of three electrodes namely working electrode, reference electrode (which controls the potential of working electrode) and an auxiliary electrode which helps to measure the current flow [24]. The working principle of this system mainly involves the measurement of variations in current results from biochemical reaction that involves oxidation and reduction of electroactive species. Usually, the current is recorded at a constant potential that facilitates the electron transfer reactions. In detail, the rate of electron transfer is measured when a molecule undergoes a redox reaction at a constant electric potential applied to the working electrode. The electrons could be transferred either form the analyte to the working electrode or from the electrode to the analyte depending positive and negative potential changes during oxidation and reduction respectively [25]. Moreover, the rate of electron flow (current) is greatly influenced by the concentration of target analyte diffusing to the surface of the working electrode. Many researchers have reported amperometric detection for various foodborne pathogens such as *Escherichia coli* O157:H7 [26], *Salmonella* [27], *Listeria monocytogenes* [28] and *Camplybacter jejuni* [29].

The bioreceptor bacteriophages can be coupled with amperometric transduction by causing rupture of cell membranes and release of intracellular enzymes from the infected bacterial species. In *E. coli*, the released β-D-galactosidase (enzyme) after cell-lysis was converted into p-aminophenol (product) with the aid of p-aminophenyl-β-D-galactopyranoside (substrate) that added externally [30]. The oxidation of the enzymes causes a change in electric current produced as a function of time at the carbon electrode and recorded in a potentiostat device. This approach enabled the detection of *E. coli* at concentrations as low as 1 cfu/100mL within 6 to 8 h and the overall sensitivity of the assay was increased with filtration and pre-incubation of the sample prior to phage infection. Similarly, Benhar et al. reported detection of *L. monocytogenes* after phage lysis (*Listeria scFvs*) using Horseradish Peroxidase conjugated to anti-M13 monoclonal antibody within 5 min and achieved detection limit of 500 cfu/mL [31].

Besides that, many researchers utilized enzyme-linked amperometric immunosensor (bioaffinity reaction) for the detection of pathogens. Antibodies are immobilized on the surface of working electrode, hence the binding of the analyte to the targeted antibody generates current signal which is quantified using labelled enzymes. Brooks et al. have reported a rapid alkaline phosphatase enzyme-linked amperometric immunosensor detection technique for *Salmonella* present in food samples within 4 hours with detection limit of 10^5 cfu/mL [32]. Croci et al. developed a sandwich ELISA assay for the detection of *Salmonella* in meat samples with detection limit of 1–10 cfu/25 g through the immobilization of horseradish peroxidase labelled monoclonal and polyclonal antibodies directly on the surface of the electrode. The 3,3′,5,5′ tetramethylbenzidine was used as substrate to electrochemically measure the binding signal [33]. Similarly, an amperometric enzyme immunosensor was developed by Zhao and Liu using monoclonal antibodies with better detection limit at levels of 0.1 cfu/g and 0.1 cfu/mL for *E. coli* present in poultry meat and pasteurized milk [34]. Besides that, the amperometric enzyme immunosensor also can be coupled with immunomagnetic separation with magnetic nanoparticle-antibody conjugates for better detection sensitivity and specificity. This approach was evaluated for the detection of *E. coli* O157:H7 in ground beef samples. The detection limits in both enriched and non-enriched samples showed promising outcome with the detection limit of 8.0×10^0 cfu/mL and 1.6×10^1 cfu/mL respectively within 15 min to 6 hours of assay time [35]. Enzyme-based amperometric biosensors have the advantage of being highly sensitive, rapid and inexpensive however it lacks of reproducibility due to the loss of enzyme activity over time [36]. Moreover, the amperometric techniques can also be coupled to DNA-based detection (genosensor). It involves the immobilization of an oligonucleotide probe on an amperometric transducer and upon the binding its complementary sequence present in the analyte solution, a measurable current signal is produced. The labelled DNA biosensors offers more sensitive and selective detection. F. Farabullini et al. developed a gold based working electrode modified with thiolated oligonucleotides that specific to *Salmonella, L. monocytogenes*, and *E. coli* [37]. Upon the binding to the target, this complex will be labeled intrinsically with biotinylated signaling probes which can be coupled with streptavidin–alkaline phosphatase conjugate for the detection purpose using α-naphthyl phosphate solution. This genosensor had successfully detected DNA samples from different bacteria at the nanomolar level without any cross-interference within 1 hour of assay time. This approach was also used for the detection of food borne hazard, *Aeromonas hydrophila* as reported by Tichoniuk et al. [38]. They proposed a self-assembled monolayer (SAM) consisting of thiolated single-stranded DNA probe selective to the targeted bacterial DNA present at the concentration of 2.5 µg cm^{-3} and methylene blue (MB) was used as electroactive indicator for the hybridization events. Besides methylene blue, horseradish peroxidase (HRP) also can be coupled with genosensor (*lol*B gene probe) developed for amperometric detection of *Vibrio cholera* genomic DNA with 100% specificity and sensitivity of 0.85 ng/uL [39]. DNA-based biosensors are highly sensitive for the detection of pathogens present at low concentrations however, the overall long assay time and high complexity in labelling process are the disadvantages of this type of biosensors [40, 41].

2.2   Impedimetric and conductometric biosensors

Impedimetric and conductometric biosensor are the most powerful electrochemical analytical systems for the detection of foodborne pathogens. The impedimetric detection system measures the conductance, capacitance and impedance of the analyte-containing medium as well as changes in the electron transfer properties on the electrode surface. In

Nanohybrids   Materials Research Forum LLC
Materials Research Foundations **87** (2021) 69-102   https://doi.org/10.21741/9781644901076-3

a technique called electrochemical impedance spectroscopy (EIS), an alternating current (AC) is applied across the electrodes. This causes electron buildup and discharge processes to occur at the electrode surface and in the electrolyte. By changing the frequency of the sinusoidal current, the electron-transfer and capacitance properties of the electrodes interacting with the analyte can be studied. During a bio-recognition event, conductance and capacitance increases while impedance decreases across a working electrode surface [42]. The increase and decrease in both conductance and impedance are depends on the analyte concentration and the changes in voltage allows the evolution of interfacial capacitance, conductance and resistance. These techniques are important for a rapid and automated detection system which allows the miniaturization of the device. Moreover, the measurement of changes in the electrical properties of bacterial cells makes the whole-cell as an attractive analytes for detection [43]. These techniques are mostly applicable to measure the bio-recognition involving DNA/nuclei acid hybridization, antibody- antigen reactions and enzyme reactions [44, 45]. Several analytical devices was developed based on conductance and impedance technology for food-borne pathogen detection however, impedimetric measurement is widely studied for culture-independent bacterial detection due to its high sensitivity regardless of sample matrix, label free, cost effective and high-throughput device [46]. The applications of conductometric biosensors are limited in food analysis due to poor sensitivity caused by variable ionic background of real food samples and requires high ionic strength media for better detection [18, 47]. Moreover, the selectivity of conductometric biosensors is relatively poor because all charge carriers could result in the change of conductivity [48].

Many researchers have reported the use of conductometric analysis with different types of biorecognition element for various foodborne pathogens detection [49-51]. Mostly, conductometric biosensor is associated with enzymatic reaction where the ionic strength is capable to produce electron flow. A conductometric sandwich immunosensor using polyaniline conductive polymers as transducer and polyclonal antibodies of *E. coli* O157:H7 and *Salmonella* spp. as biorecognition element was developed for foodborne pathogens detection [49]. They reported a specific immunosensor that target desired bacteria even in the presence of other non-targets with a lower detection limit averagely 7.9×10^1 cfu/mL within 10 min assay time after sample application. Based on this study, the further development in realm of food analysis was done by the same group of researchers in the later years. They designed a portable, sensitive and rapid electrochemical sandwich immunoassay biosensor for the detection of *E. coli* O157:H7 present in 9 different samples such as lettuce, alfalfa sprouts, and strawberries with detection limit of 81 cfu/mL in 6 min of assay time [50]. This novel approach was sensitive and specific enough for the analysis of food samples and it can be further

Materials Research Forum LLC
https://doi.org/10.21741/9781644901076-3

optimized and adapted into biomedical and bio-defense fields alternating the bulky and expensive biosensors. Pal et al. implemented this similar approach for the development of polyaniline functionalized direct-charge transfer conductometric immunosensor for the detection of *B. cereus* in food samples [51]. The conductive polyaniline was tagged with primary anti B. cereus antibodies for the direct capture of targeted analyte present in food sample without the need of pre-enrichment step and followed by conjugation with secondary antibodies functionalized on transducer surface. The detection limit of this biosensors within 6 minutes of assay time in food matrices ranges from 35–88 cfu/mL with strong discrimination between other types of bacteria (*Bacillus megaterium* and E. *coli*). Besides whole cell detection, Chen [52] have also fabricated a high accuracy, reproducible and stable conductometric immunosensors for the detection of bacterial toxin, Staphylococcal enterotoxin B (SEB). He reported the use of nanogold/chitosan-MWCNT immobilized with horseradish peroxidase (HRP)-labeled SEB antibody (HRP-anti-SEB) and the immunoreaction between this antibody-antigen results in measurement of conductometric signal with detection of toxin ranges from 0.5 to 83.5 ng/mL. When this sensor was adapted in complex food matrix, SEB of different concentration gave an average of 116% recovery of the toxin in milk samples.

In recent years, integration of impedance with biological recognition technology has received increasing attention for the application of foodborne pathogens detection. The change in impedance is results from the amount of the analyte that interacts with the bioreceptors. This interaction will cause a change in capacitance and electron transfer resistance on the surface of the working electrode, and the impendence increases as the amount of analyte binding to the bioreceptors increases [53]. Mostly, impedimetric sensors uses antibodies as bioreceptors however, whole bacterial cells is also possible to detect using this method which enable a direct detection in food analysis [44]. EIS is the most used impedimetric biosensors technique because it can monitor the physico-chemical changes results from interaction of the analyte with the bioreceptor immobilized on the transducer. These includes the information on charge transfer from the solution to the electrode surface, resistance of the solution and also diffusion rate of species to and from the bulk solution and also its formation of double layer capacitance [54]. The overall sensitivity of the impedimetric biosensor can be improved using interdigitated array microelectrode and these also allows the miniaturization of the impedimetric biosensors into a chip format. Some researchers have actively developed such platforms for rapid detection of foodborne pathogens. For instance, Varsheny and Li has developed an antibody-based interdigitated array microelectrode impedimetric biosensor for the detection of *E. coli* O157:H7 in ground beef samples within 35 minutes of assay time [55]. Moreover, the use of magnetic nanoparticle–antibody conjugates (MNAC) has improved its overall sensitivity of detection by 35% with detection limit 7.4×10^4 and 8.0×10^5 cfu/mL in pure culture and ground beef samples respectively. Yang et al. also developed similar interdigitated microelectrodes for the detection of viable *Salmonella* cells in pure culture and milk samples at different frequencies with detection limit of 0.5 cfu/mL after 10 hours of detection time [56]. The impedance reading was measured against bacterial growth time and most changes in impedance was observed at low frequency because of the increase in the double-layer capacitance.

In impedimetric biosensor, bacteriophage has also been used as cross linker between bacteria and the working electrode. Shabani et al. developed an impedimetric screen-printed carbon electrode array biosensor based bacteriophage for the detection of *E. coli* bacteria with limit of detection of approximately 10^4 cfu/mL [57]. The overall impedance in this experiment decreases as the bacteria concentration increases due to release of ionic intercellular content results from penetration of phages. Similarly, Shabani et al. also showed a sensitive detection of *E. coli* by immobilizing T4 bacteriophage coated Dynabeads on screen-printed carbon-based electrode with limit of detection 10^3 cfu/mL in milk samples [58]. They used fluorescence and flow cytometry techniques to monitor the successful binding of phages with bacteria. Besides that, enzyme-based impedimetric sensors have been developed with increasing sensitivity of sensor devices. Ruan et al. have developed an alkaline phosphatase labelled anti-*E. coli* O157:H7 antibody immobilized on indium tin oxide (ITO) electrode chips for the detection of *E. coli* with detection limit of 6×10^3 cfu/mL [59]. However, its detection limit is very high as compared to the traditional ELISA. More studies for the impedimetric detection of whole bacterial cells has emerged in recent years. Mostly the data reported is based on the change in impedance upon the analyte binding. Mantzila et al. have reported the use of faradic impedimetric immunosensor for the first time to detect *S*. Typhimurium in milk with detection limit of 10^2 cfu/mL [60]. Wang et al. also successfully developed a cost effective and robust gold nanoparticle-graphene based impedimetric immunosensor for rapid and sensitive detection of *E. coli* O157:H7 with lower detection limits of 10^4 cfu/mL and 10^3 cfu/mL in contaminated ground beef and cucumber samples, respectively with excellent specificity [61]. The use of this novel electrode material system showed high potential to be adapted in different food matrixes. Besides that, Labib et al. has developed an DNA aptasensor using aptamer designed to capture the whole-cell *S*. Typhimurium with detection limit of 600 cfu/mL with high specificity and selectivity [62].

2.3 Potentiomeric biosensors

Potentiometric is another types of electrochemical biosensor which measures the changes in pH or ion concentration occurs during interaction of analyte and working electrode. One of the advantage of potentiometric sensor is the measurement of the interaction between target analyte and biorecognition element does not involve generation of redox products however, it is still lack of overall sensitivity and prone to interferents from the sample matrix is the main disadvantage of this system [63]. Zelada-Guillén et al. developed a real-time label-free aptamer-based potentiometric biosensor for the *E. coli* O157:H7 in milk and apple juice [64]. He uses selective apatmer that immobilized on a SWCNT which acts as an ion-to-electron transducers and the interaction with targeted analyte changes the electrical potential. The detection limit of this biosensor in complex food matrix is 6 cfu/mL and 26 cfu/mL in milk apple juice respectively. Not many potentiometric biosensors were studied for the detection of foodborne pathogens, but many LAPS (light-addressable potentiometric sensor) was reported on past years. A LAPS-based immunosensor for the detection of *S.* Typhimurium has been described by Dill and co-workers [65]. This system uses biotinylated and fluorescein-labelled anti-Salmonella antibodies immobilized on a silicon chips for detection of *Salmonella* in chicken carcass with detection limit of 119 cfu/mL with 90% recovery rate from food matrix. Ercole et al. also developed potentiometric alternating biosensing based LAPS for the detection of *E. coli* cells in vegetables. This system detects the pH variations due to NH$_3$ production by a *urease-E. coli* antibody conjugate with detection limit of 10 cfu/mL within 1.5 hours of assay time [66].

Table 1. Different modes of electrochemical based foodborne pathogen detection.

Detection Technique	Analyte	Biorecognition element	Assay time	Detection limit and source	Ref
Amperometric	*L. monocytogenes*	Enzyme immunoassay	<1 hour	1.07 X10^2 CFU/mL in milk samples	[67]
Amperometric	*E. coli* O157:H7	Enzyme immunoassay	45 min	10^2 in milk sample	[68]
Amperometric	*E. coli, L. monocytogenes* and *C. jejuni*	Enzyme immunoassay	30 min	50, 10, and 50 CFU/mL in chicken extracts	[28]
Amperometric	*S.* Typhimurium	Enzyme immunoassay	2.5 h	1.09 × 10^3 in chicken carcass	[27]

Ampreometric	*Salmonella*	DNA	3 hours	0.04 ng/µL of genomic DNA	[69]
Conductometric	*E. coli*	T4 Bacteriophage	Not stated	10^2 CFU/mL of pure culture	[70]
Conductometric	*E. coli O17.H7*	Antibody-antigen	A few mins.	5 to 100 ng/mL in baby food	[71]
Conductometric	*E. coli*	Antibody-antigen	Not stated	10 CFU/mL of pure culture	[72]
Impedmetric	*L. monocytogenes*	Antibody-antigen	Not stated	4 CFU/mL in tomato extract	[73]
Impedmetric	*E. coli O157:H7*	Antibody-antigen	Not stated	10^2 cfu/mL in spiked milk, ground beef, and spinach samples.	[74]
Impedmetric	*Listeria innocua*	Bacteriophage	Not stated	10^5 CFU/mL in milk sample	[75]
Impedmetric	*Salmonella enteritidis*	DNA	Not stated	100 ng/mL in liquid food matrix	[76]
Impedmetric	*E. coli O157:H7*	Antibody-antigen	5 min	3 log CFU/mL in meat sample	[77]
Potentiometric	*E. coli O157:H7*	Antibody-antigen	30 min	10^1 to 10^2 CFU/mL in apple juice and beef meat.	[78]
Potentiometric	*E. coli O157:H7*	Antibody-antigen	30 min	10^3 CFU/mL in saline	[79]
Potentiometric	*E. coli O157:H7*	Antibody-antigen	1.5 h	10 CFU/mL in drinking water	[80]

3. Graphene properties

Graphene is a nanomaterial with a thickness of one atom and these planar sheets of carbon atoms are tightly packed in a honeycomb lattice crystal structure. This material has become the electronic material of next generation due to its physical characteristics including high thermal and current density, ballistic transport and nanometer-scaled

Nanohybrids   Materials Research Forum LLC
Materials Research Foundations **87** (2021) 69-102   https://doi.org/10.21741/9781644901076-3

hydrophobicity [81]. Graphene was initially extracted from graphite through the micromechanical cleavage method which accommodated simple production of good quality crystallites form of graphene [82]. This nanomaterial can be classified into the metallic or semi-conductor category based on its configuration; zigzag or armchair. Zigzag formation leads to graphene material with metallic property while the armchair configuration leads to either metallic or semi-conductor properties [83, 84]. Extraordinary electrical physicochemical property of graphene has led to its use such as in ballistic transistors and conducting sensors and electrodes. Increased electron mobility of graphene and minimal noise have enhanced the use of this nanomaterial as sensors. The 2-dimensional structure of graphene which are exposed to the external environment makes it an excellent candidate for detection of molecules that are being absorbed onto its surface [85].

3.1   Graphene Synthesis

Graphene sheets are known to exist in layers of single sheet, bi-layer and also few layers. The single layered graphene contains 2-D sheet of carbon in hexagonal sheets and the layer increases respective to the number of layers [86]. As graphene are known to contain hybridized sp^2 bonding where the Ơ bonds plays the role of backbone while the π bonds regulate interactions between the different layers of graphene. As all variations of graphene has uses in diverse fields, the synthesis methods of graphene play a vital role in determining the final physicochemical properties of graphene [87]. The discovery of single layered graphene was credited to Novoselov *et al.* in the year of 2004 where the group managed to discover a reproducible method of synthesizing graphene sheets consistently through the exfoliation technique [82]. Since then, several methods of synthesizing graphene have been described in the literature for the mass production of this nanomaterial.

3.2   Top-down Approach

A. Exfoliation

As graphite contained many layers of graphene sheets that are held together by weaker van der Waals forces, hence it may be possible to generate graphene sheets if the weak bond were to be broken in principle [88]. The exfoliation method uses energy either from mechanical or chemical-based source to break the interlayer bonds to produce single graphene sheets. This method employs repeated peelings of commercially available pyrolytic graphite (HOPG) which are in the thickness range of 1 mm. Scotch tape was used to finally peel off single layers of graphene from the original graphite sheets [82]. This mechanical exfoliation method is an example of top-down approach where graphene sheets are formed by taking apart the layers of graphite. As this was found to be a reliable and simple technique, this method was soon employed and modified by other researchers to produce graphene at a higher yield through manipulation of the bonding between HOPG and Si substrate [89, 90]. Exfoliation of graphite in liquid phase was conducted through dispersion of pure graphite in N-methyl-pyrrolidone solvent which managed to produce up to 12 wt% of monolayer graphene sheets. The advantage of this method is that the solvent-graphene interaction countered the energy needed to exfoliate graphite into mono-layered graphene sheets as the solvent had comparable surface energy just as the graphene. This exfoliation process shows excellent potential to mass produce graphene sheets including the liquid-phase exfoliation techniques as well [91]. Although this method has great benefits, the large number of defects due to oxidation and reduction routes highly affects the electrical property of graphene. Therefore, improvisations on this method are greatly needed to be focused on regulating the graphene layers as to be used in the industrial sector [92].

B. Sonication

Graphene synthesis through this method uses ultrasonic energy to separate layers of graphene in a precursor [91]. Graphene obtained through this method are mainly used in the sensors, polymer fillers and in transparent electrodes. Solvents such as N-methyl pyrrolidone are used mainly for the incorporation of graphene in composites for drop casting, vacuum filtration and spray coating. Similarly, solvent-aided sonication is a modification of sonication method to produce monolayers of graphene through centrifugation step. Majority of graphene produced through this method are in the thickness range of 1 nm with an electrical conductivity of 5000 $S \cdot m^{-1}$ [93]. However, to prevent the restacking of the graphene sheets after the sonication process due to the van der Waals forces, dispersing agents may be used in the solution before the sonication process itself. Additionally, this method offers graphene sheets from graphite without the use of chemicals [94]. The disadvantage of this method is that this process requires an increased amount of energy for large yield of graphene sheets and furthermore, removal of impurities from the liquid graphene solution can pose a challenge [95].

C. Reduction of graphite oxide

Graphene nanoflakes or powder can be produced through reduction of graphite oxide using chemicals. This is a preferred method as it produces lower number of graphite exfoliation. Although graphite oxide may be sonicated in water to produce graphene sheets, the resulting product has a characteristic of electrically insulating. Therefore, this method can be used to restore the conductive property of graphene [96]. Thermal or chemical reduction of graphite oxide can be performed to produce reduced graphene oxide (rGO) sheets that are strongly bonded with a tensile modulus ranging from 32 GPa [97]. While this method offers synthesis of graphene at low temperatures, but increased surface defects and low purity have been noted for the graphene sheets [92].

3.3   Bottom-up approach

A. Chemical vapor deposition

Chemical vapour deposition (CVD) through thermal routes to synthesize graphene was first found in 2006 where camphor was used in synthesizing graphene on foils made of Ni [98]. Initially, camphor was evaporated at the temperature of 180 °C then pyrolyzed in a CVD furnace in the range of 700 to 850 °C with argon gas as the carrier. Once cooling process was over, graphene synthesized through this method were found to contain up to 35 layers. Similarly, Yu *et al.* noted that 3 to 4 layers of graphene were formed on polycrystalline Ni foils where mixtures of CH_4, Ar and H_2 gas were used as the precursors [99]. Characterizations of the formed graphene sheets revealed that the graphene were only able to form under adequate cooling rates on the Ni compared to increased or reduced cooling temperatures as these temperatures were unfavorable for the formation of graphene. The authors noted that the dissimilarity in the formation of graphene under different cooling rates were attributable to carbon solubility in Ni and segregation kinetics of the carbon itself. This report provided crucial information on the growth of graphene sheets under CVD process where the carbon atoms have adequate time to grow on Ni when subjected to moderate cooling rates [99]. Further advancement of the CVD process paved way for synthesis of graphene sheets on Cu foil where the graphene sheets produced through this method are of high uniformity and quality. However, the growth of graphene on Cu was noted to be self-limiting due to the minimal solubility of C in Cu compared to the Ni [100]. Recent developments in this method have provided reproducibility of graphene of good quality and this have created the venture of graphene sheets for use in flexible and photovoltaic electronic applications. Overall, graphene synthesized through CVD method enabled the use of this nanomaterial in a variety of applications including solar cells, flexible OLED, smart windows and touch screens. However, the limitation of this method is that CVD is costly especially when scaled up for industrial use [88].

B. Epitaxial growth on SiC

Growth of graphene on SiC substrate is one of the popular techniques of synthesizing graphene. The sheets are formed when the surface containing H_2 of 6H-SiC were subjected to short period of heating at high temperatures of 1250 until 1450 °C. The graphene sheets that were epitaxially grown on the SiC are noted to have 1 to 3 layers of graphene where the number of layers are dependent on the temperature that is being used for the decomposition process [101]. Another article reported that it was possible to generate graphene films that are as thick as one atom through the use of this method where the advancement in this method have gained interest of the semiconductor industry. Continuous films of graphene were able to generated through this method where low temperature of 750 °C are used in developing graphene on SiC coated Ni thin film [102]. This method has become favorable as it produces graphene with large area such as wafer-size films for industrial applications. However, graphene produced through this method are inclined to have fragile anti-localization and additionally the SiC substrates are costly and only generates a low yield [103].

C. Growth of large-area graphene films from metal-carbon melts

This method uses graphite as a carbon source, where it is placed in contact with the transition metals and subjected to heating process at high temperatures that are suitable to melt the metal that is being used. Once the increase in the temperature induces the carbon to start dissolving into the molten metal, the temperature is then reduced to accommodate the precipitation of carbon [104]. This precipitate is then paving way for a variety of carbon types including single and few-layered graphene. Nickel has been the metal of choice in this method as it is not Raman active. Additionally, copper too has been used as the metal substrate for graphene synthesis through this method [105].

4.   Graphene nanomaterials transducer as electrochemical biosensor

Graphene and graphene-based materials are widely exploited in the field of electrochemical biosensing due to the outstanding properties such as, high charge mobility, large surface area, and ease of surface functionalization [11]. Graphene-based materials are used as transducers in electrochemical biosensors which converts the interaction of bioreceptors immobilized on its surface and target molecules into readable signal. The semi-metal graphene is a promising material for fabrication of electronic sensors that poses high carrier mobility ($15,000$ cm^2 V^{-1} s^{-1}) [106]. The bioreceptors often uses EDC/NHS chemistry and physisorption (refer with: Fig. 2) [8]. Graphene is widely engineered for targeting wide range of biomolecules. The method of graphene synthesis influences its immobilized interaction with target biomolecules. In biomedical field, the oxide-free pristine graphene is a promising candidate for biosensing because it offers high electrostatic force and infinite structure at molecular level [107]. Moreover, pristine graphene also allows stable π-π stacking, non-covalent interactions and provides abundant active sites for charge-biomolecular interactions leading to enhanced sensing properties and improved selectivity [108].

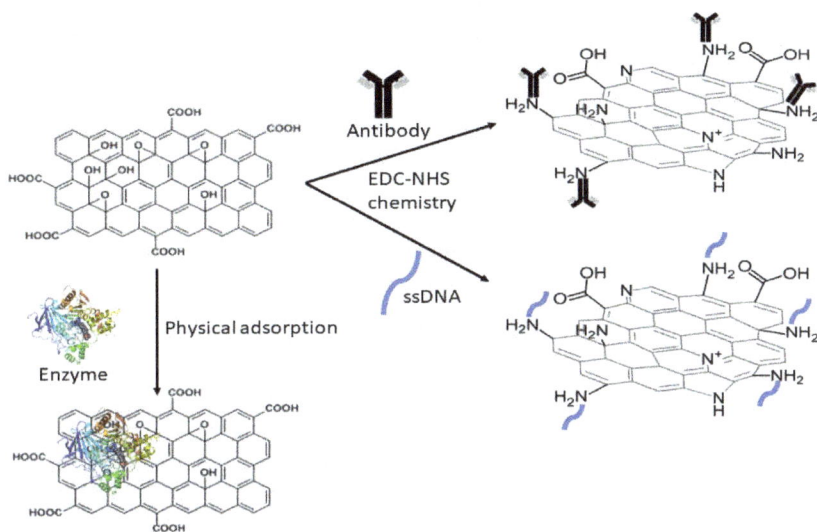

Figure 2. Schematic diagram of the attachment of bioreceptors on graphene surface.

Pure graphene can be functionalized with any charged molecules or metal ions at its charged area and vacancy defects [107]. The functionalized graphene allows binding of nanoparticles, heteroatoms, quantum dots, proteins, DNA, enzymes, antigens, antibodies, and other molecules [109, 110]. Functionalized graphene surface also enables direct detection of biomolecules due to the presence of hydroxyl, carboxyl, and epoxide groups. High C/O ratio of graphene-based materials promotes stronger interactions with biomolecules for a more sensitive detection [109]. For instance, the conjugated structure of graphene facilitates high electron transfer resulting from charge-biomolecular interactions that generates enhanced electrochemical signal. Besides unique electronic and adsorption properties, graphene also act as quencher in the transducer to generate fluorescent biosensors [111].

One of the important considerations for detection of foodborne pathogens is the limit of detection of the target molecules. The quality of graphene materials may vary from batch-to batch that lead to different properties and functionalities in the biosensors [112]. For example, the functional groups, number of layers and oxidation states of graphene will

Nanohybrids   Materials Research Forum LLC
Materials Research Foundations **87** (2021) 69-102   https://doi.org/10.21741/9781644901076-3

affect the sensing performance and interaction between the transducer and bioreceptor
[113]. Moreover, the orientation of graphene-based materials and the bioreceptors greatly
influences the selectivity and sensitivity of the biosensors. Correct orientation prevents
non-specific binding of biomolecules to the target molecules [114]. By taking into
consideration these limitations, graphene based electrochemical biosensors enable rapid
bacterial detection as compared to conventional methods potentially resulting in advances
in healthcare and diagnosis.

4.1   Graphene based electrochemical biosensors for foodborne pathogen detections

Graphene in electrochemical sensors is capable to increase the sensitivity of detection and
LOD as well as the overall performance of the sensors to to the increased rate of electron
transfer at the surface of the transducers. Recently, considerable attention has been given
for the rapid detection of foodborne pathogens due to the progressive foodborne disease
outbreaks worldwide. A wide range of graphene-based electrochemical biosensor such as
antibody, DNA and whole-cell based detection strategies have been explored for
foodborne pathogen detections (refer with: Table 2).

The versatile functional groups on graphene and modified-graphene nanomaterials allows
specific immobilization of antibody attachment for immunosensing of foodborne
pathogens [115]. The strategies of immobilization include EDC/NHS chemistry reaction,
electrostatic bonding, or using 1-pyrenebutanoic acid succinimidyl ester (PASE) linker
[114]. The adaption of graphene-based electrochemical immunosensing food foodborne
pathogen detection is popular in the field of health. Pandey et al. [116], have
experimented the electrochemical properties of graphene wrapped copper (II)-cysteine
complex for a label-free ultrasensitive electrochemical immunosensor of *E. coli* O157:
H7. Graphene modified with copper and cystine exhibited high electron transfer rate
constant (1.82×10^{-6} cm/s) and increases the overall surface area which led to to the
lower detection limit of 3.8 cfu/mL. The proposed method also showed high selectivity to
only pathogenic *E. coli* O157: H7. Moreover, a novel approach of using graphene-based
electrochemical biosensor that generates non-Faradaic impedance electrochemical signals
specific to target bacteria without using any redox molecules/mediators were reported by
Pandey et al. [117]. They explored the effect of number of graphene layers on the sensor
performance. The graphene monolayer and graphene multilayer were used to detect *E.
coli* O157:H7 by immobilizing anti-E. coli antibody using PASE in methanol solution.
The study found that, number of layers affects the charge carrier mobility, sensitivity and
biocompatibility of the sensors. Single layer graphene has better electron mobility due to

85

fewer number of defects. A lower detection limit of 10^1 cfu/mL and 10^2 cfu/mL were achieved for both mono-layer and multi-layer graphene, respectively.

Besides immunosensing, DNA based electrochemical biosensors using graphene as transducer also being widely explored. A facile and sensitive graphene oxide (GO)/chitosan electrochemical DNA biosensor was reported by Xu et al. [118]. In this study, the glassy carbon electrode (GCE) was modified with GO/chitosan for the detection of *E. coli* O157:H7 electrochemical measurements. The cyclic voltammetry was used to study the electrochemical properties of GO/chitosan electrode which showed excellent electron transfer ability. Moreover, electrochemical impedance was used to study the hybridization of ssDNA biorecognition molecule and target DNA of *E. coli.* The results showed good specificity and selectivity of the sensor with detection limit of 3.584×10^{-15} M. Besides DNA targets, whole-cell bacterial detection was made possible by using DNA aptamer which created a breakthrough in biosesning field [119]. Aptamer specifically interact with specific elements present on bacterial cell membrane and lead to novel detection of whole-cell bacteria without the need for target DNA extraction [120]. Muniandy et al. [121] recently reported a novel label-free whole-cell Salmonella Typhimurium aptasensor using a reduced GO-Titania oxide nanocomposite as the sensing platform. The ssDNA aptamer specific to the outer membrane protein of *S.* Typhimurium was immobilized on the reduced GO-titanium oxide coated GCE via covalent binding wit as]id of surface chemistry of the modified graphene. Coupling graphene with metal oxides greatly enhances the electronic properties and surface area of the graphene nanocomposite. This sensor achieved a detection limit of 10^1 cfu/mL in both pure cultures and contaminated meat samples. Similarly, Dinshaw et al. [122] reported another novel combination of graphene nanocomposite with is reduced GO-chitosan (rGO-CHI) composite as the conductive substrate for the sensing platform. The presence of chitosan increases the overall biocompatibility and electron transfer rate of the graphene. This nanocomposite was further coated with glutaraldehyde to aid the immobilization of thiolated aptamer molecule that directly binds to the Salmonella cells. The aptasensor was evaluated in comparison to a conventional PCR amplification assay and an enhanced detection limit of 10^1 cfu/mL was observed for the reduced GO-chitosan aptasensor as compared to a 10^2 cfu/mL detection limit for the PCR assay.

Table 2. Examples of some graphene-based electrochemical biosensors for foodborne pathogen detection.

Sensing Platform	Detection mode	Target analyte	Biorecognition elemenent	Assay time	Detection limit	
GO-gold nanoparticle	Electrochemical-impedimetry	*Salmonella*	Thiolatad ssDNA aptamer	Within an hour	3×10^0 cfu/mL	[
rGO–gold nanoparticle/ionic liquid	Electrochemical-Voltametry	*Enterobacter sakazakii*	HRP-anti-*E. sakazakii*	2 hours	1.19×10^2 cfu/mL	[
rGO-carbon nanotube	Electrochemical-impedimetry	*Salmonella* ATCC 50761	Amino-modified ssDNA aptamer	1 hour	2.5×10^1 cfu/mL	[
GO-silver nanoparticles	Electrochemical-Voltametry	*S.* Typhimurium	anti-*S.* Typhi	Not stated	1.0×10^1 cfu/mL	[
rGO-GO	Electrochemical-impedimetry	*S.* Typhimurium	Anti-outer membrane protein	3h	10 cfu/mL in fruit juices	[
Holey rGO	Electrochemical field-effect transistor	*E. coli*	Magainin antimirocbial peptide	30 min	8.0×10^1 cfu/mL	[
GO & rGO	Electrochemical-Potentiometric	*Staphylococcus aureus*	Aptamer	30 min	1.0×10^0 cfu/mL	[
rGO-copper (II) assisted cysteine	Electrochemical-impedimetry	*Staph. aureus*	Anti-*Staph. aureus*	2 hours	4.4 cfu/mL	[
GO-silver	Electrochemical	Sulfur reducing bacteria (SRB)	anti-SRB	Not stated	50 cfu/mL	[

Conclusion and Outlook

In this chapter, we have reported the synthesis, fabrication and recent studies of graphene and graphene-based materials with possible applications in foodborne pathogen detections. We have summarized the reported analytical performance of graphene-based sensors. The utilization of different biorecognition elements such as antibody, DNA, and enzymes with their advantages and disadvantages were also discussed. Overall, the type of sensor selected will depend on the type of application.

The development of graphene based electrochemical biosensors is vital for rapid and sensitive detections for foodborne pathogens in healthcare field. Despite the excellent sensitivity and specificity of the traditional methods of detecting bioagents involves tedious process and multi-step procedures that limits their execution at point -of-care. Moreover, the detection of low concentration targets in complex biological media can be challenging. Graphene is a most commonly exploited 2D material with outstanding physio-chemical properties such as large surface area, zero-bandgap semiconductor, high tensile strength, biocompatible, and ultra-high charge mobility. The synthesis method of graphene and graphene -based materials is important in the electrochemistry of the biosensing applications. A well-maintained graphene property throughout the synthesis and fabrication process preserves its electronic properties, biocompatibility and increases the active sites for biomolecule immobilization and recognition. The integration of graphene in electronic biosensors giving an ultra-high sensitivity and rapid detection of foodborne pathogens which is promising in healthcare settings. Besides detection limit, the economical and facile approach for sensor design and fabrication is another important element in the field of biosensing.

Although graphene is an excellent material in the field of biosensing for foodborne pathogen detection, better understanding of the physics and chemistry at the surface of graphene is crucial to ensure proper orientation of biorecognition elements which gives highly sensitive detection. In addition, cost-effective and reproducible production and miniaturization of compact electrochemical biosensors is an emergent need for a reliable diagnostic purpose. Cost-effective biosensors increase their availability in rural areas for emergency uses and miniaturization of the sensing device allows rapid and on-site detection of foodborne pathogens. These factors limit the translation of biosensors into industrial production and commercialization. Methods for producing reproducible sensor batches and scaling-up to mass production, as well as integration of biosensors into automated and miniaturized systems are yet to be developed. However, the toxicity and biocompatibility need to be evaluated for a guaranteed performance and safety of the sensors.

Acknowledgments

This work was financially supported by University of Malaya Impact-Oriented Interdisciplinary Research Grant No. IIRG018A-2019.

References

[1] J. Vidic, P. Vizzini, M. Manzano, D. Kavanaugh, N. Ramarao, M. Zivkovic, V. Radonic, N. Knezevic, I. Giouroudi, I. Gadjanski, Point-of-need dna testing for detection of foodborne pathogenic bacteria, Sensors, 19 (2019) 1100. https://doi.org/10.3390/s19051100

[2] J.W.-F. Law, N.-S. Ab Mutalib, K.-G. Chan, L.-H. Lee, Rapid methods for the detection of foodborne bacterial pathogens: principles, applications, advantages and limitations, Frontiers in microbiology, 5 (2015) 770-770. https://doi.org/10.3389/fmicb.2014.00770

[3] S.K. Vashist, Point-of-Care Diagnostics: Recent Advances and Trends, Biosensors, 7 (2017) 62. https://doi.org/10.3390/bios7040062

[4] Y. Wang, T.V. Duncan, Nanoscale sensors for assuring the safety of food products, Current Opinion in Biotechnology, 44 (2017) 74-86. https://doi.org/10.1016/j.copbio.2016.10.005

[5] F. Jia, N. Duan, S. Wu, R. Dai, Z. Wang, X. Li, Impedimetric Salmonella aptasensor using a glassy carbon electrode modified with an electrodeposited composite consisting of reduced graphene oxide and carbon nanotubes, Microchimica Acta, 183 (2016) 337-344. https://doi.org/10.1007/s00604-015-1649-7

[6] T. Hianik, J. Wang, Electrochemical Aptasensors – Recent Achievements and Perspectives, 2009. https://doi.org/10.1002/elan.200904566

[7] K.M. Abu-Salah, M.M. Zourob, F. Mouffouk, S.A. Alrokayan, M.A. Alaamery, A.A. Ansari, DNA-Based Nanobiosensors as an Emerging Platform for Detection of Disease, Sensors (Basel, Switzerland), 15 (2015) 14539-14568. https://doi.org/10.3390/s150614539

[8] J. Peña-Bahamonde, H.N. Nguyen, S.K. Fanourakis, D.F. Rodrigues, Recent advances in graphene-based biosensor technology with applications in life sciences, Journal of nanobiotechnology, 16 (2018) 75-75. https://doi.org/10.1186/s12951-018-0400-z

[9] M. Pumera, Graphene in biosensing, Materials Today, 14 (2011) 308-315. https://doi.org/10.1016/S1369-7021(11)70160-2

[10] Y. Zhang, J. Li, Z. Wang, H. Ma, D. Wu, Q. Cheng, Q. Wei, Label-free electrochemical immunosensor based on enhanced signal amplification between

Au@Pd and CoFe2O4/graphene nanohybrid, Scientific Reports, 6 (2016) 23391.
https://doi.org/10.1038/srep23391

[11]   E. bahadır, M. Kemal Sezgintürk, Applications of graphene in electrochemical
sensing and biosensing, 2015. https://doi.org/10.1016/j.trac.2015.07.008

[12]   R.A.S. Luz, R.M. Iost, F.N. Crespilho, Nanomaterials for Biosensors and
Implantable Biodevices, in: F.N. Crespilho (Ed.) Nanobioelectrochemistry: From
Implantable Biosensors to Green Power Generation, Springer Berlin Heidelberg,
Berlin, Heidelberg, 2013, pp. 27-48. https://doi.org/10.1007/978-3-642-29250-7_2

[13]   N.S. Hobson, I. Tothill, A.P. Turner, Microbial detection, Biosensors &
bioelectronics, 11 (1996) 455-477. https://doi.org/10.1016/0956-5663(96)86783-2

[14]   A.L. Ghindilis, P. Atanasov, M. Wilkins, E. Wilkins, Immunosensors:
electrochemical sensing and other engineering approaches, Biosensors &
bioelectronics, 13 (1998) 113-131. https://doi.org/10.1016/S0956-5663(97)00031-6

[15]   M. Berrettoni, D. Tonelli, P. Conti, R. Marassi, M. Trevisani, Electrochemical
sensor for indirect detection of bacterial population, Sensors and Actuators B:
Chemical, 102 (2004) 331-335. https://doi.org/10.1016/j.snb.2004.04.022

[16]   I. Tothill, Biosensors and nanomaterials and their application for mycotoxin
determination, World Mycotoxin Journal, 4 (2011) 361-374.
https://doi.org/10.3920/WMJ2011.1318

[17]   N.J. Ronkainen, H.B. Halsall, W.R. Heineman, Electrochemical biosensors,
Chemical Society reviews, 39 (2010) 1747-1763. https://doi.org/10.1039/b714449k

[18]   D.R. Thévenot, K. Toth, R.A. Durst, G.S. Wilson, Electrochemical biosensors:
recommended definitions and classification1International Union of Pure and Applied
Chemistry: Physical Chemistry Division, Commission I.7 (Biophysical Chemistry);
Analytical Chemistry Division, Commission V.5 (Electroanalytical Chemistry).1,
Biosensors and Bioelectronics, 16 (2001) 121-131. https://doi.org/10.1016/S0956-
5663(01)00115-4

[19]   R. Maalouf, C. Fournier-Wirth, J. Coste, H. Chebib, Y. Saikali, O. Vittori, A.
Errachid, J.P. Cloarec, C. Martelet, N. Jaffrezic-Renault, Label-free detection of
bacteria by electrochemical impedance spectroscopy: comparison to surface plasmon
resonance, Analytical chemistry, 79 (2007) 4879-4886.
https://doi.org/10.1021/ac070085n

[20] M.H. Abdalhai, A.M. Fernandes, X. Xia, A. Musa, J. Ji, X. Sun, Electrochemical Genosensor To Detect Pathogenic Bacteria (Escherichia coli O157:H7) As Applied in Real Food Samples (Fresh Beef) To Improve Food Safety and Quality Control, Journal of agricultural and food chemistry, 63 (2015) 5017-5025. https://doi.org/10.1021/acs.jafc.5b00675

[21] J. Wang, Electrochemical biosensors: Towards point-of-care cancer diagnostics, Biosensors and Bioelectronics, 21 (2006) 1887-1892. https://doi.org/10.1016/j.bios.2005.10.027

[22] S. Cinti, G. Volpe, S. Piermarini, E. Delibato, G. Palleschi, Electrochemical Biosensors for Rapid Detection of Foodborne Salmonella: A Critical Overview, Sensors (Basel, Switzerland), 17 (2017). https://doi.org/10.3390/s17081910

[23] A. Amine, H. Mohammadi, I. Bourais, G. Palleschi, Enzyme inhibition-based biosensors for food safety and environmental monitoring, Biosensors & bioelectronics, 21 (2006) 1405-1423. https://doi.org/10.1016/j.bios.2005.07.012

[24] Y. Wang, H. Xu, J. Zhang, G. Li, Electrochemical sensors for clinic analysis, sensors (Basel, Switzerland), 8 (2008) 2043-2081. https://doi.org/10.3390/s8042043

[25] F.-G. Bănică, Amperometric Enzyme Sensors, in: Chemical Sensors and Biosensors, John Wiley & Sons, Ltd, 2012, pp. 314-331. https://doi.org/10.1002/9781118354162.ch14

[26] A. Singh, S. Poshtiban, S. Evoy, Recent advances in bacteriophage based biosensors for food-borne pathogen detection, Sensors (Basel), 13 (2013) 1763-1786. https://doi.org/10.3390/s130201763

[27] L. Yang, C. Ruan, Y. Li, Rapid Detection Of Salmonella Typhimurium In Food Samples Using A Bienzyme Electrochemical Biosensor With Flow Injection, Journal of Rapid Methods & Automation in Microbiology, 9 (2001) 229-240. https://doi.org/10.1111/j.1745-4581.2001.tb00249.x

[28] S. Chemburu, E. Wilkins, I. Abdel-Hamid, Detection of pathogenic bacteria in food samples using highly-dispersed carbon particles, Biosensors & bioelectronics, 21 (2005) 491-499. https://doi.org/10.1016/j.bios.2004.11.025

[29] X. Yang, J. Kirsch, A. Simonian, Campylobacter spp. detection in the 21st century: A review of the recent achievements in biosensor development, Journal of

Nanohybrids
Materials Research Forum LLC
Materials Research Foundations **87** (2021) 69-102
https://doi.org/10.21741/9781644901076-3

microbiological methods, 95 (2013) 48-56.
https://doi.org/10.1016/j.mimet.2013.06.023

[30]   T. Neufeld, A. Schwartz-Mittelmann, D. Biran, E.Z. Ron, J. Rishpon, Combined phage typing and amperometric detection of released enzymatic activity for the specific identification and quantification of bacteria, Analytical chemistry, 75 (2003) 580-585. https://doi.org/10.1021/ac026083e

[31]   I. Benhar, I. Eshkenazi, T. Neufeld, J. Opatowsky, S. Shaky, J. Rishpon, Recombinant single chain antibodies in bioelectrochemical sensors, Talanta, 55 (2001) 899-907. https://doi.org/10.1016/S0039-9140(01)00497-0

[32]   J.L. Brooks, B. Mirhabibollahi, R.G. Kroll, Experimental enzyme-linked amperometric immunosensors for the detection of salmonellas in foods, Journal of Applied Bacteriology, 73 (1992) 189-196. https://doi.org/10.1111/j.1365-2672.1992.tb02977.x

[33]   L. Croci, E. Delibato, G. Volpe, G. Palleschi, A RAPID ELECTROCHEMICAL ELISA FOR THE DETECTION OF SALMONELLA IN MEAT SAMPLES, Analytical Letters, 34 (2001) 2597-2607. https://doi.org/10.1081/AL-100108407

[34]   Z.J. Zhao, X.M. Liu, Preparation of monoclonal antibody and development of enzyme-linked immunosorbent assay specific for Escherichia coli O157 in foods, Biomedical and environmental sciences : BES, 18 (2005) 254-259.

[35]   M. Varshney, L. Yang, X.L. Su, Y. Li, Magnetic nanoparticle-antibody conjugates for the separation of Escherichia coli O157:H7 in ground beef, Journal of food protection, 68 (2005) 1804-1811. https://doi.org/10.4315/0362-028X-68.9.1804

[36]   M.D. Gouda, M.A. Kumar, M.S. Thakur, N.G. Karanth, Enhancement of operational stability of an enzyme biosensor for glucose and sucrose using protein based stabilizing agents, Biosensors & bioelectronics, 17 (2002) 503-507. https://doi.org/10.1016/S0956-5663(02)00021-0

[37]   F. Farabullini, F. Lucarelli, I. Palchetti, G. Marrazza, M. Mascini, Disposable electrochemical genosensor for the simultaneous analysis of different bacterial food contaminants, Biosensors and Bioelectronics, 22 (2007) 1544-1549. https://doi.org/10.1016/j.bios.2006.06.001

[38]   M. Tichoniuk, D. Gwiazdowska, M. Ligaj, M. Filipiak, Electrochemical detection of foodborne pathogen Aeromonas hydrophila by DNA hybridization biosensor,

Materials Research Forum LLC
https://doi.org/10.21741/9781644901076-3

Biosensors and Bioelectronics, 26 (2010) 1618-1623.
https://doi.org/10.1016/j.bios.2010.08.030

[39] K.-F. Low, K. Chuenrangsikul, P. Rijiravanich, W. Surareungchai, Y.-Y. Chan, Electrochemical genosensor for specific detection of the food-borne pathogen, Vibrio cholerae, World Journal of Microbiology and Biotechnology, 28 (2012) 1699-1706. https://doi.org/10.1007/s11274-011-0978-x

[40] F. Lucarelli, G. Marrazza, A.P.F. Turner, M. Mascini, Carbon and gold electrodes as electrochemical transducers for DNA hybridisation sensors, Biosensors and Bioelectronics, 19 (2004) 515-530. https://doi.org/10.1016/S0956-5663(03)00256-2

[41] M.I. Pividori, A. Merkoçi, S. Alegret, Electrochemical genosensor design: immobilisation of oligonucleotides onto transducer surfaces and detection methods, Biosensors and Bioelectronics, 15 (2000) 291-303. https://doi.org/10.1016/S0956-5663(00)00071-3

[42] D. Ivnitski, I. Abdel-Hamid, P. Atanasov, E. Wilkins, S. Stricker, Application of electrochemical biosensors for detection of food pathogenic bacteria, Electroanalysis, 12 (2000) 317-325. https://doi.org/10.1002/(SICI)1521-4109(20000301)12:5<317::AID-ELAN317>3.0.CO;2-A

[43] L. Yang, R. Bashir, Electrical/electrochemical impedance for rapid detection of foodborne pathogenic bacteria, Biotechnology advances, 26 (2008) 135-150. https://doi.org/10.1016/j.biotechadv.2007.10.003

[44] E. Katz, I. Willner, Probing Biomolecular Interactions at Conductive and Semiconductive Surfaces by Impedance Spectroscopy: Routes to Impedimetric Immunosensors, DNA-Sensors, and Enzyme Biosensors, Electroanalysis, 15 (2003) 913-947. https://doi.org/10.1002/elan.200390114

[45] P. D'Orazio, Biosensors in clinical chemistry, Clinica chimica acta; international journal of clinical chemistry, 334 (2003) 41-69. https://doi.org/10.1016/S0009-8981(03)00241-9

[46] D. Grieshaber, R. MacKenzie, J. Vörös, E. Reimhult, Electrochemical Biosensors - Sensor Principles and Architectures, Sensors (Basel, Switzerland), 8 (2008) 1400-1458. https://doi.org/10.3390/s8031400

[47]   R. Rogers K, M. Mascini, Biosensors for field analytical monitoring, Field analytical chemistry and technology, 2 (1998) 317-331. https://doi.org/10.1002/(SICI)1520-6521(1998)2:6<317::AID-FACT2>3.0.CO;2-5

[48]   S.R. Mikkelsen, G.A. Rechnitz, Conductometric tranducers for enzyme-based biosensors, Analytical chemistry, 61 (1989) 1737-1742. https://doi.org/10.1021/ac00190a029

[49]   Z. Muhammad-Tahir, E.C. Alocilja, A conductometric biosensor for biosecurity, Biosensors and Bioelectronics, 18 (2003) 813-819. https://doi.org/10.1016/S0956-5663(03)00020-4

[50]   Z. Muhammad-Tahir, E.C. Alocilja, A Disposable Biosensor for Pathogen Detection in Fresh Produce Samples, Biosystems Engineering, 88 (2004) 145-151. https://doi.org/10.1016/j.biosystemseng.2004.03.005

[51]   S. Pal, W. Ying, E.C. Alocilja, F.P. Downes, Sensitivity and specificity performance of a direct-charge transfer biosensor for detecting Bacillus cereus in selected food matrices, Biosystems Engineering, 99 (2008) 461-468. https://doi.org/10.1016/j.biosystemseng.2007.11.015

[52]   Z.-G. Chen, Conductometric immunosensors for the detection of staphylococcal enterotoxin B based bio-electrocalytic reaction on micro-comb electrodes, Bioprocess and Biosystems Engineering, 31 (2008) 345-350. https://doi.org/10.1007/s00449-007-0168-2

[53]   J.S. Daniels, N. Pourmand, Label-Free Impedance Biosensors: Opportunities and Challenges, Electroanalysis, 19 (2007) 1239-1257. https://doi.org/10.1002/elan.200603855

[54]   V.F. Lvovich, Impedance Spectroscopy: Applications to Electrochemical and Dielectric Phenomena, Wiley, 2015.

[55]   M. Varshney, Y. Li, Interdigitated array microelectrode based impedance biosensor coupled with magnetic nanoparticle–antibody conjugates for detection of Escherichia coli O157:H7 in food samples, Biosensors and Bioelectronics, 22 (2007) 2408-2414. https://doi.org/10.1016/j.bios.2006.08.030

[56] L. Yang, Y. Li, G.F. Erf, Interdigitated Array microelectrode-based electrochemical impedance immunosensor for detection of Escherichia coli O157:H7, Analytical chemistry, 76 (2004) 1107-1113. https://doi.org/10.1021/ac0352575

[57]   A. Shabani, M. Zourob, B. Allain, C.A. Marquette, M.F. Lawrence, R. Mandeville, Bacteriophage-Modified Microarrays for the Direct Impedimetric Detection of Bacteria, Analytical chemistry, 80 (2008) 9475-9482. https://doi.org/10.1021/ac801607w

[58]   A. Shabani, C.A. Marquette, R. Mandeville, M.F. Lawrence, Magnetically-assisted impedimetric detection of bacteria using phage-modified carbon microarrays, Talanta, 116 (2013) 1047-1053. https://doi.org/10.1016/j.talanta.2013.07.078

[59]   C. Ruan, L. Yang, Y. Li, Immunobiosensor Chips for Detection of Escherichia coli O157:H7 Using Electrochemical Impedance Spectroscopy, Analytical chemistry, 74 (2002) 4814-4820. https://doi.org/10.1021/ac025647b

[60]   A.G. Mantzila, V. Maipa, M.I. Prodromidis, Development of a faradic impedimetric immunosensor for the detection of Salmonella typhimurium in milk, Analytical chemistry, 80 (2008) 1169-1175. https://doi.org/10.1021/ac071570l

[61]   Y. Wang, J. Ping, Z. Ye, J. Wu, Y. Ying, Impedimetric immunosensor based on gold nanoparticles modified graphene paper for label-free detection of Escherichia coli O157:H7, Biosensors & bioelectronics, 49 (2013) 492-498. https://doi.org/10.1016/j.bios.2013.05.061

[62]   M. Labib, A.S. Zamay, O.S. Kolovskaya, I.T. Reshetneva, G.S. Zamay, R.J. Kibbee, S.A. Sattar, T.N. Zamay, M.V. Berezovski, Aptamer-based viability impedimetric sensor for bacteria, Analytical chemistry, 84 (2012) 8966-8969. https://doi.org/10.1021/ac302902s

[63]   I. Palchetti, M. Mascini, Electroanalytical biosensors and their potential for food pathogen and toxin detection, Analytical and Bioanalytical Chemistry, 391 (2008) 455-471. https://doi.org/10.1007/s00216-008-1876-4

[64]   G.A. Zelada-Guillén, S.V. Bhosale, J. Riu, F.X. Rius, Real-Time Potentiometric Detection of Bacteria in Complex Samples, Analytical chemistry, 82 (2010) 9254-9260. https://doi.org/10.1021/ac101739b

[65]   K. Dill, L.H. Stanker, C.R. Young, Detection of salmonella in poultry using a silicon chip-based biosensor, Journal of biochemical and biophysical methods, 41 (1999) 61-67. https://doi.org/10.1016/S0165-022X(99)00027-5

[66]   C. Ercole, M. Del Gallo, L. Mosiello, S. Baccella, A. Lepidi, Escherichia coli detection in vegetable food by a potentiometric biosensor, Sensors and Actuators B: Chemical, 91 (2003) 163-168. https://doi.org/10.1016/S0925-4005(03)00083-2

[67]   Y. Lu, Y. Liu, Y. Zhao, W. Li, L. Qiu, L. Li, A Novel and Disposable Enzyme-Labeled Amperometric Immunosensor Based on MWCNT Fibers for Listeria monocytogenes Detection, Journal of Nanomaterials, 2016 (2016) 8. https://doi.org/10.1155/2016/3895920

[68]   Y. Li, P. Cheng, J. Gong, L. Fang, J. Deng, W. Liang, J. Zheng, Amperometric immunosensor for the detection of Escherichia coli O157:H7 in food specimens, Analytical Biochemistry, 421 (2012) 227-233. https://doi.org/10.1016/j.ab.2011.10.049

[69]   M.I. Pividori, A. Merkoçi, J. Barbé, S. Alegret, PCR-Genosensor Rapid Test for Detecting Salmonella, Electroanalysis, 15 (2003) 1815-1823. https://doi.org/10.1002/elan.200302764

[70]   L. Yao, P. Lamarche, N. Tawil, R. Khan, A.M. Aliakbar, M.H. Hassan, V.P. Chodavarapu, R. Mandeville, CMOS Conductometric System for Growth Monitoring and Sensing of Bacteria, IEEE Transactions on Biomedical Circuits and Systems, 5 (2011) 223-230. https://doi.org/10.1109/TBCAS.2010.2089794

[71]   M. Yang, S. Sun, H.A. Bruck, Y. Kostov, A. Rasooly, Electrical percolation-based biosensor for real-time direct detection of staphylococcal enterotoxin B (SEB), Biosensors and Bioelectronics, 25 (2010) 2573-2578. https://doi.org/10.1016/j.bios.2010.04.019

[72]   Y. Huang, X. Dong, Y. Liu, L.-J. Li, P. Chen, Graphene-based biosensors for detection of bacteria and their metabolic activities, Journal of Materials Chemistry, 21 (2011) 12358-12362. https://doi.org/10.1039/c1jm11436k

[73]   R. Radhakrishnan, M. Jahne, S. Rogers, I.I. Suni, Detection of Listeria Monocytogenes by Electrochemical Impedance Spectroscopy, Electroanalysis, 25 (2013) 2231-2237. https://doi.org/10.1002/elan.201300140

[74]   D. Li, Y. Feng, L. Zhou, Z. Ye, J. Wang, Y. Ying, C. Ruan, R. Wang, Y. Li, Label-free capacitive immunosensor based on quartz crystal Au electrode for rapid and sensitive detection of Escherichia coli O157:H7, Analytica Chimica Acta, 687 (2011) 89-96. https://doi.org/10.1016/j.aca.2010.12.018

[75]   M. Tolba, M.U. Ahmed, C. Tlili, F. Eichenseher, M.J. Loessner, M. Zourob, A bacteriophage endolysin-based electrochemical impedance biosensor for the rapid detection of Listeria cells, Analyst, 137 (2012) 5749-5756. https://doi.org/10.1039/c2an35988j

[76] S.A. Vetrone, M.C. Huarng, E.C. Alocilja, Detection of Non-PCR Amplified S. enteritidis Genomic DNA from Food Matrices Using a Gold-Nanoparticle DNA Biosensor: A Proof-of-Concept Study, Sensors (Basel, Switzerland), 12 (2012) 10487-10499. https://doi.org/10.3390/s120810487

[77]   S.M. Radke, E.C. Alocilja, A high density microelectrode array biosensor for detection of E. coli O157:H7, Biosensors & bioelectronics, 20 (2005) 1662-1667. https://doi.org/10.1016/j.bios.2004.07.021

[78]   Z. Qiao, C. Lei, Y. Fu, Y. Li, Rapid and sensitive detection of E. coli O157:H7 based on antimicrobial peptide functionalized magnetic nanoparticles and urease-catalyzed signal amplification, Analytical Methods, 9 (2017) 5204-5210. https://doi.org/10.1039/C7AY01643C

[79]   A.G. Gehring, D.L. Patterson, S.-I. Tu, Use of a Light-Addressable Potentiometric Sensor for the Detection ofEscherichia coliO157:H7, Analytical Biochemistry, 258 (1998) 293-298. https://doi.org/10.1006/abio.1998.2597

[80]   C. Ercole, M.D. Gallo, M. Pantalone, S. Santucci, L. Mosiello, C. Laconi, A. Lepidi, A biosensor for Escherichia coli based on a potentiometric alternating biosensing (PAB) transducer, Sensors and Actuators B: Chemical, 83 (2002) 48-52. https://doi.org/10.1016/S0925-4005(01)01027-9

[81]   A. D Ghuge, A. R Shirode, V. J Kadam, Graphene: a comprehensive review, Current drug targets, 18 (2017) 724-733. https://doi.org/10.2174/1389450117666160709023425

[82]   K.S. Novoselov, A.K. Geim, S.V. Morozov, D. Jiang, Y. Zhang, S.V. Dubonos, I.V. Grigorieva, A.A. Firsov, Electric field effect in atomically thin carbon films, science, 306 (2004) 666-669. https://doi.org/10.1126/science.1102896

[83]   M. Pumera, Graphene-based nanomaterials and their electrochemistry, Chemical Society Reviews, 39 (2010) 4146-4157. https://doi.org/10.1039/c002690p

[84]   N. Gorjizadeh, Y. Kawazoe, Chemical functionalization of graphene nanoribbons, J. Nanomaterials, 2010 (2010) 1-4. https://doi.org/10.1155/2010/513501

Materials Research Forum LLC
https://doi.org/10.21741/9781644901076-3

[85] A. Nag, A. Mitra, S.C. Mukhopadhyay, Graphene and its sensor-based applications: A review, Sensors and Actuators A: Physical, 270 (2018) 177-194. https://doi.org/10.1016/j.sna.2017.12.028

[86] J. Hass, W. De Heer, E. Conrad, The growth and morphology of epitaxial multilayer graphene, Journal of Physics: Condensed Matter, 20 (2008) 323202. https://doi.org/10.1088/0953-8984/20/32/323202

[87] S. Bharech, R. Kumar, A review on the properties and applications of graphene, J Mater Sci Mechan Eng, 2 (2015) 70.

[88] W. Choi, I. Lahiri, R. Seelaboyina, Y.S. Kang, Synthesis of graphene and its applications: a review, Critical Reviews in Solid State and Materials Sciences, 35 (2010) 52-71. https://doi.org/10.1080/10408430903505036

[89] S.S. Datta, D.R. Strachan, S.M. Khamis, A.C. Johnson, Crystallographic etching of few-layer graphene, Nano letters, 8 (2008) 1912-1915. https://doi.org/10.1021/nl080583r

[90] J.C. Meyer, C. Girit, M. Crommie, A. Zettl, Hydrocarbon lithography on graphene membranes, Applied Physics Letters, 92 (2008) 123110. https://doi.org/10.1063/1.2901147

[91] Y. Hernandez, V. Nicolosi, M. Lotya, F.M. Blighe, Z. Sun, S. De, I. McGovern, B. Holland, M. Byrne, Y.K. Gun'Ko, High-yield production of graphene by liquid-phase exfoliation of graphite, Nature nanotechnology, 3 (2008) 563. https://doi.org/10.1038/nnano.2008.215

[92] S.S. Shams, R. Zhang, J. Zhu, Graphene synthesis: a Review, Materials Science-Poland, 33 (2015) 566-578. https://doi.org/10.1515/msp-2015-0079

[93] M. Zhou, T. Tian, X. Li, X. Sun, J. Zhang, P. Cui, J. Tang, L.-C. Qin, Production of graphene by liquid-phase exfoliation of intercalated graphite, Int. J. Electrochem. Sci, 9 (2014) 810-820.

[94] D. Nuvoli, L. Valentini, V. Alzari, S. Scognamillo, S.B. Bon, M. Piccinini, J. Illescas, A. Mariani, High concentration few-layer graphene sheets obtained by liquid phase exfoliation of graphite in ionic liquid, Journal of Materials Chemistry, 21 (2011) 3428-3431. https://doi.org/10.1039/C0JM02461A

[95]   B. Jayasena, S. Subbiah, A novel mechanical cleavage method for synthesizing few-layer graphenes, Nanoscale research letters, 6 (2011) 95. https://doi.org/10.1186/1556-276X-6-95

[96]   D.R. Dreyer, S. Park, C.W. Bielawski, R.S. Ruoff, The chemistry of graphene oxide, Chemical society reviews, 39 (2010) 228-240. https://doi.org/10.1039/B917103G

[97]   D.A. Dikin, S. Stankovich, E.J. Zimney, R.D. Piner, G.H. Dommett, G. Evmenenko, S.T. Nguyen, R.S. Ruoff, Preparation and characterization of graphene oxide paper, Nature, 448 (2007) 457. https://doi.org/10.1038/nature06016

[98]   P.R. Somani, S.P. Somani, M. Umeno, Planer nano-graphenes from camphor by CVD, Chemical Physics Letters, 430 (2006) 56-59. https://doi.org/10.1016/j.cplett.2006.06.081

[99]   Q. Yu, J. Lian, S. Siriponglert, H. Li, Y.P. Chen, S.-S. Pei, Graphene segregated on Ni surfaces and transferred to insulators, Applied Physics Letters, 93 (2008) 113103. https://doi.org/10.1063/1.2982585

[100] X. Li, W. Cai, J. An, S. Kim, J. Nah, D. Yang, R. Piner, A. Velamakanni, I. Jung, E. Tutuc, Large-area synthesis of high-quality and uniform graphene films on copper foils, science, 324 (2009) 1312-1314. https://doi.org/10.1126/science.1171245

[101] Z.-S. Wu, W. Ren, L. Gao, B. Liu, C. Jiang, H.-M. Cheng, Synthesis of high-quality graphene with a pre-determined number of layers, Carbon, 47 (2009) 493-499. https://doi.org/10.1016/j.carbon.2008.10.031

[102] E. Rollings, G.-H. Gweon, S. Zhou, B. Mun, J. McChesney, B. Hussain, A. Fedorov, P. First, W. De Heer, A. Lanzara, Synthesis and characterization of atomically thin graphite films on a silicon carbide substrate, Journal of Physics and Chemistry of Solids, 67 (2006) 2172-2177. https://doi.org/10.1016/j.jpcs.2006.05.010

[103] S.V. Morozov, K.S. Novoselov, M. Katsnelson, F. Schedin, L. Ponomarenko, D. Jiang, A.K. Geim, Strong suppression of weak localization in graphene, Physical review letters, 97 (2006) 016801. https://doi.org/10.1103/PhysRevLett.97.016801

[104] S. Amini, J. Garay, G. Liu, A.A. Balandin, R. Abbaschian, Growth of large-area graphene films from metal-carbon melts, Journal of Applied Physics, 108 (2010) 094321. https://doi.org/10.1063/1.3498815

[105] I. Pletikosić, M. Kralj, P. Pervan, R. Brako, J. Coraux, A. N'diaye, C. Busse, T. Michely, Dirac cones and minigaps for graphene on Ir (111), Physical Review Letters, 102 (2009) 056808. https://doi.org/10.1103/PhysRevLett.102.056808

[106] K. Bolotin, K.J. Sikes, Z. Jiang, M. Klima, G. Fudenberg, J. Hone, P. Kim, H.L. Stormer, Ultrahigh Electron Mobility in Suspended Graphene, 2008. https://doi.org/10.1016/j.ssc.2008.02.024

[107] P. Suvarnaphaet, S. Pechprasarn, Graphene-Based Materials for Biosensors: A Review, 2017.

[108] V. Georgakilas, J.N. Tiwari, K.C. Kemp, J.A. Perman, A.B. Bourlinos, K.S. Kim, R. Zboril, Noncovalent Functionalization of Graphene and Graphene Oxide for Energy Materials, Biosensing, Catalytic, and Biomedical Applications, Chemical reviews, 116 (2016) 5464-5519. https://doi.org/10.1021/acs.chemrev.5b00620

[109] P. Suvarnaphaet, S. Pechprasarn, Graphene-Based Materials for Biosensors: A Review, Sensors, 17 (2017). https://doi.org/10.3390/s17102161

[110]   A. Ambrosi, C.K. Chua, A. Bonanni, M. Pumera, Electrochemistry of Graphene and Related Materials, Chemical reviews, 114 (2014) 7150-7188. https://doi.org/10.1021/cr500023c

[111] Z. Zhu, An Overview of Carbon Nanotubes and Graphene for Biosensing Applications, Nano-Micro Letters, 9 (2017) 25. https://doi.org/10.1007/s40820-017-0128-6

[112] D.A.C. Brownson, P.J. Kelly, C.E. Banks, In situ electrochemical characterisation of graphene and various carbon-based electrode materials: an internal standard approach, RSC Advances, 5 (2015) 37281-37286. https://doi.org/10.1039/C5RA03049H

[113] T. Kuila, S. Bose, P. Khanra, A.K. Mishra, N.H. Kim, J.H. Lee, Recent advances in graphene-based biosensors, Biosensors & bioelectronics, 26 (2011) 4637-4648. https://doi.org/10.1016/j.bios.2011.05.039

[114] J. Peña-Bahamonde, H.N. Nguyen, S.K. Fanourakis, D.F. Rodrigues, Recent advances in graphene-based biosensor technology with applications in life sciences, Journal of Nanobiotechnology, 16 (2018) 75. https://doi.org/10.1186/s12951-018-0400-z

[115] M. Pan, Y. Gu, Y. Yun, M. Li, X. Jin, S. Wang, Nanomaterials for Electrochemical Immunosensing, 2017.

[116] C.M. Pandey, I. Tiwari, V.N. Singh, K.N. Sood, G. Sumana, B.D. Malhotra, Highly sensitive electrochemical immunosensor based on graphene-wrapped copper oxide-cysteine hierarchical structure for detection of pathogenic bacteria, Sensors and Actuators B: Chemical, 238 (2017) 1060-1069. https://doi.org/10.1016/j.snb.2016.07.121

[117] A. Pandey, Y. Gurbuz, V. Ozguz, J.H. Niazi, A. Qureshi, Graphene-interfaced electrical biosensor for label-free and sensitive detection of foodborne pathogenic E. coli O157:H7, Biosensors & bioelectronics, 91 (2017) 225-231. https://doi.org/10.1016/j.bios.2016.12.041

[118] S. Xu, Electrochemical DNA Biosensor Based on Graphene Oxide- Chitosan Hybrid Nanocomposites for Detection of Escherichia Coli O157:H7, 2017. https://doi.org/10.20964/2017.04.16

[119] A. Ahmed, J.V. Rushworth, N.A. Hirst, P.A. Millner, Biosensors for Whole-Cell Bacterial Detection, Clinical Microbiology Reviews, 27 (2014) 631-646. https://doi.org/10.1128/CMR.00120-13

[120] Y.-X. Wang, Z.-Z. Ye, C.-Y. Si, Y.-B. Ying, Application of Aptamer Based Biosensors for Detection of Pathogenic Microorganisms, 2012. https://doi.org/10.1016/S1872-2040(11)60542-2

[121] S. Muniandy, S.J. Teh, J.N. Appaturi, K.L. Thong, C.W. Lai, F. Ibrahim, B.F. Leo, A reduced graphene oxide-titanium dioxide nanocomposite based electrochemical aptasensor for rapid and sensitive detection of Salmonella enterica, Bioelectrochemistry (Amsterdam, Netherlands), 127 (2019) 136-144. https://doi.org/10.1016/j.bioelechem.2019.02.005

[122] I.J. Dinshaw, S. Muniandy, S.J. Teh, F. Ibrahim, B.F. Leo, K.L. Thong, Development of an aptasensor using reduced graphene oxide chitosan complex to detect Salmonella, Journal of Electroanalytical Chemistry, 806 (2017) 88-96. https://doi.org/10.1016/j.jelechem.2017.10.054

[123] X. Ma, Y. Jiang, F. Jia, Y. Yu, J. Chen, Z. Wang, An aptamer-based electrochemical biosensor for the detection of Salmonella, Journal of Microbiological Methods, 98 (2014) 94-98. https://doi.org/10.1016/j.mimet.2014.01.003

[124] X. Hu, W. Dou, G. Zhao, Electrochemical immunosensor for Enterobacter sakazakii detection based on electrochemically reduced graphene oxide–gold nanoparticle/ionic liquid modified electrode, Journal of Electroanalytical Chemistry, 756 (2015) 43-48. https://doi.org/10.1016/j.jelechem.2015.08.009

[125] C. Sign, G. Sumana, Antibody conjugated graphene nanocomposites for pathogen detection, 704 (2016) 012014. https://doi.org/10.1088/1742-6596/704/1/012014

[126] R. Mutreja, M. Jariyal, P. Pathania, A. Sharma, D.K. Sahoo, C.R. Suri, Novel surface antigen based impedimetric immunosensor for detection of Salmonella typhimurium in water and juice samples, Biosensors and Bioelectronics, 85 (2016) 707-713. https://doi.org/10.1016/j.bios.2016.05.079

[127] Y. Chen, Z. P Michael, G. Kotchey, Y. Zhao, A. Star, Electronic Detection Of Bacteria Using Holey Reduced Graphene Oxide, 2014. https://doi.org/10.1021/am500364f

[128] R. Hernández, C. Vallés, A.M. Benito, W.K. Maser, F. Xavier Rius, J. Riu, Graphene-based potentiometric biosensor for the immediate detection of living bacteria, Biosensors and Bioelectronics, 54 (2014) 553-557. https://doi.org/10.1016/j.bios.2013.11.053

[129] Y. Wu, H. Chai, Development of an Electrochemical Biosensor for Rapid Detection of Foodborne Pathogenic Bacteria, 2017. https://doi.org/10.20964/2017.05.09

[130] Y. Wan, Y. Wang, J. Wu, D. Zhang, Graphene Oxide Sheet-Mediated Silver Enhancement for Application to Electrochemical Biosensors, Analytical Chemistry, 83 (2011) 648-653. https://doi.org/10.1021/ac103047c

Materials Research Forum LLC
https://doi.org/10.21741/9781644901076-4

Chapter 4

Hybrid Materials based on Silica Nanostructures for Biomedical Scaffolds (Bone Regeneration) and Drug Delivery

Mojdeh Rahnama Ghahfarokhi[1], Jhaleh Amirian[2,3]*

[1]Department of Nanotechnology Engineering Faculty of Advanced Sciences and Technologies, University of Isfahan, 8174673441, Iran

[2] College of Materials Science and Engineering, Shenzhen Key Laboratory of Polymer Science and Technology, Guangdong Research Center for Interfacial Engineering of Functional Materials, Nanshan District Key Laboratory for Biopolymers and Safety Evaluation, Shenzhen University, Shenzhen, 518055, PR China

[3] Key Laboratory of Optoelectronic Devices and Systems of Ministry of Education and Guangdong Province, College of Optoelectronic Engineering, Shenzhen University, Shenzhen 518060, P. R. China

*jalehamirian@gmail.com

Abstract

Silica nanoparticles with nanoporous nature are introduced as thermally and chemically stable nanomaterials with controllable porosity and morphology. The nanoparticles can be divided into three groups: microporous, mesoporous, and macroporous based on the porous size. The use of these materials for different applications is associated with their unique properties as disinfectants. This chapter discusses different synthesis methodologies to prepare well-dispersed mesoporous silica nanoparticles (MSNs) and hollow silica nanoparticles (HSNs) with tunable dimensions ranging from a few to hundreds of nanometers with different mesostructures. Several good characteristics of the MSNs, best biocompatibility and low toxicity, are proposed as the basis of the carrier for the controlled release of drugs, genes into living cells and bone regeneration.

Keywords

Silica, Sol-Gel, Mesoporous, Biomedical Applications, Drug Delivery

Materials Research Forum LLC
https://doi.org/10.21741/9781644901076-4

Contents

1.   Introduction

1.1   Synthesis of nanoporous silica

Silica nanoparticles have been synthesized with different pore size, morphology and structure. These can be designed by controlling synthesis parameters and conditions such as pH, surfactants, and different concentrations of silica and its various sources, among others. Thus, in the synthesis of "ideal" silica nanoparticles, the properties such as uniformity in particle size, pore size, and large volume pores should be taken into consideration in terms of homogeneity [1]. There are many obtention methods for silica nanoparticles like Sol-Gel, soft template, hard template, and core-shell structure. Nevertheless, soft template is one of the most important methods. On the other hand, porous materials, especially mesoporous silica nanoparticles (MSNs) are depended on the type of material used for forming the cavities. In General, mesoporous scilica materials categorized into the two classes such as silica gel (is an amorphous and porous form of silicon dioxide) and silica glass (fused quartz or fused silica is glass consisting of silica in the amorphous (non-crystalline form) and non-silica such as zeolite (Zeolites are microporous, aluminosilicate minerals commonly used as commercial adsorbents and catalysts).

Generally, the synthesis of these materials includes the formation of liquid-crystalline mesophases of surfactants which act as a pattern for polymerization of orthosilicic acid. This synthesis can be carried out under acidic or basic conditions, and using different sources of silica as fumed silica [2], sodium silicate [3], tetra-alkyl oxide of silane [4], quartz [5], rice husk ash [6], bagasse ash [7], and tetramethoxysilane [8]. The great balance of properties of these materials is basically associated with a good biocompatibility and low toxicity. In addition, sodium silicate and any ash are inexpensive sources. There are many ways to synthesize nanoparticles for different applications, and the most important of them will be discussed below.

Sol-Gel:

The Sol-Gel method (shown in figure 1) was first introduced by Graham when he worked on silica sol [9]. The Sol-Gel method is a process that uses a colloidal source for ceramic synthesis with an intermediate, which includes Sol and a Gel state.

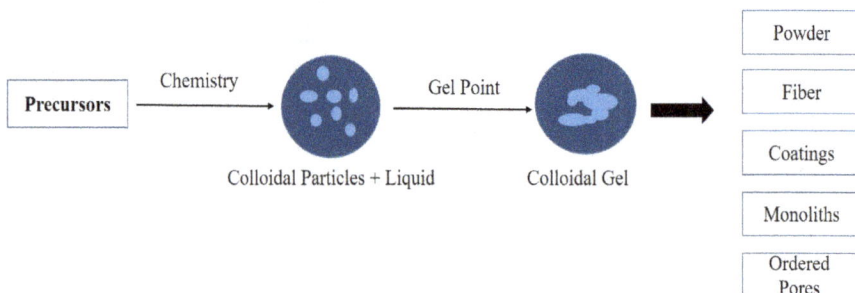

Figure 1. Simplified Chart of Sol-Gel Processes.

In Figure 1, we see the main process of Sol-Gel, the first step involves the preparation of a homogeneous solution. To do this, the precursor are dissolved in the solvent (e.g water or alcohol) to obtain a homogeneous solution. The second stage is the conversion of a homogeneous solution to Sol, where the hydrolysis reaction (decomposition by water)

Nanohybrids
Materials Research Foundations **87** (2021) 103-120

Materials Research Forum LLC
https://doi.org/10.21741/9781644901076-4

takes place, actually breaking down the molecular water and converting it to simple molecules. To form soluble tubes, some water is added to the reaction medium to hydrolyze reaction. Subsequently, the fine particles are dispersed in the solvent, forming a colloidal mixture, which is a solid particle dispersed in the solvent phase. The third step is the formation of the Gel. It is enough to stimulate the solution so that the dispersed fine particles begin to form, and the reaction is exactly the opposite of the hydrolysis reaction. At the end, drying and sintering are performed to achieve the final products [10].

The temperature required for Sol-Gel steps is low and is often close to room temperature. Precursors such as metal alkaloids and alkaline solutions are easily purified in very high percentages. In the Sol-Gel method, a sharp decrease in pH can be prevented. Hydrolysis speed can be controlled with the use of appropriate modified chemical precursors[11].

Stober method:

In 1968, Stober *et al.* reported the first method for the synthesis of nanoporous silica particles. This method briefly includes hydrolysis of tetra-alkyl silicates in a mixture of alcohol and water using ammonia as a catalyst [12].

The "Stober method" was examined, and consequently, particles with diameters from nanometers to a few microns have been obtained. By controlling the parameters of the Stober method, silica materials and non-silica ones can be synthesized [13, 14].

Grun *et al.* [15] modified the Stober method by adding a cationic surfactant to the reaction mixture. Subsequently, mixtures of alcohol, water, and ammonia with different systems were used to synthesize mesoporous silica nanoparticles with different sizes, porosities, and structures. Also, Yano *et al.* [16] synthesized mono-dispersed MSNs, with pore diameter in the range of tens to several hundred nanometers by changing the methanol-water ratio. In this method, we can control the size of MSNs by controlling the pH. Mou and co-workers [17] by changing the pH in the range of 10.86-11.52, they were able to control the particle size. A lower pH gave a smaller size of MSNs but in contrast, Qiao *et al.* [18] reported that the size of MSNs increases from 30 nm to 80 nm with the decrease of pH from 10.0 to 6.0.

Core-shell structure (Binary surfactants):

The use of binary surfactants of different molecular weights during the synthesis can lead to the formation of twin material structures that are dual-mesoporous silica spheres (DMSS) called particles. Core-shell structures consist of smaller pores (2.0 nm) in the shell and larger tunable pores (12.8-18.5 nm) in the core [19]. For this method pointing to the research of Suteewong *et al.* [20], they synthesized highly aminated MSNs and successfully controlled the sizes of particles.

In the above Figure 2, we see the schematic illustration for the formation of dual-mesoporous silica spheres. Surfactants are surrounded by existing polymers to form the desired molds and the precursor particles are coated on the surface of these molds or nuclei. Then the germination begins to grow and the crust is formed. After calcination, the nuclei are removed and the porous structure is obtained.

Soft template:

It is a simple synthesis method to prepare hollow mesoporous silica nanoparticles (HMSNs). Hollow silica nanoparticles (HSNs) are a subclass of mesoporous silica nanoparticles (MSNs) with a lot of recent research activity[21]. Hollow silica particles made by soft templates such as micelles[22], microemulsion droplets[23], or vesicular structures[24]. Very small HSNs are formed by single-micelle templating. Favors the formation of hollow nanoparticles templated are synthesized by single-micelle nanoparticles by lowering of the framework-precursor/surfactant ratio. In vesicle templating, the size of HSNs increases with the addition of a cationic surfactant, and anionic co-surfactant, which can be used to lower the curvatures as mesostructure templates. With careful control of silica framework, hollow silica nanospheres have been successfully synthesized in microemulsion-templating. The particles synthesized by this method are widely used in many applications such as drug delivery and bone regeneration because these particles have a big size pore volume to improve the loading capacity.

Hard template:

The synthesis of polymer lattices, silica colloids, and metal oxides with uniform sizes requires specific conditions. These materials are synthesized by hard templates for monodispersed mesoporous silica nanoparticle products. The hard template should respond to three basic criteria for the achievement of inorganic silica replica. One of them is the silicification at the surface of the template and is faster compared to silicification at the self-condensation of the silica. Second, the template must be stable at the silicate deposition and condensation throughout the processes. Finally, the template should be easily removed without fracturing the inorganic silica cast.

1.2   Hybrid materials based on silica nanostructures

Many natural materials have an inorganic and organic structure at the nanoscale. In most cases, the mineral part provides mechanical strength to the bodies, while the organic part provides the bond between soft tissues. Thousands of years ago, the first synthetic materials were dyes made from organic and inorganic materials. The Sol-Gel process was

Materials Research Forum LLC
https://doi.org/10.21741/9781644901076-4

developed in the 1930s as one of the reasons for the development of organic and inorganic compounds.

As stated above, mesoporous silica materials are one of a variety of nanoporous materials. The discovery of MSNs by scientists of Mobil Corporation in 1992, was recognized as a breakthrough that could lead to a variety of important applications, such as catalyst, adsorption, polymer filler, optical devices, bio-imaging, drug delivery, and biomedical applications [25].

After the discovery of the above-mentioned materials, many research efforts have been underway to achieve control over the final characteristics. This research allows getting new families of mesoporous silica materials, such as SBA [26], MSU [27] and FSM [28] (are mesoporous materials with a hierarchical structure from a family of silicate and alumosilicate solids that were first developed by researchers at Mobil Oil Corporation and that can be used as catalysts or catalyst supports), where particle shapes and characteristic porosities can be controlled. The differences of these particles in their morphology, the used materials, and their synthesis method. On the other hand, mesoporous organic-inorganic hybrid materials are a class of materials that present good pore sizes and large specific surface areas.

1.2.1 Classification of materials

Porous materials can be divided into two groups of silica materials and non-silica materials.

Solid materials that can exhibit cavities in their structure are porous. Porous materials have two kinds of cavities including open and close cavities as shown in Figure 4. Open cavities have access to the surface of the material, but closed cavities are contained within the material[29].

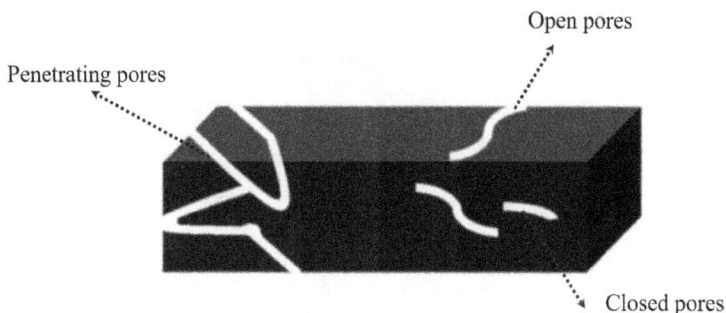

Figure 2. Schematic illustration of different morphology of porous.

According to the definition of (IUPAK)[1], porous materials are divided into three general categories[30], as shown in Table 1.

Table 1. Classification of porous materials.

Name	Size (nm)
Microporous materials	smaller than 2
Mesoporous materials	2 to 50
Macroporous materials	larger than 50

Nanoporous materials are a subset of porous materials. They have the largest porosity of 0.4nm and the size is in the range of 1-100 nm.

Nanoporous silica materials have become an important field of biomaterials, which are usually applied as nanoparticles or thin films. They have a large surface area, large pore volume, high reactivity on surface, and good control of the morphology, the aspect ratio between pore size and porosity, particle size, dispersity, and uniformity. The tunability of these characteristics gives MSNs advantages over other nanoparticles used for the polymer filler, optical devices, drug delivery, film and coating, and biomedical applications [31, 32].

1.2.2 Methods of preparation

In general, the preparation of hybrid materials is divided into two: direct and indirect methods. For example, in the below, Figure 3 shown the Sol-Gel method for the preparation of hybrid material-based silica.

Hydrolysis

$$Si(OR)_4 \longrightarrow (RO)_3 Si\text{-}OH + H_2O$$

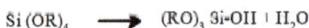

Condesation

$$\equiv Si\text{-}OH + HO\text{-}Si \equiv \longrightarrow \equiv Si\text{-}O\text{-}Si \equiv + H_2O$$

$$\equiv Si\text{-}OH + RO\text{-}Si \equiv \longrightarrow \equiv Si\text{-}O\text{-}Si \equiv + ROH$$

Figure 3. Sol-Gel process.

---

[1] International Union of Pure and Applied Chemistry

The Sol-Gel process is the best technique for preparing organic-inorganic hybrid materials. In these materials, the inorganic phase is formed inside an organic polymer matrix by the Sol-Gel process. The Sol-Gel process makes available for use methods for preparing a variety of organic-inorganic hybrid materials[33].

For example, the preparation of polyimide-silica hybrid films is shown in figure 4. This figure depicts that after the addition of TEOS and water to the acid, a homogeneous solution of silica is created as a Sol, then dried at the appropriate temperature and the Gel obtained, and then finally, calcined at high temperature.

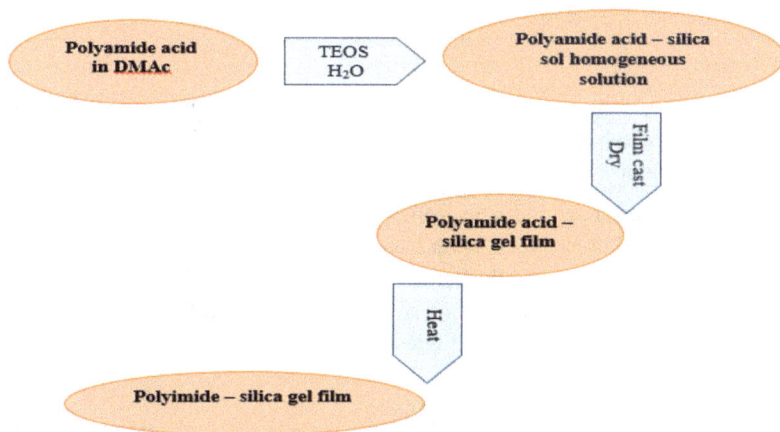

Figure 4. Preparation of polyimide-silica hybrid films.

2.   Characteristics

Organic-inorganic composites are a new class of materials that have large specific regions and pore sizes between 2 and 15 nm, obtained by the method of linking mineral and organic components with pattern synthesis. Functions can be investigated in three ways: by subsequent binding of the organic components to a pure silica matrix, by simultaneous reaction of compressible inorganic silica species and silicified organic compounds, using the bi-acetylated organic precursors leading to the period. They are called periodic mesoporous organic (PMO). The compatibility between the organic (polymer) and the inorganic phases (silica) powerfully affects on the properties of materials [34].

They are highly biocompatible and have no harm to the human body. The cytotoxicity of these substances can be checked by various tests on human fibroblasts. These materials have good mechanical properties and have shown excellent thermal and chemical resistance. X-ray diffraction and transmission electron microscopy are used for measuring pore of MSNs, and nitrogen sorption is used to measure the pore width. Large pore size 6-20 nm were recorded [26].

3. Properties

The unique properties of these materials include the grain size, controllable cavity size, the high specific surface area, the volume of the cavities, and the other properties, which are briefly described in below.

Particle size:

Unger, Stucky, and Zhao were among the first groups that made micrometer-sized mesoporous silica spheres with narrow size distribution [35-37]. Their goal was to control the particle size of mesoporous silica spheres.

When preparing the nanoporous silica particle, we can regulate the size and morphology by controlling the pH and/or using several templates [38]. On the other hand, the stirring rate is important for controlling the particle size. When the rate is slow, long fibers are produced and when the rate is fast, fine powder is formed[39]. Particle sizes can range from hundreds of microns to several millimeters.

Pore size:

Porosity is obtained by dividing the volume of cavities to the total volume of matter. By controlling the source of the silica type, and amount of applied surfactant, the size of the cavities can be regulated. Environmental conditions of synthesis and *in vitro* conditions are one of the factors for cavity resizing [40].

Surface area:

The surface area of the MSNs is an important factor for most of the applications. The value of the surface area is measured by the BET method. When the pore size is low, the surface area is high and when the surface area is high, this material is best candidate for applying in drug delivery applications.

Pore volume:

As mentioned before, the volume of cavities plays an important role in porous materials and drug delivery applications. Highly porous materials, possessing the high volume of pore, can load more drug, and eventually, the drug release can be increased desponded on the size of volume and other factors such as hydrophilicity, functionality of materials and other factors. One of the drug delivery goals is to achieve the controllable drug release for the specific application. [41].

4. Applications

In contrast to the mesoporous silica in the micrometer or larger dimensions, MSNs have a great number of special characteristics useful for many applications. For instance, their short channels can be used as solid supports for active sites in catalysis[42, 43].

Well-dispersed mesoporous silica nanoparticles with high porosity, thermal, mechanical, and chemical stability can be used as nano-fillers which can merge in silica or polymer matrixes to produce nanocomposite films. Surface-modified MSNs in the 20-30 nm range can be applied to various substrates and used as a single layer antireflection coating by using simple wet deposition techniques at ambient temperature [44].

MSNs possess superior features compared with organic and inorganic nanostructures such as their pore size, particle size, tunable porosity, biocompatibility, and high specific surface area. These unique characteristics have resulted in using these particles in many applications.

Since the discovery of the synthesis of mesoporous silica materials in 1992, many researchers have explored the functionalization of these materials for various applications such as high-performance catalysis, antireflection coating, transparent polymer-MSNs nanocomposites, drug-release, and theranostic systems, separation, sensors, etc.

4.1 Drug delivery

In 2001, the first reported mesoporous silica material as a drug delivery system [45]. Since then, silica nanoparticles have become widely used as drug carriers because of their high biocompatibility and convenient drug formulation. After about ten years, silica nanoparticles became a popular drug delivery system. A wide range of drugs such as small molecules, light-sensitive molecules, and proteins. have been used in the treatment of diseases such as cancer, Parkinson's, and heart diseases. Prolonged blood circulation provided improved disease targeting and reduced side effects [46-48].

Mesoporous silica materials can be considered as carriers for drug delivery because of their textural properties which increase the loading amount of drugs inside the pore channels. Similarly, drug diffusion kinetics can be controlled due to the functionalization of the silanol group [49].

Nanohybrids   Materials Research Forum LLC
Materials Research Foundations **87** (2021) 103-120   https://doi.org/10.21741/9781644901076-4

4.2   Tissue regeneration

MSNs have the potential to serve as vehicles to carry growth factors, peptides, or stem cells to a tissue-engineered scaffold in order to enhance tissue regeneration. The mechanical and nanotopographical features of the scaffolds can also be tuned by modulating of the surrounding cells microenvironments [50].

In bone regenerative medicine, MSNs exhibit superior osteoconductivity and osteoinductivity in comparison to the solid microparticle. Furthermore, the bioactivity of the MSNs improved compared to the conventional bioglass particles due to faster releasing of Si ions.

4.3   Application of mesoporous silica nanoparticles in gene transfer

Along with the transfer of small molecules of drugs and proteins, transcription is another important application of silica nanoparticle [51]. The use of silica nanoparticles for gene delivery has been extensively studied. The surface of the mesoporous silica nanoparticles is easily altered by cationic molecules due to the negatively charged groups and makes them resistant to nuclease formation for stable interactions with nucleotides.

In early 2000, the Saltzman Group bonded silica nanoparticles through co-incubation with the DNA complex, and it was observed that beta-galactosidase gene expression increased by 750% as the concentration of the DNA complex increased at the cell surface [52, 53].

5.   New steps

In the past two decades, great progress has been made in the synthesis of mesoporous silica nanoparticles, but we can see further improvements in the future.

By using these nanoparticles, optimizing their synthesis conditions, consumables, as well as modulating of the structral, physical, and chemical properties match with the applications, a brighter future can be envisioned for these materials. Using these substances will certainly make useful changes to the Earth's biological and health conditions. Also, with the advancement of these substances and their applications in the medical field, we can talk about improving the conditions of diseases all over the world and less vulnerability of human beings. Furthermore, these materials could reduce the impact of some dangerous substances over the Earth's environment.

As we now observe more and more, nanoparticles have become part of human daily life, and in the future, they will certainly dominate and influence the whole of human life.

Conclusions

In summary, in the past two decades, great progress has been made in the synthesis of mesoporous silica and hybrid materials. The discovery of this novel drug delivery system has shown potential for researchers to work on it. Morphological changes help in modifying these materials to produce diversified from the material. Diversified nature of MSNs has an approach for applications like drug/ gene delivery, targeted drug delivery for cancer drugs, bone regeneration, as diagnostic and imaging agent and cell tracing, and more and more. Many of these methods are based on the high dilution method, and the concentration of resulting products is usually low, typically in the millimolar range. Scaling-up the quantity to a commercial scale (kilograms) for practical applications is expected to present major challenges with regard to collection, uniformity, and reproducibility. In addition, the environmental impact of mesoporous silica nanoparticles should be also considered. MSNs with high surface areas could have a significant binding affinity toward proteins, and thus cause denaturation. The strong bonding of negatively charged mesoporous silica nanoparticles particularly to the surface of particular cells is due to the presence of quaternary ammonium ions in the membrane in the physiological pH range. Especially in powder form, inhalation of MSNs could have serious toxic effects in the same way as aerogels. To date, both *in vitro* and *in vivo* tests results suggest that MSNs are not immunogenic or toxic, but their impact on human health and the environment still requires further investigation. Studies on the potential hazards of MSNs will be helpful in establishing new guidelines on their use.

References

[1] S.H. Wu, C.Y. Mou, H.P. Lin, Synthesis of mesoporous silica nanoparticles, Chem Soc Rev, 42 (2013) 3862-3875. https://doi.org/10.1039/c3cs35405a

[2] W. Zhu, Y. zhou, W. Ma, M. Li, J. Yu, K. Xie, Using silica fume as silica source for synthesizing spherical ordered mesoporous silica, Materials Letters, 92 (2013) 129-131. https://doi.org/10.1016/j.matlet.2012.10.044

[3] P.F. Fulvio, S. Pikus, M. Jaroniec, Short-time synthesis of SBA-15 using various silica sources, J Colloid Interface Sci, 287 (2005) 717-720. https://doi.org/10.1016/j.jcis.2005.02.045

[4] F. Lu, S.H. Wu, Y. Hung, C.Y. Mou, Size effect on cell uptake in well-suspended, uniform mesoporous silica nanoparticles, Small, 5 (2009) 1408-1413. https://doi.org/10.1002/smll.200900005

[5] X. Dong, Y. Wang, H. Dan, Z. Hong, K. Song, Q. Xian, Y. Ding, A facile route to synthesize mesoporous SBA-15 silica spheres from powder quartz, Materials Letters, 204 (2017) 97-100. https://doi.org/10.1016/j.matlet.2017.05.115

[6] T. Witoon, M. Chareonpanich, J. Limtrakul, Synthesis of bimodal porous silica from rice husk ash via sol–gel process using chitosan as template, Materials Letters, 62 (2008) 1476-1479. https://doi.org/10.1016/j.matlet.2007.09.004

[7] N.A. Rahman, I. Widhiana, S.R. Juliastuti, H. Setyawan, Synthesis of mesoporous silica with controlled pore structure from bagasse ash as a silica source, Colloids and Surfaces A: Physicochemical and Engineering Aspects, 476 (2015) 1-7. https://doi.org/10.1016/j.colsurfa.2015.03.018

[8] D. Liu, M. Sasidharan, K. Nakashima, Micelles of poly(styrene-b-2-vinylpyridine-b-ethylene oxide) with blended polystyrene core and their application to the synthesis of hollow silica nanospheres, J Colloid Interface Sci, 358 (2011) 354-359. https://doi.org/10.1016/j.jcis.2011.03.004

[9] K.J. Klabunde, J. Stark, O. Koper, C. Mohs, D.G. Park, S. Decker, Y. Jiang, I. Lagadic, D. Zhang, Nanocrystals as stoichiometric reagents with unique surface chemistry, The Journal of Physical Chemistry, 100 (1996) 12142-12153. https://doi.org/10.1021/jp960224x

[10] R. Li, A. Clark, L. Hench, An investigation of bioactive glass powders by sol-gel processing, Journal of Applied Biomaterials, 2 (1991) 231-239. https://doi.org/10.1002/jab.770020403

[11] S.-A. Wallington, T. Labayen, A. Poppe, N.A. Sommerdijk, J.D. Wright, Sol-Gel entrapped materials for optical sensing of solvents and metal ions, Sensors and Actuators B: Chemical, 38 (1997) 48-52. https://doi.org/10.1016/S0925-4005(97)80170-0

[12] W. Stöber, A. Fink, E. Bohn, Controlled growth of monodisperse silica spheres in the micron size range, Journal of colloid and interface science, 26 (1968) 62-69. https://doi.org/10.1016/0021-9797(68)90272-5

[13] J. Liu, S.Z. Qiao, H. Liu, J. Chen, A. Orpe, D. Zhao, G.Q. Lu, Extension of the Stöber method to the preparation of monodisperse resorcinol–formaldehyde resin polymer and carbon spheres, Angewandte Chemie International Edition, 50 (2011) 5947-5951. https://doi.org/10.1002/anie.201102011

[14] J. Choma, D. Jamioła, K. Augustynek, M. Marszewski, M. Gao, M. Jaroniec, New opportunities in Stöber synthesis: preparation of microporous and mesoporous carbon spheres, Journal of Materials Chemistry, 22 (2012) 12636-12642. https://doi.org/10.1039/c2jm31678a

[15] M. Grün, I. Lauer, K.K. Unger, The synthesis of micrometer-and submicrometer-size spheres of ordered mesoporous oxide MCM-41, Advanced Materials, 9 (1997) 254-257. https://doi.org/10.1002/adma.19970090317

[16] K. Yano, Y. Fukushima, Synthesis of mono-dispersed mesoporous silica spheres with highly ordered hexagonal regularity using conventional alkyltrimethylammonium halide as a surfactant, Journal of Materials Chemistry, 14 (2004) 1579-1584. https://doi.org/10.1039/b313712k

[17] F. Lu, S.H. Wu, Y. Hung, C.Y. Mou, Size effect on cell uptake in well-suspended, uniform mesoporous silica nanoparticles, Small, 5 (2009) 1408-1413. https://doi.org/10.1002/smll.200900005

[18] Z.-A. Qiao, L. Zhang, M. Guo, Y. Liu, Q. Huo, Synthesis of mesoporous silica nanoparticles via controlled hydrolysis and condensation of silicon alkoxide, Chemistry of Materials, 21 (2009) 3823-3829. https://doi.org/10.1021/cm901335k

[19] D. Niu, Z. Ma, Y. Li, J. Shi, Synthesis of core– shell structured dual-mesoporous silica spheres with tunable pore size and controllable shell thickness, Journal of the American Chemical Society, 132 (2010) 15144-15147. https://doi.org/10.1021/ja1070653

[20] T. Suteewong, H. Sai, R. Cohen, S. Wang, M. Bradbury, B. Baird, S.M. Gruner, U. Wiesner, Highly aminated mesoporous silica nanoparticles with cubic pore structure, Journal of the American Chemical Society, 133 (2010) 172-175. https://doi.org/10.1021/ja1061664

[21] J. Hu, M. Chen, X. Fang, L. Wu, Fabrication and application of inorganic hollow spheres, Chemical Society Reviews, 40 (2011) 5472-5491. https://doi.org/10.1039/c1cs15103g

[22] J. Liu, Q. Yang, L. Zhang, H. Yang, J. Gao, C. Li, Organic– inorganic hybrid hollow nanospheres with microwindows on the shell, Chemistry of Materials, (2008). https://doi.org/10.1021/cm800192f

[23] J. Liu, S.Z. Qiao, J.S. Chen, X.W.D. Lou, X. Xing, G.Q.M. Lu, Yolk/shell nanoparticles: new platforms for nanoreactors, drug delivery and lithium-ion

batteries, Chemical Communications, 47 (2011) 12578-12591.
https://doi.org/10.1039/c1cc13658e

[24] Y.-Q. Yeh, B.-C. Chen, H.-P. Lin, C.-Y. Tang, Synthesis of hollow silica spheres with mesostructured shell using cationic− anionic-neutral block copolymer ternary surfactants, Langmuir, 22 (2006) 6-9. https://doi.org/10.1021/la052129y

[25] C. Kresge, M. Leonowicz, W.J. Roth, J. Vartuli, J. Beck, Ordered mesoporous molecular sieves synthesized by a liquid-crystal template mechanism, nature, 359 (1992) 710. https://doi.org/10.1038/359710a0

[26] D. Zhao, J. Feng, Q. Huo, N. Melosh, G.H. Fredrickson, B.F. Chmelka, G.D. Stucky, Triblock copolymer syntheses of mesoporous silica with periodic 50 to 300 angstrom pores, science, 279 (1998) 548-552.

[27] S.A. Bagshaw, E. Prouzet, T.J. Pinnavaia, Templating of mesoporous molecular sieves by nonionic polyethylene oxide surfactants, Science, 269 (1995) 1242-1244. https://doi.org/10.1126/science.269.5228.1242

[28] S. Inagaki, Y. Fukushima, K. Kuroda, Synthesis of highly ordered mesoporous materials from a layered polysilicate, Journal of the Chemical Society, Chemical Communications, (1993) 680-682. https://doi.org/10.1039/c39930000680

[29] K. Ishizaki, S. Komarneni, M. Nanko, Porous Materials: Process technology and applications, Springer science & business media, 2013.

[30] J.L. Vivero-Escoto, Y.-D. Chiang, K.C. Wu, Y. Yamauchi, Recent progress in mesoporous titania materials: adjusting morphology for innovative applications, Science and technology of advanced materials, 13 (2012) 013003. https://doi.org/10.1088/1468-6996/13/1/013003

[31] D.D. Borawake, A. Gupta, S.S. Thorat, Recent Trends for the Formulation of Theranostic Mesoporous Silica Nanoparticles: A Review.

[32] S. Williams, A. Neumann, I. Bremer, Y. Su, G. Dräger, C. Kasper, P. Behrens, Nanoporous silica nanoparticles as biomaterials: evaluation of different strategies for the functionalization with polysialic acid by step-by-step cytocompatibility testing, Journal of Materials Science: Materials in Medicine, 26 (2015) 125. https://doi.org/10.1007/s10856-015-5409-3

Materials Research Forum LLC
https://doi.org/10.21741/9781644901076-4

[33] N. Khimich, A study of the possibility of transforming monolithic silica gel into silica glass upon low-temperature sintering, Glass physics and chemistry, 29 (2003) 513-515.

[34] F. Hoffmann, M. Cornelius, J. Morell, M. Fröba, Silica-based mesoporous organic–inorganic hybrid materials, Angewandte Chemie International Edition, 45 (2006) 3216-3251. https://doi.org/10.1002/anie.200503075

[35] L. Qi, J. Ma, H. Cheng, Z. Zhao, Micrometer-sized mesoporous silica spheres grown under static conditions, Chemistry of materials, 10 (1998) 1623-1626. https://doi.org/10.1021/cm970811a

[36] K. Kosuge, P.S. Singh, Rapid Synthesis of Al-Containing Mesoporous Silica Hard Spheres of 30− 50 µm Diameter, Chemistry of Materials, 13 (2001) 2476-2482. https://doi.org/10.1021/cm000623b

[37] M. Grün, K.K. Unger, A. Matsumoto, K. Tsutsumi, Novel pathways for the preparation of mesoporous MCM-41 materials: control of porosity and morphology, Microporous and mesoporous materials, 27 (1999) 207-216. https://doi.org/10.1016/S1387-1811(98)00255-8

[38] Q. Huo, D.I. Margolese, G.D. Stucky, Surfactant control of phases in the synthesis of mesoporous silica-based materials, Chemistry of Materials, 8 (1996) 1147-1160. https://doi.org/10.1021/cm960137h

[39] S. Schacht, Q. Huo, I. Voigt-Martin, G. Stucky, F. Schüth, Oil-water interface templating of mesoporous macroscale structures, Science, 273 (1996) 768-771. https://doi.org/10.1126/science.273.5276.768

[40] J. Vartuli, K. Schmitt, C. Kresge, W. Roth, M. Leonowicz, S. McCullen, S. Hellring, J. Beck, J. Schlenker, Effect of surfactant/silica molar ratios on the formation of mesoporous molecular sieves: inorganic mimicry of surfactant liquid-crystal phases and mechanistic implications, Chemistry of Materials, 6 (1994) 2317-2326. https://doi.org/10.1021/cm00048a018

[41] T. Azaïs, C. Tourné-Péteilh, F. Aussenac, N. Baccile, C. Coelho, J.-M. Devoisselle, F. Babonneau, Solid-state NMR study of ibuprofen confined in MCM-41 material, Chemistry of Materials, 18 (2006) 6382-6390. https://doi.org/10.1021/cm061551c

[42]   D.J. Mihalcik, W. Lin, Mesoporous silica nanosphere-supported chiral ruthenium
        catalysts: synthesis, characterization, and asymmetric hydrogenation studies,
        ChemCatChem, 1 (2009) 406-413. https://doi.org/10.1002/cctc.200900188

[43]   Y. Huang, W. Deng, E. Guo, P.W. Chung, S. Chen, B.G. Trewyn, R.C. Brown,
        V.S.Y. Lin, Mesoporous Silica Nanoparticle-Stabilized and Manganese-Modified
        Rhodium Nanoparticles as Catalysts for Highly Selective Synthesis of Ethanol and
        Acetaldehyde from Syngas, ChemCatChem, 4 (2012) 674-680.
        https://doi.org/10.1002/cctc.201100460

[44]   Y. Hoshikawa, H. Yabe, A. Nomura, T. Yamaki, A. Shimojima, T. Okubo,
        Mesoporous silica nanoparticles with remarkable stability and dispersibility for
        antireflective coatings, Chemistry of Materials, 22 (2009) 12-14.
        https://doi.org/10.1021/cm902239a

[45]   M. Vallet-Regi, A. Ramila, R. Del Real, J. Pérez-Pariente, A new property of
        MCM-41: drug delivery system, Chemistry of Materials, 13 (2001) 308-311.
        https://doi.org/10.1021/cm0011559

[46]   K. Unger, H. Rupprecht, B. Valentin, W. Kircher, The use of porous and surface
        modified silicas as drug delivery and stabilizing agents, Drug Development and
        Industrial Pharmacy, 9 (1983) 69-91. https://doi.org/10.3109/03639048309048546

[47]   B. Qian, Y. Deng, J.H. Im, R.J. Muschel, Y. Zou, J. Li, R.A. Lang, J.W. Pollard, A
        distinct macrophage population mediates metastatic breast cancer cell extravasation,
        establishment and growth, PloS one, 4 (2009) e6562.
        https://doi.org/10.1371/journal.pone.0006562

[48]   T. López, M. Alvarez, S. Arroyo, A. Sánchez, D. Rembao, R. López, Obtaining of
        SiO2 nanostructured materials for local drug delivery of methotrexate, J Biotechnol
        Biomaterial S, 4 (2011) 2. https://doi.org/10.4172/2155-952X.S4-001

[49]   E. Tasciotti, X. Liu, R. Bhavane, K. Plant, A.D. Leonard, B.K. Price, M.M.-C.
        Cheng, P. Decuzzi, J.M. Tour, F. Robertson, Mesoporous silicon particles as a
        multistage delivery system for imaging and therapeutic applications, Nature
        nanotechnology, 3 (2008) 151. https://doi.org/10.1038/nnano.2008.34

[50]   D. Böcking, O. Wiltschka, J. Niinimäki, H. Shokry, R. Brenner, M. Lindén, C.
        Sahlgren, Mesoporous silica nanoparticle-based substrates for cell directed delivery
        of Notch signalling modulators to control myoblast differentiation, Nanoscale, 6
        (2014) 1490-1498. https://doi.org/10.1039/C3NR04022D

[51]   I.I. Slowing, J.L. Vivero-Escoto, C.-W. Wu, V.S.-Y. Lin, Mesoporous silica nanoparticles as controlled release drug delivery and gene transfection carriers, Advanced drug delivery reviews, 60 (2008) 1278-1288. https://doi.org/10.1016/j.addr.2008.03.012

[52]   D. Luo, W.M. Saltzman, Enhancement of transfection by physical concentration of DNA at the cell surface, Nature biotechnology, 18 (2000) 893. https://doi.org/10.1038/78523

[53]   D. Luo, E. Han, N. Belcheva, W.M. Saltzman, A self-assembled, modular DNA delivery system mediated by silica nanoparticles, Journal of Controlled Release, 95 (2004) 333-341. https://doi.org/10.1016/j.jconrel.2003.11.019

Nanohybrids   Materials Research Forum LLC
Materials Research Foundations **87** (2021) 121-133   https://doi.org/10.21741/9781644901076-5

Chapter 5

Nanohybrids for Wound Healing Application

A. Saravanan[1], P. Senthil Kumar[2,3*], R. Jayasree[1], S. Jeevanantham[1]

[1]Department of Biotechnology, Rajalakshmi Engineering College, Chennai 602105, India

[2]Department of Chemical Engineering, SSN College of Engineering, Chennai 603110, India

[3]SSN-Centre for Radiation, Environmental Science and Technology (SSN-CREST), SSN College of Engineering, Chennai 603110, India

senthilkumarp@ssn.edu.in*

Abstract

The microfluidics-delivered nanohybrids invest the framework with an arranged course from wound discovery, receptive oxygen species rummaging and drug release. The drug release conduct mirrors the dynamic wound healing process, hence rendering an upgraded bio-mimetic regeneration. The properties of nanomaterials that are 1– 100 nm in size can be controlled to influence their capacities while connecting with biomaterials and biomedicines. Among the different sorts of nanomaterials, the clay minerals are universal in soils and viewed as protected materials for use in medicinal applications. Anti-bacterial activity is a vital factor for wound healing. Re-epithelization happens amid wound healing and includes the expansion of keratinocytes and the separation of fibroblasts. Ongoing improvements in nanotechnology for blending nanometer-estimate materials may give a chance to empowering viable wound healing because of material surface collaboration with cells and tissue.

Keywords

Nanomaterials, Wound Healing, Fibroblasts, Keratinocytes, Anti-Bacterial

Contents

1.   Introduction

Different sorts of wounds are driving reasons for sombreness and mortality, and each year they affect a huge number of people the world over. The nonhealing wounds causes several problems to the patients as well as the healthcare professionals such as agony, lessened personal satisfaction, expanded dreariness, requires more hospitalization and mortality which results in chronic ulcer [1]. Healing is the process by which the integrity of the wounded tissue could be restored. The wounds are majorly classified into two types such as acute and chronic skin wounds. Acute skin wounds were healing within a prescribed period of time through a certain continuous process including inflammation, tissue formation and remodelling but the chronic wounds do not heal through continuous process within particular period of time. Examples of acute and chronic skin wounds are skin burn and venous or arterial ulcer and diabetic foot ulcer (DFU) respectively. There are several factors including age, chronic diseases such as vascular disorders and diabetics, obesity, smoking, change in collagen formation, inflammatory response, basement membrane flattening and the rate of blood supply could affect or minimize the rate of wound healing process [2]. Because of the unique characteristics features of wounds and mending stages, distinctive wound dressings can be utilized to meet most of the requirements in a specific wound healing process.

There are several therapeutic approaches that have been utilised for the controlling of the wound healing process which reduces the requirement of secondary complications such as; biomaterial and nanotechnology based methods using hydrogel, hybrids, films, foams, hydrocolloids, nanoparticles, nanotubes, nanocomposites, nanoclays, nanofillers, collagen, chitosan, gelatin, skin fibroin and nanofibers. Some advanced dressing methods like biobrane dressings and transplantation methods such as split-thickness autografts, cultured epithelial autografts and use of donor keratinocytes are also used for wound healing process [3]. The remarkable properties of nanoparticles (NPs) are the consequences of their nanometer size and cause different changes in their physical, chemical and antimicrobial properties. There are several organic and inorganic materials

Nanohybrids   Materials Research Forum LLC
Materials Research Foundations **87** (2021) 121-133   https://doi.org/10.21741/9781644901076-5

were used for the synthesis of nanohybrids, among them insoluble polysaccharides are utilized more due to its unique properties such as biocompatibility, biodegradability, availability, renewable nature and high specific strength. Currently there are different studies focussed on the various low cost organic supporting matrixes including cellulose and chitosan based network for the impregnation of different nanoparticles such as MnO, platinum, Fe_3O_4, silver, magnetic Fe_2O_3 and gold nanoparticles.

Various methods have been utilized for the construction of nanohybrids including in-situ and ex-situ co-precipitation, sonochemical method and direct mixing method. Among the various preparation methods, the direct mixing method is effectively utilized for the synthesis of nanohybrids due to certain advantages like low cost method, required lower chemicals and do not requires any precursors for the nanohybrid synthesis [4]. From the literature survey, it was observed that the in-situ co-precipitation strategy offers better control of development of uniform size and dissemination nanoparticles throughout the polymer matrix. Preparation of nanohybrids by immobilizing the nanoparticles onto the organic and inorganic matrices has several advantages such as; reusability, absence of molecule conglomeration, uniform dissemination of nanoparticles, low cost production and high catalytic activity [5]. Various analytical methods including Fourier transform infra-red (FTIR) spectroscopy, X-ray diffraction (XRD), Field emission scanning electron microscopy (FE-SEM) and thermogravimetric analysis were used to characterise the properties and mechanism of the nanohybrids. FTIR analysis is used to study the functional group arrangements on the surface of the nanohybrids and their possible interactions with organic and inorganic substances. XRD and SEM analysis were employed to study the structural properties and morphological characteristics of the nanohybrids. Thermogravimetric analysis was utilized to determine the thermal degradation nature of the nanohybrids. Nanohybrids have various applications in different fields, in biology its most commonly used for wound healing and tissue engineering applications.

This chapter mainly focuses on the various nanotechnology advancements and advancement for wound healing process. It describes the nanohybrids and their preparation, principle, multifunctional and their major application in the wound healing process major application of nanohybrids in wound healing and tissue engineering research. The current status of the nanotechnology research for the development of nanohybrids and future expectations of them are also discussed in this chapter.

Nanohybrids   Materials Research Forum LLC
Materials Research Foundations **87** (2021) 121-133   https://doi.org/10.21741/9781644901076-5

2.   Wound healing

Wound is depicted as harm and any confusion in the healthy structure and capacity of the skin. The wound occurs because of damage, hereditary issue, intense injury, warm injury, and surgical mediations. Commonly, wound is classified into two types based on their nature such as open and close wounds. Close wounds incorporate injury, hematoma, and scraped spots for example, harm of delicate tissues, little veins or profound tissue layers. While open injuries by and large comprise of gashes, cutting pricing apparatus injury, surgical injuries, creepy crawly nibbles, stings, radionecrosis, vascular neurological and metabolic injuries. Wounds can be clinically partitioned into acute and chronic wounds based on their healing nature. In acute wounds, the damage could be actuated by various components, for example, radiation, extraordinary temperature changes, or contact with synthetics. The acute wounds take 8-12 weeks for complete healing but the chronic wounds require months for the healing process because of its delayed inflammation. Wound healing is a multi-factorial physiological procedure. However, its multifaceted nature can bring about a few anomalies. Specifically, it should have the option to evacuate exorbitant exudate, improve autolytic debridement, and keep the dampness sufficient for recuperating. A general consideration technique incorporates swabbing for disease, cleaning of the injury bed from tissue trash, regardless of whether it needs to the split-thickness skin autografts or allografts, at last applying the injury dress.

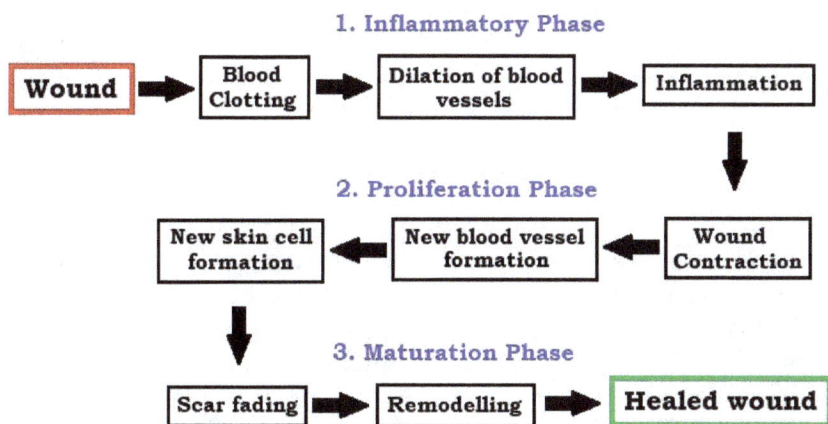

Figure 1. Stages of wound healing.

Nanohybrids   Materials Research Forum LLC
Materials Research Foundations **87** (2021) 121-133   https://doi.org/10.21741/9781644901076-5

Wound healing process consisting of three different stages by which the injured tissues could be regenerated. The wound healing process via three stages was shown in figure 1. The first stage of wound healing is named as inflammation phase. At the inflammation phase, initially the hemostasis occurs to avoid the excessive release of blood through the injury. Blood dilation takes place followed by the blood clotting which wallows the necessary inflammatory cells and metabolites including proteolytic enzymes, pro-inflammatory cytokines, macrophages and neutrophils to the wounded area which causes inflammation. After the inflammation phase the migrated inflammatory cells induce the beginning of the re-epithelialisation phase within a short period of injury. At the proliferation phase the new blood vessels were starting to form as a result of re-epithelialisation. The angiogenesis to inaugurate the synthesis and deposition of extra cellular protein fragments. The new skin cells start to form due to the fibroblast cells which support the skin cell growth by utilizing ECM protein fragments including collagen as building squares. Thereby, the fibroblast cells were playing an important role in tissue regeneration for the wound healing process at the proliferation phase. Final stage of wound healing is known as maturation phase where the remodelling of cells occurs to fade the wound scares [6]. The process of wound healing does not just include the reclamation of skin hindrance respectability and inner homeostasis, yet in addition, lessens the danger of disease and optional intricacies. There are various inconveniences that could hinder the ordinary injury recuperating procedures and lead to the change of minor injuries to the non-linear injuries. Dressing and ointments were utilized to maintain the moist environment for the wound healing process which will minimize the external influences that could reduce the wound healing process. There are several biomaterials effectively used for restoration tissue characteristics after the injury including hydrogels, nanohybrids, impregnated or immobilized nanomaterials, nanoclays, chitosan, skin fibroin, collagen and gelatin.

There are different internal and external factors which affects the wound healing process. The major factors which delay the wound healing process are shown in figure 2.

The factors are aging, obesity, pressure, trauma and edema, chronic diseases such as diabetes and anemia, stress, malnutrition, necrosis, smoking, immunodeficiency, vascular insufficiency, skin moisture, medication and lack of oxygen delivery to the tissue delay the wound healing process [7]. Aging was considered as important factor for the delayed wound healing process because it affects the structure and function of the skin by reducing the inflammatory responses and making the skins even thinner. Skin needs a satisfactory measure of liquid and wetness to be practical. In case you're inclined to dry skin you might be in danger for skin sores, disease, and thickening, which will all weaken the wound healing process. On the other side, if the skin is excessively wet, you're in danger of creating maceration as well as contaminations, so keeping up an ideal degree of skin dampness is

Nanohybrids   Materials Research Forum LLC
Materials Research Foundations **87** (2021) 121-133   https://doi.org/10.21741/9781644901076-5

basic for healing of wounds. Cardiovascular conditions are among the most adverse, however diabetes and immunodeficiency conditions can likewise slow twisted the wound healing process. Physician endorsed meds can negatively affect recuperating. For example, as indicated by the American Academy of Orthopedic Surgeons, nonsteroidal mitigating drugs frequently recommended for arthritis and found over the counter as headache medicine and ibuprofen, can meddle with the aggravation phase of the recuperating procedure. Anticoagulants have the ability to alter blood thickening, while immunosuppressant may debilitate the insusceptible framework and upgrade the danger of contamination.

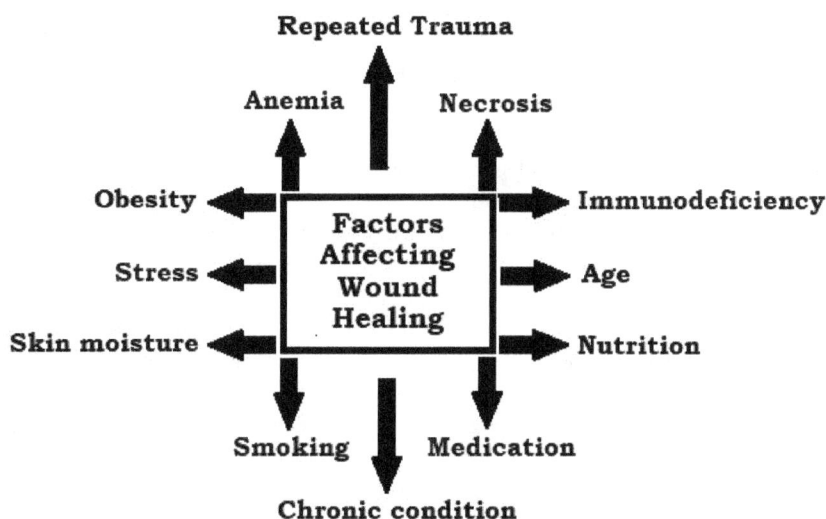

Figure 2. Factors affecting the wound healing process.

3.   Nanotechnology advancements and approaches

Recently, researchers have focused upon the improvement of nanotechnology-based conveyance frameworks because of the more noteworthy relevance to accomplishing target-explicit conveyance of restorative load while limiting their off-target impacts. Nanoencapsulation is a practical and invaluable methodology that empowers target-explicit conveyance of curcumin and improves its restorative adequacy. Nanotechnology-based conveyance frameworks increased far-reaching acknowledgment in view of their promising potential and points of interest over the traditional methodologies. Polymeric nanoparticles have been considered by numerous scientists to assess their capacity to improve pharmaceutical essentialness and remedial feasibility of the embodied medications payload. The accomplishment of this nanocarrier conveyance framework is because of their ultra-little size, high epitome proficiency, ideal zeta potential, biodegradability and biocompatibility. Moreover, these conveyance frameworks can forestall untimely corruption of epitomized drugs, give controlled/continued conveyance, accomplish target-situated conveyance of the medications and forestall foundational lethality [8].

There is a huge improvement in wound healing adequacy of the curcumin was developed recently by stacking in poly(ε-caprolactone) (PCL)/gum tragacanth (GT) electrospun nanofibers [9]. They formed PCL/GT/CUR-stacked nanofibers by electrospinning system. Nanoencapsulation of the curcumin utilizing polymeric nanoparticles has likewise indicated more prominent potential in recuperating consume wounds. Nanoencapsulation of curcumin as poly (lactic-co-glycolic corrosive (PLGA)- NPs has likewise indicated promising potential in improving injury recuperating of cutaneous excisional wounds. Treatment of burned wounds is frequently convoluted because of bacterial disease; therefore, there is an absence of outright decision because of deficient microbial inclusion, improvement of bacterial obstruction, and insufficient entrance into the objective tissues. Subsequently, to improve wound recuperating movement, microbial action, solvency, and steadiness of the curcumin, Krausz and associates endeavoured exemplification of curcumin as polymeric nanoparticles. They dissected the readied curcumin-stacked nanoparticles which mean molecule size, zeta potential, in vitro discharge qualities, cell cytotoxicity, zebrafish cytotoxicity, and antibacterial movement against MRSA and P. aeruginosa. Further, they additionally assessed the injury mending adequacy of the curcumin-stacked nanoparticle in both murine and contaminated murine consumption models [10].

The nanoencapsulation of curcumin as nanohybrid frameworks has likewise been explored for wound healing viability in diabetic incessant injuries. Authors endeavoured combination of a novel nanohybrid support by typifying curcumin in CSNPs to improve fluid dissolvability and steadiness of curcumin, trailed by impregnation of curcumin-stacked curcumin scaffold nanoparticles into collagen platform or nanohybrid support. Polymeric micelles are nanoscopic center/shell structures framed by amphiphilic square copolymers. Both the inalienable and modifiable properties of the polymeric micelles render them appropriate for the medication conveyance purposes. Micelles and hydrogel-based conveyance network have additionally been broadly used in improving the pharmaceutical hugeness and helpful adequacy of a wide scope of dynamic specialists including development factors have been utilized to embody a wide scope of pharmacological moieties including silver sulfadiazine, cyclosporine A, quercetin (QUE), hyaluronic corrosive, and nanohybrid to improve their wound recuperating efficacies for the administration of cutaneous, diabetic, and excisional wounds [11-13].

Currently, a novel nanocomposite hydrogel made of nano-curcumin, chitosan and oxidized alginate has demonstrated a more noteworthy capacity to quicken dermal injury regeneration. In this examination, specialists assessed the in vitro and in vivo viability of curcumin-encapsulated nanocomposite hydrogel which was recently created by dainty film vanishing strategy by utilizing methoxy poly(ethylene glycol)- b-poly(ε caprolactone) copolymer (MPEG-PCL) as a bearer/vehicle trailed by consolidation into the N,O-carboxymethyl CS/oxidized alginate hydrogel (CCS-OA hydrogel). The created nanocomposite hydrogel was assessed for in vitro dependability, discharge profile, cell reinforcement movement and in vivo wound recuperating adequacy [14].

4. Nanohybrids as a promising wound healing agent

Biomaterial based organic and inorganic nanohybrids have been effectively utilized for various purposes due to its unique characteristics such as bioavailability and improved features. The heterostructure layered nanohybrids were prepared by intercalation reaction which means reversible insertion of required materials into the two-dimensional (2D) compound. Different varieties of biomaterials and drug molecules with two different charges (cationic and anionic) have been consolidated into the conveyance transporters for creating heterostructured layered nanohybrids with chemo-remedial and quality restorative capacities [15]. There are different kinds of biomaterials such as hydrogels, clays, collagen and chitosan based polymers, silica, nanofibers, nanofillers and certain nanoparticles including silver, gold, platinum, MnO, etc. have been used for the hybridization of required materials. The advancement of biomaterial-based hybrid bearers with various useful constituents synergizes the proficient medication epitome and conveyance, subsequently lessening the half-maximal inhibitory fixation esteems with the improved restorative capability of the concerned medication. In such manner, hybridization of the bearer utilizing inorganic metal/metal oxide nanoparticles is of an incredible remedial advantage. The nanofiber based nanohybrids prepared through the fabrication of organic and inorganic nanofiber and it could be effectively utilized for various applications including as sensing agents in electrical devices, as a part of energy conservation devices, as catalyst, as adsorbent for the heavy metal remediation and as an effective supporting matrix for wound healing.

Wounds can be recuperated by covering a slender snare of nanofiber nanohybrids due to their unique features such as having enough pores guaranteeing the exchange of fluids and gases with the environment and also prevent microorganism's entry from the environment. Compared with commercially used skin substitutes, electrospun nanofibers are substantially more invaluable to decrease the bleakness. In diabetic conditions, this parity is upset bringing about wounds with more elevated levels of incendiary mediators and oxidative stress conditions causing collagen corruption. There are different collagen based nanohybrid scaffolds been utilized to convey anti-toxins, growth factors, cytokines as regenerative medication in diabetic injury recuperating. The nanofibrous structure of the nanowebs can advance skin development; if appropriate can be coordinated with therapeutic mixes inside the fibres. Polymer/Ag electrospun hybridized nanofibers have been viably used for wound dressing since Ag NPs and Ag salts have been utilized as antimicrobial specialists for quite a long time. In addition, Ag is additionally ready to encourage the generation of oxygen which thus decimates the membranes of microscopic organisms [16].

5.   Nanohybrids: Design principle, preparation, multifunction

A nanohybrid is a blend of nanometric organometallic or a natural/inorganic blend including clay, polymers, hydrogels and different nanoparticles. The principle behind the nanohybrid preparation is fabrication, through fabrication the biologically active molecules were conjugated with support matrix and most commonly biomaterials are used as supporting matrix. Nanohybrids have been prepared by the following methods; direct method, in-situ co-precipitation, ex-situ co-precipitation, immobilization, fabrication and sonochemical methods. For synthesis of nanohybrid, initially the biologically active compounds and the supporting matrix should be synthesized separately. Two types of nanohybrids have been prepared using graphene oxide (GO) and silver (Ag) nanoparticle [17]. For the preparation of AgNPs-GO nanohybrid the graphene oxide sheets were formed via oxidation of graphite by utilizing graphite powder, sodium nitrate, potassium permanganate and concentrated sulphuric acid. By utilizing the synthesized graphene oxide solution, silver nanoparticles graphene oxide (AgNPs) nanohybrids were prepared. For the preparation of AgNPs-GO nanohybride, the GO solution was thoroughly mixed with $AgNO_3$ solution. The wound healing properties was included into the prepared AgNPs-GO nanohybride by grafting the chitosan biopolymer onto the AgNPs-GO through self-assembly method (Marta et al., 2015). The preparation of nanofiber based nanohybrid was prepared by fabricating the synthesized nanofiber via electrospinning method. The nanoparticles based nanohybrids were prepared by fabricating the immobilized nanoparticles. Zero-valent metal nanoparticles such as silver, iron, platinum, gold, etc. are the advanced materials currently used for the synthesis of nanoparticle based hybrids due to their specific properties including catalytic, electronic, magnetic, thermal and optical properties.

Materials Research Forum LLC
https://doi.org/10.21741/9781644901076-5

The prepared nanohybrid has several applications including environmental remediation, catalysis, electronic and sensing device, energy-conservation devices and wound dressing were shown in figure 3.

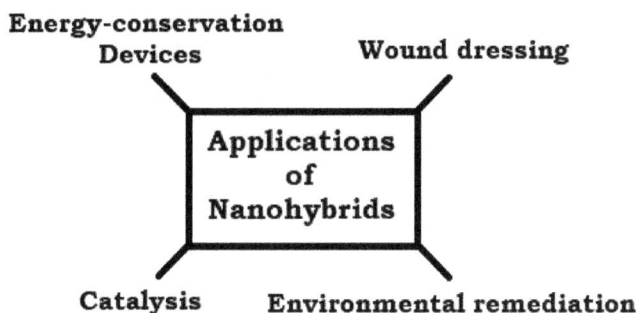

Figure 3. Applications of Nanohybrids.

Nanohybrids have been utilized effectively to remove the toxic pollutants from the environmental resources because of their higher surface area, cost effectiveness and porous structure. They could remove toxic heavy metals, electrical wastes, and organic contaminants from the water resources. An essential worry in the advancement of impetus is the recyclability of the catalyst. The immobilization of homogeneous or heterogeneous nanoscale catalysts in electrospun nanofibers shows an intriguing arrangement inferable from their huge surface area and high porosity that permit the substrate particles to survey the surface of the immobilized nanoscale catalyst. Several nanoparticles including silver, gold and platinum and nanofiber based nanohybrids have been used as sensing agents in various electronical and sensing devices. To increase the electricity production the prepared nanofibers and nanoparticle based nanohybrids are used in energy-conservation and energy storage devices. Wound healing is the process of regeneration of tissues in the injured area; during regeneration the cells required a specific supporting material for their nutrition and support. The electrospun nanofiber based nanohybrids have enough pores that enhance the wound healing process by providing suitable moisture environment through exchanging liquids and gases and prevent the microbial contamination.

Conclusion and future perspectives

Though there are several techniques available for wound dressing and healing, the nanohybrid based wound dressing process considered as effective and rapid healing process due to their features; providing higher surface area, maintaining proper moisture content and preventing the microbial contamination. A wide variety of biomaterials have been used for the synthesis of nanofiber nanohybrids through electrospinning method by immobilizing with different nanoparticles. Though the researchers have developed advanced many techniques for the synthesis of nanohybrids by utilizing various biomaterials and polymers, they have certain difficulties during as well as after the formation such as they required high initial economical investment. Though an assortment of electrospun hybrid nanofibers have been created by means of various in situ techniques for various applications, this region of research despite everything stays open and there are as yet numerous issues staying unexplored. For example, progressively essential examinations despite everything stay to be completed and the determination of polymer frameworks from the perspective of structure, nanotexture control, and mechanical solidness of the shaped nanofibers, enhancement of electrospinning conditions to more readily control the morphology of the filaments, and investigating new in situ responses to form new nanohybrid materials.

References

[1]   N.B. Menke, K.R. Ward, T.M. Witten, D.G. Bonchev, R.F. Diegelmann, Impaired wound healing, Clinics in Dermatology, 25 (2007) 19-25. https://doi.org/10.1016/j.clindermatol.2006.12.005

[2]   A. Bhatia, K. O'Brien, M. Chen, A. Wong, W. Garner, D.T. Woodley, W. Li, Dual therapeutic functions of F-5 fragment in burn wounds: preventing wound progression and promoting wound healing in pigs, Molecular Therapy - Methods & Clinical Development, 3 (2016) 16041. https://doi.org/10.1038/mtm.2016.41

[3]   Z. Hussain, H.E. Thu, S.-F. Ng, S. Khan, H. Katas, Nanoencapsulation, an efficient and promising approach to maximize wound healing efficacy of curcumin: A review of new trends and state-of-the-art, Colloids and Surfaces B: Biointerfaces, 150 (2017) 223-241. https://doi.org/10.1016/j.colsurfb.2016.11.036

[4]   M. Moritz, M. Geszke-Moritz, The newest achievements in synthesis, immobilization and practical applications of antibacterial nanoparticles, Chemical Engineering Journal, 228 (2013) 596–613. https://doi.org/10.1016/j.cej.2013.05.046

[5] S.S. Mirtalebi, H. Almasi, M.A. Khaledabad, Physical, morphological, antimicrobial and release properties of novel MgO-bacterial cellulose nanohybrids prepared by in-situ and ex-situ methods, International Journal of Biological Macromolecules, 128 (2019) 848-857. https://doi.org/10.1016/j.ijbiomac.2019.02.007

[6] M. Kuhlmann, W. Wigger-Alberti, Y.V. Mackensen, M. Ebbinghaus, R. Williams, F. Krause-Kyora, R. Wolber, Wound healing characteristics of a novel wound healing ointment in an abrasive wound model: a randomised, intra-individual clinical investigation, Wound Medicine, 24 (2019) 24-32. https://doi.org/10.1016/j.wndm.2019.02.002

[7] S. Guo, L.A. DiPietro, Factors Affecting Wound Healing, Journal of Dental Research, 89 (2010) 219–229. https://doi.org/10.1177/0022034509359125

[8] Z. Hussain, H. Katas, M.C.I.M. Amin, E. Kumolosasi, S. Sahudin, Downregulation of immunological mediators in 2,4-dinitrofluorobenzene-induced atopic dermatitis-like skin lesions by hydrocortisone-loaded chitosan nanoparticles, International Journal of Nanomedicine, 9 (2014) 5143–5156. https://doi.org/10.2147/IJN.S71543

[9] M. Ranjbar-Mohammadi, S.H. Bahrami, Electrospun curcumin loaded poly(ε caprolactone)/gum tragacanth nanofibers for biomedical application, International Journal of Biological Macromolecules, 84 (2016) 448–456. https://doi.org/10.1016/j.ijbiomac.2015.12.024

[10] A.E. Krausz, B.L. Adler, V. Cabral, M. Navati, J. Doerner, R.A. Charafeddine, D. Chandra, H. Liang, L. Gunther, A. Clendaniel, S. Harper, J.M. Friedman, J.D. Nosanchuk, A.J. Friedman, Curcumin-encapsulated nanoparticles as innovative antimicrobial and wound healing agent, Nanomedicine: Nanotechnology, Biology and Medicine, 11 (2015) 195–206. https://doi.org/10.1016/j.nano.2014.09.004

[11] C. Gong, Q. Wu, Y. Wang, D. Zhang, F. Luo, X. Zhao, Y. Wei, Z. Qian, A biodegradable hydrogel system containing curcumin encapsulated in micelles for cutaneous wound healing, Biomaterials, 34 (2013) 6377–6387. https://doi.org/10.1016/j.biomaterials.2013.05.005

[12] E. Dellera, M.C. Bonferoni, G. Sandri, S. Rossi, F. Ferrari, C. Del Fante, C. Perotti, P. Grisoli, C. Caramella, Development of chitosan oleate ionic micelles loaded with silver sulfadiazine to be associated with platelet lysate for application in wound healing, European Journal of Pharmaceutics and Biopharmaceutics, 88 (2014) 643–650. https://doi.org/10.1016/j.ejpb.2014.07.015

[13]   V.V.S.R. Karri, G. Kuppusamy, S.V. Talluri, S.S. Mannemala, R. Kollipara, A.D. Wadhwani, S. Mulukutla, K.R.S. Raju, R. Malayandi, Curcumin loaded chitosan nanoparticles impregnated into collagen-alginate scaffolds for diabetic wound healing, International Journal of Biological Macromolecules, 93 (2016) 1519–1529. https://doi.org/10.1016/j.ijbiomac.2016.05.038

[14]   X. Li, S. Chen, B. Zhang, M. Li, K. Diao, Z. Zhang, J. Li, Y. Xu, X. Wang, H. Chen, In situ injectable nanocomposite hydrogel composed of curcumin, N, Ocarboxymethyl chitosan and oxidized alginate for wound healing application, International Journal of Pharmaceutics, 437 (2012) 110–119. https://doi.org/10.1016/j.ijpharm.2012.08.001

[15]   J.Y. Kim, J.H. Yang, J.H. Lee, G. Choi, M.R. Jo, S.J. Choi, J.H. Choy, 2D inorganic-antimalaria drug–polymer hybrid with pH-responsive solubility, Chemistry: An Asian Journal, 10 (2015) 2264–2271. https://doi.org/10.1002/asia.201500347

[16]   Z. Qiao, M. Shen, Y. Xiao, M. Zhu, S. Mignani, J.P. Majoral, X. Shi, Organic/inorganic nanohybrids formed using electrospun polymer nanofibers as nanoreactors, Coordination Chemistry Reviews, 372 (2018) 31–51. https://doi.org/10.1016/j.ccr.2018.06.001

[17]   B. Marta, M. Potara, M. Iliut, E. Jakab, T. Radu, F. Imre-Lucaci, G. Katona, O. Popescu, S. Astilean, Designing chitosan–silver nanoparticles–graphene oxide nanohybrids with enhanced antibacterial activity against Staphylococcus aureus, Colloids and Surfaces A: Physicochemical and Engineering Aspects, 487 (2015) 113–120. https://doi.org/10.1016/j.colsurfa.2015.09.046

Nanohybrids
Materials Research Foundations **87** (2021) 134-155

Materials Research Forum LLC
https://doi.org/10.21741/9781644901076-6

Chapter 6

Bioresorbable Metals for Cardiovascular and Fracture Repair Implants

S.C. Cifuentes[1,*], V. San-Miguel[2], Y. Wang[3], A. García-Peñas[2]

[1]Área de Ciencia e Ingeniería de Materiales, ESCET, Universidad Rey Juan Carlos, C/Tulipán s/n, 28933 Móstoles, Madrid, Spain

[2]Departamento de Ciencia e Ingeniería de Materiales e Ingeniería Química (IAAB), Universidad Carlos III de Madrid, 28911 Leganés, Madrid, Spain

[3]College of Materials Science and Engineering, Shenzhen Key Laboratory of Polymer Science and Technology, Guangdong Research Center for Interfacial Engineering of Functional Materials, Nanshan District Key Laboratory for Biopolymers and Safety Evaluation, Shenzhen University, Shenzhen, 518055, P. R. China

sandra.cifuentes@urjc.es*

Abstract

The potential of bioresorbable metals to revolutionize current and future medical devices fascinates researchers. Magnesium, iron, and zinc have been thoroughly studied for the treatment of cardiovascular diseases or to repair fractures. Iron was the first type of metal being researched and introduced in biomedical applications. Magnesium is the most studied one, it has been tested by clinical trials and commercially available products have been already developed. The interest in zinc has recently emerged and is continuously growing. This chapter offers an overview of the role that Mg, Fe, and Zn are playing advancing the evolution of bioresorbable implants.

Keywords

Biodegradable Metals, Cardiovascular Implants, Medical Devices for Fracture Repair, Magnesium, Iron, Zinc

Contents

1.   Introduction

The continuous and rapid progress in global life expectancy, together with the high rate of world population growth, are closely linked to population aging. This demographic change has resulted in higher proportions of older people than younger people. People who are over 60 are more likely to suffer fractures or coronary heart diseases. In fact, cardiovascular disease is the main cause of death worldwide.

The European Society of Cardiology states that each year cardiovascular disease causes over 3.9 million deaths in Europe [1]. According to the Global Trauma Devices Market Analysis, Size and Trends, more than 1.9 million fracture repair procedures are performed only in the United States annually [2]. In the next decades, the incidence of fractures and cardiovascular diseases will not slow. Therefore, doctors and scientists are working together gathering efforts to improve the quality of life of an aging population. One of the approached strategies is the development of innovative medical devices that fulfil their main function and at the same time reduce the costs of healthcare spending.

This chapter is focused on bioresorbable medical devices based mainly on metallic materials. The first part presents an overview of both, cardiovascular implants and medical devices for fracture repair, showing the importance and the mission of bioresorbable implants in both applications. The second part is focused on specific materials, mainly magnesium (Mg), iron (Fe), and zinc (Zn), that are either being used or under study for cardiovascular implants or fracture repair.

2.   Cardiovascular implants

The principal mission of a cardiovascular implant is to support the inner wall of blood vessels in the treatment of obstructive lesions. The stent has to improve blood flow and prevent the contraction of the veins and arteries reducing the recurrence of a blockage (restenosis) [3]. To fulfil its function, the stent has to: (1) withstand external forces, for

Nanohybrids   Materials Research Forum LLC
Materials Research Foundations **87** (2021) 134-155   https://doi.org/10.21741/9781644901076-6

which it must have an adequate radial resistance; (2) be flexible, to negotiate tortuous vessels; (3) be compatible with blood, and (4) show non-toxic reactions [4].

The implantation of bare metal stents (BMS), fabricated from stainless steel, cobalt-chromium alloy, and Nitinol (nickel-titanium alloy), has shown great improvements compared to simple balloon angioplasty. Nevertheless, permanent implants present some limitations that narrow their applicability. The long-term presence of a BMS causes in-stent restenosis due to the growth of scar tissue within the vessels (neointimal), chronic inflammatory local reactions, mismatches in mechanical behaviour between stented and non-stented vessel areas, interference of remodelling, and inability to adapt to growth in young patients [5].

Since the role of a coronary implant is temporary and limited to a period of 6 months, [6] there is no need to carry the permanent implant together with its long-term complications. For this reason, stenting technology has moved toward the development of a new generation of biodegradable stents able to "fulfil the mission and step away" [3]. A biodegradable stent must be biocompatible, as well as its degradation products, it has to maintain the radial force while the vessel is healing during the period of high risk for recoil and complete biodegrade afterwards [7].

3. Medical devices for fracture repair

Metallic prosthetic devices commonly used to repair, and regenerate injured bones have noticeably improved the quality of life of patients. Stainless steel (316L) is one of the most preferred biomaterials for plates and screws due to the combination of good mechanical properties, corrosion resistance, and cost-effectiveness. Titanium alloys (Ti-6Al-4V among others) are preferred for intramedullary rods, spinal clamps, self-drilling bone screws, and other implants, because of their high strength and lower elastic modulus [8].

However, permanent implants have some drawbacks, which imply serious nuisances for the patients, especially for the old ones. A second surgical procedure is often required to remove the implant after the tissue has sufficiently healed. This is needed in order to avoid detrimental side effects. For instance, ion release or stress shielding, are associated with the reduction of the stimulation of new bone growth and remodelling [9]. Complications related to thermal conductivity, metal hypersensitivity, chemical carcinogenesis, and/or infection can also stand out [10]. Implant removal procedure can be complex and lead to medical problems, such as neurovascular injury, refracture, or recurrence of deformity [11]. Moreover, the use of permanent implants is restricted for young patients, because permanent devices would inhibit the bone growth.

Nanohybrids   Materials Research Forum LLC
Materials Research Foundations **87** (2021) 134-155   https://doi.org/10.21741/9781644901076-6

To overcome the disadvantages implicit in the use of permanent implants for fracture repair, the development of bioresorbable devices was proposed. Degradable materials for osteosynthesis must fulfil an additional requirement: to be metabolized by the human body without leaving trace, and gradually lose their mechanical strength while the bone tissue is regenerated. In this way the system Bone+Implant can maintain its mechanical strength.

4.   Bioresorbable metals

Implants that can be metabolized by the human body have appeared as one of the most attractive and promising solutions for the treatment of cardiovascular diseases and fractures. The main objective of degradable implants is to overcome the limitations and improve the features of current implantable devices. Biodegradable metals have played an important role writing the history of resorbable implants. However, the use of these materials implies many new challenges that scientific community is facing and researching on. There are three main challenges to be addressed: achieve good mechanical properties, control of degradation times, and fulfil the biocompatibility of degradation products [12-15].

The potential of biodegradable metals to improve applicability of medical devices attracts a great interest from both fields: medical science and materials science and engineering. Iron (Fe), magnesium (Mg), and zinc (Zn) are some of the most thoroughly studied biodegradable metals that are either used or researched for the treatment of cardiovascular diseases or to repair fractures. Fe, Mg, and Zn are essential elements in the human body, playing an important role in physiological processes. Table 1 shows for Mg, Fe and Zn, and some of the most common alloying elements (Ca, Mn), their average amount in human body, their level in serum blood, and the daily allowance.

Fe and its alloys were the first type of degradable metals in being researched and introduced in biomedical applications. However, the slow *in-vivo* degradation of these alloys still remains a challenge for researchers. Mg alloys are the most studied ones, they have been tested by clinical trials and commercially available products have been already developed. Companies like Syntellix (Germany), Biotronik (Germany), or K-Met (Korea) have launched their own biodegradable metals products based on Mg alloys. By the side of Zn, the interest of its alloys for medical devices has very recently emerged and is continuously growing. The striking importance of biodegradable metallic medical devices has led to the development by the ASTM subcommittee F04.15 of new guides for testing procedures and processes of these metals [15].

Table 1. Amount in human organism, blood serum level, and daily allowance of Mg, Fe and Zn and the alloying elements Ca and Mn [16, 17].

Element	Amount in human organism	Blood serum level	Daily allowance
Mg	25 g	0.73 - 1.06 mN	0.7 g
Fe	4 - 5 g	5.0 - 17.6 g/L	10 - 20 mg
Zn	2 g	12.4 - 17.4 mM	12 - 15 mg
Alloying elements			
Ca	1100 g	0.919 - 0.933 mM	0.8 g
Mn	12 mg	< 0.8 mg/l	4 mg

4.1 Magnesium - Mg

Magnesium and its alloys are an interesting option to develop biodegradable materials with appropriate mechanical properties and good biocompatibility. Magnesium possesses exceptional mechanical and physiological features that make it attractive for bone repair and cardiovascular applications. The mechanical features of this metal are compatible with those of human bone. Mg density of 1.74 g/cm^3 and Young's modulus of 41 – 45 GPa are close to that of natural bone (1.8 – 2.1 g/cm^3 and 7 – 30 GPa) [16]. Physiologically, Mg is an essential mineral for bone formation and plays an important role in the regulation of calcium homeostasis inducing the mineralization process of bone tissue [18]. Its beneficial role enhancing the osteoblastic response and reducing bacterial adhesion has been pointed out, [19, 20] as well as Mg beneficial antithrombotic, antiarrhythmic, and antiproliferative properties [21]. Therefore, Mg and its alloys have been thoroughly studied for both, fracture repair and cardiovascular applications. This research has led to the development of commercially available resorbable devices based on Mg. Table 2 summarizes the evolution of bioresorbable Mg alloys research for cardiovascular and fracture repair applications.

Fracture repair:

Lambotte in 1907 attempted for the first time the use of Mg to stabilize a fractured leg [22]. Error! Bookmark not defined.However, pure magnesium corroded too rapidly *in vivo*, implant dissolved only 8 days after surgery and released a high amount of hydrogen gas. Nowadays, more than a century after Lambotte's failed attempt, the fast degradation rate in physiological environment and the evolution of hydrogen hinders magnesium successful application for bone repair.

Strategies to slow down Mg corrosion rate maintaining good mechanical properties have been addressed by tailoring the composition and microstructure of the base alloys.

Commercial alloys with aluminium (Al) and rare earth (RE) elements (AZ31, AZ91, WE43, WE54, ZM21, ZEK100) originally designed for automotive industry have been investigated *in vitro* and *in-vivo* [23, 24]. Although the alloys provide a high initial stability and positive biodegradation, the contents of Al and RE may be harmful. Neurotoxic effects of aluminium and the unknown biological effects of REs in the human body require further investigation to prove their biosafety [25, 26]. Moreover, some of the alloys containing REs have shown pathological effects on the host tissue [23]. Consequently, above certain levels, these alloying elements could be unsuitable for biomedical magnesium materials.

The design of alloys intended to repair bone fractures includes the incorporation of biologically important elements like Ca and Zn. Binary alloys Mg-Ca, Mg-Zn, ternary alloys Mg-Zn-Ca, and bulk metallic glasses have been thoroughly studied and show slower corrosion rate and good biocompatibility [27, 28]. These alloys have made great progress in the development of magnesium osteosynthesis implants. Nevertheless, an initially low degradation rate to avoid local alkalization, due to the formation of OH^- ions, has not been reached; as a result, it is still desirable to prevent further deterioration of the adjacent tissue and ensure proper tissue healing and growth [29]. The way to mitigate degradation in the initial stage is by surface treatments or coatings [30]. Different coating processes are reviewed in the literature, but among these processes, polymeric coatings showed a higher improvement in the initial corrosion protection and a better initial cytocompatibility. As polymers can play the role of a protective environment for Mg alloys, an alternative material consisting in the incorporation of Mg particles into a biodegradable polymeric matrix has been also proposed [31].

Pre-clinical studies have been performed in animals (mouse, rabbit, sheep, or pig) in order to assess the *in-vivo* performance of pure-Mg [32], ternary MgZnCa alloys, [33] and commercially available Mg alloys [34]. A MgCaZn alloy (RESOMET) has been evaluated in clinical trials for the treatment of hand fractures. This alloy system has received the approval of the Food and Drug Administration in Korea [35]. Nowadays, companies like Syntellix (Germany) or K-Met (Korea) offer their own products for fracture repair based on bioresorbable Mg alloys.

Cardiovascular:

The first approach for cardiovascular resorbable stents was the use of biodegradable polymers. However, in the circulatory system, polymer implants elicit more severe reactions than that of osteosynthesis devices. Willem J van der Giessen *et al.* [36] tested five different biodegradable polymers (polyglycolic acid/polylactic acid [PGLA], polycaprolactone [PCL], polyhydroxybutyrate valerate [PHBV], polyorthoester [POE],

Nanohybrids   Materials Research Forum LLC
Materials Research Foundations 87 (2021) 134-155   https://doi.org/10.21741/9781644901076-6

and polyethyleneoxide/polybutylene terephthalate [PEO/PBTP]) in porcine coronary arteries, finding that these polymers induced marked inflammatory changes during the absorption process, leading to thrombotic occlusion.

Encouraging results were obtained with poly-L-lactic acid and the first clinical implantation in humans of a fully biodegradable stent made of PLLA was performed by Tamai *et al.*[37] in 2000. However, to approximate the mechanical performance required in arteries, the implant was bulky, and this prevented the stent from becoming a viable alternative. It also induced some inflammatory responses, growth of neointimal scar tissue, and restenosis [37]. To prevent these adverse reactions, some drugs designed to prevent neointimal formation were incorporated into the stents [38]. The stent acts as a vehicle that elutes the drug and treats the problem right in the site of interest.

The failure of biodegradable polymeric stents has been centred on three major hurdles, namely: the lack of radio-opacity, a radial force lower than that of metallic stents which leads to thicker stents, and the reduced ability to be deformed. These needs spurred research on metal-based fully resorbable stents.

Given that magnesium is a natural body component with beneficial antithrombotic, antiarrhythmic, and antiproliferative properties, it is a promising material for cardiovascular implants. However, Mg rapid degradation results in tissue overload with degradation products and causes neointima formation. Additionally, the increase of OH^- ions leads to high local pH and can produce chronic inflammatory and thrombogenic reactions [21].

Commercial magnesium alloys that combine good mechanical properties and corrosion resistance were selected for further research on biodegradable coronary implants. This is the reason why magnesium alloys, with small contents of aluminium, zinc, and rare earth elements (WE43, AZ31, AE21), originally intended for engineering applications, have emerged as the cardiovascular stent material. Heublein *et.al.* were the pioneers of implantation in animals of a stent made of a magnesium alloy. The stent prototype consisted of an alloy containing 2% aluminium and 1% rare earth elements (AE21) [39].

An absorbable metal stent (AMS) made of WE43 alloy was implanted in more than 60 patients during a pilot feasibility trial showing promising results. The study highlighted the need to extend the degradation time and enable drug elution to prevent in-stent restenosis and neointima formation [40]. These findings raised the possibility to incorporate a bioresorbable matrix for controlled release of an antiproliferative drug as a plausible solution [38]. With this in mind, efforts have been gathered to develop a new generation of biodegradable stents, made by combining polymers and Mg properties, with improved radial strength, better ductility, controlled drug delivery, and prolonged degradation times. Most recently, a new approach consisting on bioresorbable Polymer/Mg composites has been proposed for cardiovascular applications [41].

Currently, efforts are focused on long-term assessments of Mg based stents in pre-clinical and clinical trials [42]. The main goal is to gather representative and reliable data in order to validate resorbable stents for clinical use [43]. The company Biotronik (Germany) offers resorbable Mg stents for vascular intervention.

Table 2. Evolution of the research on bioresorbable Mg alloys for cardiovascular and fracture repair applications.

MAGNESIUM			
CARDIOVASCULAR	Ref.	FRACTURE REPAIR	Ref.
Mg and coronary disease	[21]	First attempt of Mg for osteosynthesis	[22]
Mg alloys in cardiovascular implant technology	[39]	In vivo studies (AZ31, AZ91, WE43, LAE442)	[44]
		Review of Mg alloys as biomaterials (AZ31, AZ91, WE43)	[16]
Drug eluting bioresorbable Mg stent	[38]	Toxicity of alloying elements (rare earth, Al)	[25]
		Binary Mg-Zn alloy coated with PLGA	[29]
Absorbable Metal Stent insight (WE31 alloy)	[40]	Polymer/Mg composites for orthopedic applications	[31]
		Review of coatings on Mg alloys	[30]
Preclinical evaluation of a Mg stent (Mg scaffold Magmaris)	[42]	Pd alloying of Mg-Zn-Ca alloys	[27]
		ZEK100 alloy	[23]
Drug eluting fully bioresorbable Mg scaffold (Magmaris)	[43]	Ternary Mg-Zn-Ca alloys	[28]
		Long-term clinical trial of alloys MgZnCa	[45]
Polymer/Mg as potential bioresorbable/biodegradable stent (BRS)	[41]	*In vivo* studies Pure Mg	[32]
		Small and large animal models (MgZnCa alloys)	[33]

4.2. Iron Fe

Iron, one of the indispensable elements in human body, is involved in the transport of vital oxygen and is a component of metalloproteins such as catalase, hydrogenase, iron responsive element-binding proteins (IRE-BP), or aconitase. The daily allowance level of this metal is up to 20 mg [46].

Fe has better mechanical properties compared with metallic Mg. Its Young's modulus is more than six times larger than that of cortical bone (Fe: 200 GPa, Bone: 30 GPa), and its elongation is more than ten times that of Mg (Fe: 40%, Mg: 3%) [47]. These mechanical properties are very attractive for implants that require high strength and high ductility, such as stents that require a high radial strength and be easily deployed into the arteries

[14]. The high strength values of Fe alloys could seem unsuitable for osteosynthesis applications, but, in fact, these values enable the development of slender wires to design very thin implants with sufficient mechanical strength [48].

Fe corrodes in the presence of dissolved oxygen, forming a protective oxide layer over the surface. This is the main reason why the corrosion rate of iron is the lowest among the biodegradable metals [14]. The suitability of iron as a degradable implant for the treatment of cardiovascular diseases has been verified in pre-clinical *in-vivo* studies. Stents based on Fe have been investigated in animal tests. Results were satisfactory as there were neither obstruction due to inflammation, neointimal proliferation nor thrombotic events. Nonetheless, a faster degradation rate in physiologic media is desirable to improve the suitability of Fe for cardiovascular implants, given that larger portions of the stents remained intact after 1 year implanted *in-vivo* [49]. The considerably low degradation rate lead to reactions that are similar to those found in permanent applications [50]. The suitability of Fe-based alloys for biodegradable implants for fracture repair remains questionable. The presence of oxygen is a mandatory prerequisite to induce Fe corrosion, and the availability of this gas is very limited in bony tissue [48]. Table 3 shows an insight of the iron based alloys that have been studied for both cardiovascular and osteosynthesis applications.

Cardiovascular:

The design approach for biodegradable Fe-based stents is focused towards the increment of iron degradation rate. Attempts have been done following the next three strategies: (1) Alloying [51], (2) Surface modification [52], and (3) Novel fabrication methods [53]. Among these strategies, the one that has been most comprehensively studied is alloying.

Hermawan *et al.* investigated for the first time the effect of alloying elements on the biodegradability of Fe-based stents [54, 55]. The selection of Mn was related with the objective to obtain a higher degradation rate than that of pure Fe and an antiferromagnetic alloy, compatible with magnetic resonance imaging (MRI). The alloys developed (Fe30Mn and Fe35Mn) showed degradation rates more than two times higher than that of pure iron and contained single austenitic phase. Further *in-vitro* and *in-vivo* studies were developed [56, 57] by designing a strategy based on alloying with manganese and palladium. The first one lowers the electrode potential and the last one forms noble (Fe, Mn) Pd intermetallics that act as cathodic sites. As a result, the Fe-Mn-Pd alloys exhibited a degradation rate that was one order of magnitude higher than the one observed in pure iron. Mechanical behaviour was adjustable by heat treatment procedures. Fe-Mn-Pd alloys have also been investigated by Kraus *et al.* [48] for osteosynthesis applications. The outcome was that dense layers of degradation products were formed on the surfaces of the implants and therefore the degradation proceeded rather slowly. Comparing the behaviour of pure Fe and the alloys with Mn and Pd, no differences were observed. The dissimilarities between Schinhammer *et al.* and Kraus *et al.* results, highlight the importance of considering the specific application and implantation site when studying a specific biodegradable material. B. Liu and Y.F Zheng [51] have explored the effect of Mn, Co, Al, W, B, C, and S on the biodegradability and biocompatibility of Fe-X binary alloys. They studied the mechanical properties, corrosion behaviour, cytotoxicity, and hemocompatibility of Fe-X binary alloys, compared to that of pure Fe. The elements Co, W, C, and S were selected as suitable alloying elements for iron biomaterials given the improvement on mechanical properties, adequate corrosion rates, and good biocompatibility.

Moravej *et al.* [53] investigated the relationship between microstructural modification and the increment of the degradation rate of pure iron. They found that electroformed iron had a highly oriented fine microstructure, with columnar grains which influences the degradation behaviour of the material [58]. Zhu *et al.* [52] have modified the surface of pure Fe with FeO thin films improving its corrosion resistance together with the biocompatibility of pure iron.

Table 3. Iron based alloys for cardiovascular and fracture repair applications.

IRON			
CARDIOVASCULAR	Ref.	FRACTURE REPAIR	Ref.
FeMn alloy	[54]		
FeMn alloy	[55]		
In vitro studies FeMn	[56]	Fe-Mn-Pd	[48]
Fe-Mn-Pd	[57]		
Fe-X alloys (X: Mn, Co, Al, W, Sn, B, C and S)	[51]		

4.3 Zinc -Zn

Zinc is also an essential trace element in the human body. It plays a crucial role in the structural, catalytic, or regulatory action of several metalloenzymes, including alkaline phosphatase (ALP), proteins engaged in replication of nucleic acid, or RNA polymer [59]. The daily level of allowance of zinc is between 12 – 15 mg [60]. This metal is being explored for both bone and cardiovascular implants.

The advantage of Zn over Mg is its easier processing by classical routes. This is possible because Zn exhibits a lower melting point, lower chemical reactivity, better machinability, in comparison with Mg, and its processing can be performed in air. In addition to the ease of processability, Zn is more noble than Mg, therefore its corrosion

Nanohybrids   Materials Research Forum LLC
Materials Research Foundations **87** (2021) 134-155   https://doi.org/10.21741/9781644901076-6

rate is lower than that of pure Mg and higher than the one of Fe [61]. The biocompatibility of this metal has been already proved for bone and vascular applications [62]. However, Zn mechanical properties are less attractive than Mg or Fe mechanical stability. Pure Zn is soft and brittle, its ultimate tensile strength is approximately 30 MPa and the elongation is smaller than 0.25%. These values are far to be suitable for most medical applications. The main challenge with this metal, is, therefore, the design of alloys with improved mechanical properties (higher strength and larger elongation) but maintaining its biocompatibility. The main strategies to improve Zn mechanical behaviour are two: (1) modifying the chemical composition by the addition of alloying elements and (2) modifying the microstructure by thermomechanical refinement of grain size [63].

Binary Zn-X and ternary Zn-X-Y alloys have been developed in order to obtain materials with higher strength and better ductility. Researchers have explored the following "X" alloying elements: Magnesium, [61, 64, 65] Calcium, and Strontium, [66] Aluminum [67,68] Copper [69] and in less extent Silver [70] and Lithium [71]. However, the addition of Al and Li is restricted given the known toxicity of these elements. Most common "Y" alloying elements in ternary Zn alloys are: Calcium, Strontium, and Manganese for Zn-Mg-*Y* ternary alloys [72, 73], Strontium for Zn-Ca-*Y* alloys [74], and Magnesium for Zn-Cu-*Y* ternary alloys [75]. Table 3 summarizes the Zn based alloys that have been studied through in-vitro and in-vivo tests in order to assess their potential for cardiovascular or fracture repair applications.

Osteosynthesis

Zn^{2+} ions stimulate osteogenesis in bone and play an essential role in bone formation, mineralization, and preservation of bone mass. In fact, a decrement of zinc in bone matrix is linked to the alteration of bone metabolism associated with skeletal diseases and aging [76]. Human bone marrow mesenchymal stem cells (hMSC) in the presence of Zn ions, are able to grow and differentiate. They form an extracellular matrix (ECM) that is able to mineralize, and induce osteogenesis [77].

Alloying Zn with other essential elements in bone regeneration like Mg, was the first step to design suitable Zn alloys for bone fixation. Alloying with Mg lead to the increment of the overall corrosion rate, and to gain adequate ultimate and yield strengths for osteosynthesis applications [61, 78]. Zn-Mg binary alloys have demonstrated acceptable cytotoxic effects on human osteoblasts cells [79].

More recently, pre-clinical animal trials have been performed by Li *et al.* using binary Zn-1X alloys (X: Mg, Ca, and Sr). Promising outcomes resulted from these *in-vivo* tests. Pins implanted into mouse femurs led to new bone formation without the occurrence of inflammation. The best behaviour was shown by Zn-1Sr alloy. The corrosion of the alloys studied released zinc ions at a permissible concentration, below the daily allowance of Zn [66].

Cardiovascular

The response of primary human coronary artery endothelial cells (HCECs) and human smooth muscle cells (SMCs) to different concentrations (0 - 140μM) of extracellular Zn ions has been investigated by Ma *et al.* [80] Low concentrations of Zn promoted cell viability, proliferation, spreading, and migration. Whereas, cells subjected to high Zn concentrations experienced harmful effects [81]. Apart from *in-vitro* tests, *in-vivo* animal studies have also demonstrated the biocompatibility of Zn with the arterial tissue. Pure Zn wires have been implanted into the abdominal aorta of adult rats, demonstrating excellent biocompatibility and regeneration of the tissue. Zn wires degrade at a suitable rate far below the daily allowance of zinc (15 mg/day). It was also observed that Zn ions suppressed restenosis pathways [82, 83]. The degradation of pure Zinc stents implanted in a rabbit abdominal model has been also investigated. The attempt was successful as the stent retained its mechanical integrity for 6 months and did not exert inflammation, platelet aggregation, thrombosis formation, or obvious intimal hyperplasia. The attractiveness of Zn for vascular implants lies on its potential to reduce the problem of in-stent restenosis and its anti-inflammatory and anti-proliferative properties [82, 84, 85].

Table 4. Zn based alloys for cardiovascular and fracture repair applications.

ZINC			
CARDIOVASCULAR	Ref.	FRACTURE REPAIR	Ref.
Effect of Zn on cells from carotid arteries	[85]	Zn-Mg	[78]
In vivo tests Zn	[82]	Zn-Mg	[61]
In vivo tests Zn	[83]	Zn-Mg Cytotoxicity	[79]
Zn-Mg	[64]	Zn-X (X:Mg, Ca and Sr)	[66]

Concluding remarks

Bioresorbable metals have gained a relevant place in the field of medical devices. Significant efforts have been undertaken in order to understand the behavior of Mg, Fe, and Zn within a physiological environment. This has led to meet an important milestone in the scientific research, the development of commercially available implants for the treatment of cardiovascular diseases and fracture repair.

In order to move forward in this field, it is necessary to broaden the range of applications, look for more versatile manufacturing processes, and extent the spectra of available materials. The scientific community faces a big challenge in order to obtain a material that is easy to process, exhibits an adequate mechanical stability for the required application, has a controlled degradation rate which is well synchronized with the loss of mechanical properties, and more important, is completely bioresorbable and biocompatible.

Mg, Fe, and Zn are the colors available in the palette that paints the future of the generation of biodegradable metals. The alloying elements and surface modifications are the special tones that will give the special and essential touch to the painting in order to obtain the perfect harmony and balance. The artists are all those people involved in the development of these bioresorbable materials, such as engineers in the field of materials, processing or biomedicine, biologists, chemists, physicists, and medical specialists. It is necessary to continue working as a team to improve current available materials and provide solutions to unmet needs.

References

[1]   E. Wilkins, L. Wilson, K. Wickramasinghe, P. Bhatnagar, J. Leal, R. Luengo-Fernandez, R. Burns, M. Rayner, N. Townsend, European cardiovascular disease statistics (2017). European Heart Network, Brussels http://www.ehnheart.org/images/CVD-statistics-report-August-2017.pdf

[2]   Trauma Devices Market Analysis, Size, Trends – Global-2019-2025 MedSuite (2017) 1-187.

[3]   A. Colombo, E. Karvouni, Biodegradable Stents: "Fulfilling the Mission and Stepping Away", in, Am Heart Assoc, 102 (2000) 371-373. https://doi.org/10.1161/01.CIR.102.4.371

[4]   S.H. Duda, J. Wiskirchen, G. Tepe, M. Bitzer, T.W. Kaulich, D. Stoeckel, C.D. Claussen, Physical properties of endovascular stents: an experimental comparison, Journal of Vascular and Interventional Radiology, 11 (2000) 645-654. https://doi.org/10.1016/S1051-0443(07)61620-0

[5]   M. Moravej, D. Mantovani, Biodegradable metals for cardiovascular stent application: interests and new opportunities, International journal of molecular sciences, 12 (2011) 4250-4270. https://doi.org/10.3390/ijms12074250

[6]   T. Kimura, H. Yokoi, Y. Nakagawa, T. Tamura, S. Kaburagi, Y. Sawada, Y. Sato, H. Yokoi, N. Hamasaki, H. Nosaka, Three-year follow-up after implantation of

metallic coronary-artery stents, New England Journal of Medicine, 334 (1996) 561-567. https://doi.org/10.1056/NEJM199602293340903

[7] D.Y. Kwon, J.I. Kim, H.J. Kang, B. Lee, K.W. Lee, M.S. Kim, Biodegradable stent,5 (2012) 208-216. https://doi.org/10.4236/jbise.2012.54028

[8] K. Pawelec, J.A. Planell, Bone Repair Biomaterials: Regeneration and Clinical Applications, Woodhead Publishing, 12 (2018) 35-39.

[9] J. Nagels, M. Stokdijk, P.M. Rozing, Stress shielding and bone resorption in shoulder arthroplasty, Journal of shoulder and elbow surgery, 12 (2003) 35-39. https://doi.org/10.1067/mse.2003.22

[10] I. Matthew, J. Frame, Policy of consultant oral and maxillofacial surgeons towards removal of miniplate components after jaw fracture fixation: pilot study, British Journal of Oral and Maxillofacial Surgery, 37 (1999) 110-112. https://doi.org/10.1054/bjom.1997.0084

[11] M.L. Busam, R.J. Esther, W.T. Obremskey, Hardware removal: indications and expectations, JAAOS-Journal of the American Academy of Orthopaedic Surgeons, 14 (2006) 113-120. https://doi.org/10.5435/00124635-200602000-00006

[12] Y. Yun, Z. Dong, N. Lee, Y. Liu, D. Xue, X. Guo, J. Kuhlmann, A. Doepke, H.B. Halsall, W. Heineman, Revolutionizing biodegradable metals, Materials Today, 12 (2009) 22-32. https://doi.org/10.1016/S1369-7021(09)70273-1

[13] C. Shuai, S. Li, S. Peng, P. Feng, Y. Lai, C. Gao, Biodegradable metallic bone implants, Materials Chemistry Frontiers, 3 (2019) 544-562. https://doi.org/10.1039/C8QM00507A

[14] D. Zindani, K. Kumar, J.P. Davim, Metallic biomaterials—A review, in: Mechanical Behaviour of Biomaterials, Elsevier, (2019) 83-99.

[15] D. Paramitha, M. Ulum, A. Purnama, D. Wicaksono, D. Noviana, H. Hermawan, Monitoring degradation products and metal ions in vivo, in: Monitoring and Evaluation of Biomaterials and their Performance In Vivo, Elsevier, (2017) 19-44. https://doi.org/10.1016/B978-0-08-100603-0.00002-X

[16] M.P. Staiger, A.M. Pietak, J. Huadmai, G. Dias, Magnesium and its alloys as orthopedic biomaterials: a review, Biomaterials, 27 (2006) 1728-1734. https://doi.org/10.1016/j.biomaterials.2005.10.003

[17] E. Underwood, Trace elements in human and animal nutrition, Academic Press, INC. (1977). New York. https://doi.org/10.1016/B978-0-12-709065-8.50006-7

[18] P.A. Revell, E. Damien, X. Zhang, P. Evans, C.R. Howlett, The effect of magnesium ions on bone bonding to hydroxyapatite coating on titanium alloy implants, in: Key Engineering Materials, Trans Tech Publ, (2004) 447-450. https://doi.org/10.4028/www.scientific.net/KEM.254-256.447

[19] C. Janning, E. Willbold, C. Vogt, J. Nellesen, A. Meyer-Lindenberg, H. Windhagen, F. Thorey, F. Witte, Magnesium hydroxide temporarily enhancing osteoblast activity and decreasing the osteoclast number in peri-implant bone remodelling, Acta biomaterialia, 6 (2010) 1861-1868. https://doi.org/10.1016/j.actbio.2009.12.037

[20] D.A. Robinson, R.W. Griffith, D. Shechtman, R.B. Evans, M.G. Conzemius, In vitro antibacterial properties of magnesium metal against Escherichia coli, Pseudomonas aeruginosa and Staphylococcus aureus, Acta biomaterialia, 6 (2010) 1869-1877. https://doi.org/10.1016/j.actbio.2009.10.007

[21] P. Delva, Magnesium and coronary heart disease, Molecular aspects of medicine, 24 (2003) 63-78. https://doi.org/10.1016/S0098-2997(02)00092-4. https://doi.org/10.1016/S0098-2997(02)00092-4

[22] A. Lambotte, L'utilisation du magnesium comme materiel perdu dans l'osteosynthèse, Bull Mem Soc Nat Chir, 28 (1932) 1325-1334.

[23] D. Dziuba, A. Meyer-Lindenberg, J.M. Seitz, H. Waizy, N. Angrisani, J. Reifenrath, Long-term in vivo degradation behaviour and biocompatibility of the magnesium alloy ZEK100 for use as a biodegradable bone implant, Acta biomaterialia, 9 (2013) 8548-8560. https://doi.org/10.1016/j.actbio.2012.08.028

[24] M. Easton, A. Beer, M. Barnett, C. Davies, G. Dunlop, Y. Durandet, S. Blacket, T. Hilditch, P. Beggs, Magnesium alloy applications in automotive structures, Jom, 60 (2008) 57-62. https://doi.org/10.1007/s11837-008-0150-8

[25] F. Feyerabend, J. Fischer, J. Holtz, F. Witte, R. Willumeit, H. Drücker, C. Vogt, N. Hort, Evaluation of short-term effects of rare earth and other elements used in magnesium alloys on primary cells and cell lines, Acta biomaterialia, 6 (2010) 1834-1842. https://doi.org/10.1016/j.actbio.2009.09.024

Materials Research Forum LLC
https://doi.org/10.21741/9781644901076-6

[26] S.S.A. El-Rahman, Neuropathology of aluminum toxicity in rats (glutamate and GABA impairment), Pharmacological Research, 47 (2003) 189-194. https://doi.org/10.1016/S1043-6618(02)00336-5

[27] S. González, E. Pellicer, J. Fornell, A. Blanquer, L. Barrios, E. Ibánez, P. Solsona, S. Surinach, M. Baró, C. Nogués, Improved mechanical performance and delayed corrosion phenomena in biodegradable Mg–Zn–Ca alloys through Pd-alloying, Journal of the mechanical behavior of biomedical materials, 6 (2012) 53-62. https://doi.org/10.1016/j.jmbbm.2011.09.014

[28] E. Pellicer, S. Gonzalez, A. Blanquer, S. Suriñach, M.D. Baró, L. Barrios, E. Ibáñez, C. Nogués, J. Sort, On the biodegradability, mechanical behaviour and cytocompatibility of amorphous Mg72Zn23Ca5 and crystalline Mg70Zn23Ca5Pd2 alloys as temporary implant materials, Journal of Biomedical Materials Research, 101A (2013) 502 - 517.

[29] J. Li, P. Cao, X. Zhang, S. Zhang, Y. He, In vitro degradation and cell attachment of a PLGA coated biodegradable Mg–6Zn based alloy, Journal of materials science, 45 (2010) 6038-6045. https://doi.org/10.1007/s10853-010-4688-9

[30] H. Hornberger, S. Virtanen, A. Boccaccini, Biomedical coatings on magnesium alloys–a review, Acta biomaterialia, 8 (2012) 2442-2455. https://doi.org/10.1016/j.actbio.2012.04.012

[31] S.C. Cifuentes, E. Frutos, J.L. González-Carrasco, M. Muñoz, M. Multigner, J. Chao, R. Benavente, M. Lieblich, Novel PLLA/magnesium composite for orthopedic applications: A proof of concept, Materials Letters, 74 (2012) 239-242. https://doi.org/10.1016/j.matlet.2012.01.134

[32] L. Tian, Y. Sheng, L. Huang, D.H.-K. Chow, W.H. Chau, N. Tang, T. Ngai, C. Wu, J. Lu, L. Qin, An innovative Mg/Ti hybrid fixation system developed for fracture fixation and healing enhancement at load-bearing skeletal site, Biomaterials, 180 (2018) 173-183. https://doi.org/10.1016/j.biomaterials.2018.07.018

[33] N.G. Grün, P. Holweg, S. Tangl, J. Eichler, L. Berger, J.J. Van den Beucken, J.F. Löffler, T. Klestil, A.M. Weinberg, Comparison of a resorbable magnesium implant in small and large growing-animal models, Acta biomaterialia, 78 (2018) 378-386. https://doi.org/10.1016/j.actbio.2018.07.044

[34] F. Witte, V. Kaese, H. Haferkamp, E. Switzer, Meyer-Lindenberg, a., Wirth, CJ, & Windhagen, H.(2005), vivo corrosion of four magnesium alloys and the associated

bone response. Biomaterials, 26 (2004) 3557-3563.
https://doi.org/10.1016/j.biomaterials.2004.09.049

[35] J. Lee, M. Xue, J.S. Wzorek, T. Wu, M. Grabowicz, L.S. Gronenberg, H.A.
Sutterlin, R.M. Davis, N. Ruiz, T.J. Silhavy, Characterization of a stalled complex on
the β-barrel assembly machine, Proceedings of the National Academy of Sciences,
113 (2016) 8717-8722. https://doi.org/10.1073/pnas.1604100113

[36] W.J. Van der Giessen, A.M. Lincoff, R.S. Schwartz, H.M. Van Beusekom, P.W.
Serruys, D.R. Holmes, S.G. Ellis, E.J. Topol, Marked inflammatory sequelae to
implantation of biodegradable and nonbiodegradable polymers in porcine coronary
arteries, Circulation, 94 (1996) 1690-1697. https://doi.org/10.1161/01.CIR.94.7.1690

[37] H. Tamai, K. Igaki, E. Kyo, K. Kosuga, A. Kawashima, S. Matsui, H. Komori, T.
Tsuji, S. Motohara, H. Uehata, Initial and 6-month results of biodegradable poly-l-
lactic acid coronary stents in humans, Circulation, 102 (2000) 399-404.
https://doi.org/10.1161/01.CIR.102.4.399

[38] C. Di Mario, H. Griffiths, O. Goktekin, N. Peeters, J. Verbist, M. Bosiers, K.
Deloose, B. Heublein, R. Rohde, V. Kasese, Drug-eluting bioabsorbable magnesium
stent, Journal of interventional cardiology, 17 (2004) 391-395.
https://doi.org/10.1111/j.1540-8183.2004.04081.x

[39] B. Heublein, R. Rohde, V. Kaese, M. Niemeyer, W. Hartung, A. Haverich,
Biocorrosion of magnesium alloys: a new principle in cardiovascular implant
technology?, Heart, 89 (2003) 651-656. https://doi.org/10.1136/heart.89.6.651

[40] M. Bosiers, A.I. Investigators, AMS INSIGHT—absorbable metal stent
implantation for treatment of below-the-knee critical limb ischemia: 6-month analysis,
Cardiovascular and interventional radiology, 32 (2009) 424-435.
https://doi.org/10.1007/s00270-008-9472-8

[41] A. Srivastava, R. Ahuja, P. Bhati, P. Singh, P. Chauhan, P. Vashisth, A. Kumar, N.
Bhatnagar, Fabrication and Characterization of PLLA/Mg Composite Tube as the
potential Bioresorbable/biodegradable stent (BRS), Materialia, (2020) 1-12.
https://doi.org/10.1016/j.mtla.2020.100661

[42] R. Waksman, P. Zumstein, M. Pritsch, E. Wittchow, M. Haude, C. Lapointe-
Corriveau, G. Leclerc, M. Joner, Second-generation magnesium scaffold Magmaris:
device design and preclinical evaluation in a porcine coronary artery model,
EuroIntervention: journal of EuroPCR in collaboration with the Working Group on

Interventional Cardiology of the European Society of Cardiology, 13 (2017) 440-449. https://doi.org/10.4244/EIJ-D-16-00915

[43] C. Rapetto, M. Leoncini, Magmaris: a new generation metallic sirolimus-eluting fully bioresorbable scaffold: present status and future perspectives, Journal of thoracic disease, 9 (2017) 903-913. https://doi.org/10.21037/jtd.2017.06.34

[44] F. Witte, V. Kaese, H. Haferkamp, E. Switzer, A. Meyer-Lindenberg, C. Wirth, H. Windhagen, In vivo corrosion of four magnesium alloys and the associated bone response, Biomaterials, 26 (2005) 3557-3563. https://doi.org/10.1016/j.biomaterials.2004.09.049

[45] J.-W. Lee, H.-S. Han, K.-J. Han, J. Park, H. Jeon, M.-R. Ok, H.-K. Seok, J.-P. Ahn, K.E. Lee, D.-H. Lee, Long-term clinical study and multiscale analysis of in vivo biodegradation mechanism of Mg alloy, Proceedings of the National Academy of Sciences, 113 (2016) 716-721. https://doi.org/10.1073/pnas.1518238113

[46] M. Fontecave, J. Pierre, Iron: metabolism, toxicity and therapy, Biochimie, 75 (1993) 767-773. https://doi.org/10.1016/0300-9084(93)90126-D

[47] G. Poologasundarampillai, A. Nommeots-Nomm, Materials for 3D printing in medicine: Metals, polymers, ceramics, hydrogels, in: 3D Printing in Medicine, Elsevier, 2017, pp. 43-71. https://doi.org/10.1016/B978-0-08-100717-4.00002-8

[48] T. Kraus, F. Moszner, S. Fischerauer, M. Fiedler, E. Martinelli, J. Eichler, F. Witte, E. Willbold, M. Schinhammer, M. Meischel, Biodegradable Fe-based alloys for use in osteosynthesis: Outcome of an in vivo study after 52 weeks, Acta biomaterialia, 10 (2014) 3346-3353. https://doi.org/10.1016/j.actbio.2014.04.007

[49] R. Waksman, R. Pakala, R. Baffour, R. Seabron, D. Hellinga, F.O. Tio, Short-term effects of biocorrodible iron stents in porcine coronary arteries, Journal of interventional cardiology, 21 (2008) 15-20. https://doi.org/10.1111/j.1540-8183.2007.00319.x

[50] M. Peuster, C. Hesse, T. Schloo, C. Fink, P. Beerbaum, C. von Schnakenburg, Long-term biocompatibility of a corrodible peripheral iron stent in the porcine descending aorta, Biomaterials, 27 (2006) 4955-4962. https://doi.org/10.1016/j.biomaterials.2006.05.029

[51] B. Liu, Y. Zheng, Effects of alloying elements (Mn, Co, Al, W, Sn, B, C and S) on biodegradability and in vitro biocompatibility of pure iron, Acta biomaterialia, 7 (2011) 1407-1420. https://doi.org/10.1016/j.actbio.2010.11.001

[52] S. Zhu, N. Huang, L. Xu, Y. Zhang, H. Liu, Y. Lei, H. Sun, Y. Yao, Biocompatibility of Fe–O films synthesized by plasma immersion ion implantation and deposition, Surface and Coatings Technology, 203 (2009) 1523-1529. https://doi.org/10.1016/j.surfcoat.2008.11.033

[53] M. Moravej, F. Prima, M. Fiset, D. Mantovani, Electroformed iron as new biomaterial for degradable stents: Development process and structure–properties relationship, Acta biomaterialia, 6 (2010) 1726-1735. https://doi.org/10.1016/j.actbio.2010.01.010

[54] H. Hermawan, D. Dubé, D. Mantovani, Development of degradable Fe-35Mn alloy for biomedical application, in: Advanced Materials Research, Trans Tech Publ, (2007) 107-112. https://doi.org/10.4028/0-87849-429-4.107

[55] H. Hermawan, D. Dubé, D. Mantovani, Degradable metallic biomaterials: design and development of Fe–Mn alloys for stents, Journal of Biomedical Materials Research Part A: An Official Journal of The Society for Biomaterials, The Japanese Society for Biomaterials, and The Australian Society for Biomaterials and the Korean Society for Biomaterials, 93 (2010) 1-11. https://doi.org/10.1002/jbm.a.32224

[56] H. Hermawan, A. Purnama, D. Dube, J. Couet, D. Mantovani, Fe–Mn alloys for metallic biodegradable stents: degradation and cell viability studies, Acta biomaterialia, 6 (2010) 1852-1860.

[57] M. Schinhammer, A.C. Hänzi, J.F. Löffler, P.J. Uggowitzer, Design strategy for biodegradable Fe-based alloys for medical applications, Acta biomaterialia, 6 (2010) 1705-1713. https://doi.org/10.1016/j.actbio.2009.07.039

[58] M. Moravej, S. Amira, F. Prima, A. Rahem, M. Fiset, D. Mantovani, Effect of electrodeposition current density on the microstructure and the degradation of electroformed iron for degradable stents, Materials Science and Engineering: B, 176 (2011) 1812-1822. https://doi.org/10.1016/j.mseb.2011.02.031

[59] J.E. Coleman, Zinc proteins: enzymes, storage proteins, transcription factors, and replication proteins, Annual review of biochemistry, 61 (1992) 897-946. https://doi.org/10.1146/annurev.bi.61.070192.004341

[60] G.J. Fosmire, Zinc toxicity, The American journal of clinical nutrition, 51 (1990) 225-227. https://doi.org/10.1093/ajcn/51.2.225

[61] D. Vojtěch, J. Kubásek, J. Šerák, P. Novák, Mechanical and corrosion properties of newly developed biodegradable Zn-based alloys for bone fixation, Acta Biomaterialia, 7 (2011) 3515-3522. https://doi.org/10.1016/j.actbio.2011.05.008

[62] J. Venezuela, M. Dargusch, The influence of alloying and fabrication techniques on the mechanical properties, biodegradability and biocompatibility of zinc: A comprehensive review, Acta biomaterialia, 87 (2019) 1-40. https://doi.org/10.1016/j.actbio.2019.01.035

[63] G. Katarivas Levy, J. Goldman, E. Aghion, The prospects of zinc as a structural material for biodegradable implants—a review paper, Metals, 7 (2017) 402-420. https://doi.org/10.3390/met7100402

[64] E. Mostaed, M. Sikora-Jasinska, A. Mostaed, S. Loffredo, A. Demir, B. Previtali, D. Mantovani, R. Beanland, M. Vedani, Novel Zn-based alloys for biodegradable stent applications: design, development and in vitro degradation, Journal of the mechanical behavior of biomedical materials, 60 (2016) 581-602. https://doi.org/10.1016/j.jmbbm.2016.03.018

[65] G.K. Levy, A. Leon, A. Kafri, Y. Ventura, J.W. Drelich, J. Goldman, R. Vago, E. Aghion, Evaluation of biodegradable Zn-1% Mg and Zn-1% Mg-0.5% Ca alloys for biomedical applications, Journal of Materials Science: Materials in Medicine, 28 (2017) 174-185. https://doi.org/10.1007/s10856-017-5973-9

[66] H. Li, X. Xie, Y. Zheng, Y. Cong, F. Zhou, K. Qiu, X. Wang, S. Chen, L. Huang, L. Tian, Development of biodegradable Zn-1X binary alloys with nutrient alloying elements Mg, Ca and Sr, Scientific reports, 5 (2015) 1-13. https://doi.org/10.1038/srep12190

[67] H. Bakhsheshi-Rad, E. Hamzah, H. Low, M. Kasiri-Asgarani, S. Farahany, E. Akbari, M. Cho, Fabrication of biodegradable Zn-Al-Mg alloy: mechanical properties, corrosion behavior, cytotoxicity and antibacterial activities, Materials Science and Engineering: C, 73 (2017) 215-219. https://doi.org/10.1016/j.msec.2016.11.138

[68] H. Bakhsheshi-Rad, E. Hamzah, H. Low, M. Cho, M. Kasiri-Asgarani, S. Farahany, A. Mostafa, M. Medraj, Thermal characteristics, mechanical properties, in vitro degradation and cytotoxicity of novel biodegradable Zn–Al–Mg and Zn–Al–Mg–xBi

alloys, Acta Metallurgica Sinica (English Letters), 30 (2017) 201-211. https://doi.org/10.1007/s40195-017-0534-2

[69] Z. Tang, J. Niu, H. Huang, H. Zhang, J. Pei, J. Ou, G. Yuan, Potential biodegradable Zn-Cu binary alloys developed for cardiovascular implant applications, Journal of the mechanical behavior of biomedical materials, 72 (2017) 182-191. https://doi.org/10.1016/j.jmbbm.2017.05.013

[70] M. Sikora-Jasinska, E. Mostaed, A. Mostaed, R. Beanland, D. Mantovani, M. Vedani, Fabrication, mechanical properties and in vitro degradation behavior of newly developed ZnAg alloys for degradable implant applications, Materials Science and Engineering: C, 77 (2017) 1170-1181. https://doi.org/10.1016/j.msec.2017.04.023

[71] S. Zhao, C.T. McNamara, P.K. Bowen, N. Verhun, J.P. Braykovich, J. Goldman, J.W. Drelich, Structural characteristics and in vitro biodegradation of a novel Zn-Li alloy prepared by induction melting and hot rolling, Metallurgical and Materials Transactions A, 48 (2017) 1204-1215. https://doi.org/10.1007/s11661-016-3901-0

[72] X. Liu, J. Sun, K. Qiu, Y. Yang, Z. Pu, L. Li, Y. Zheng, Effects of alloying elements (Ca and Sr) on microstructure, mechanical property and in vitro corrosion behavior of biodegradable Zn–1.5 Mg alloy, Journal of Alloys and Compounds, 664 (2016) 444-452. https://doi.org/10.1016/j.jallcom.2015.10.116

[73] X. Liu, J. Sun, F. Zhou, Y. Yang, R. Chang, K. Qiu, Z. Pu, L. Li, Y. Zheng, Micro-alloying with Mn in Zn–Mg alloy for future biodegradable metals application, Materials & Design, 94 (2016) 95-104. https://doi.org/10.1016/j.matdes.2015.12.128

[74] H. Li, H. Yang, Y. Zheng, F. Zhou, K. Qiu, X. Wang, Design and characterizations of novel biodegradable ternary Zn-based alloys with IIA nutrient alloying elements Mg, Ca and Sr, Materials & Design, 83 (2015) 95-102. https://doi.org/10.1016/j.matdes.2015.05.089

[75] Z. Tang, H. Huang, J. Niu, L. Zhang, H. Zhang, J. Pei, J. Tan, G. Yuan, Design and characterizations of novel biodegradable Zn-Cu-Mg alloys for potential biodegradable implants, Materials & Design, 117 (2017) 84-94. https://doi.org/10.1016/j.matdes.2016.12.075

[76] N.R. Calhoun, J.C. Smith Jr, K.L. Becker, The role of zinc in bone metabolism, Clinical Orthopaedics and Related Research, 103 (1974) 212-234. https://doi.org/10.1097/00003086-197409000-00084

[77] Y. Qiao, W. Zhang, P. Tian, F. Meng, H. Zhu, X. Jiang, X. Liu, P.K. Chu, Stimulation of bone growth following zinc incorporation into biomaterials, Biomaterials, 35 (2014) 6882-6897. https://doi.org/10.1016/j.biomaterials.2014.04.101

[78] T. Prosek, A. Nazarov, U. Bexell, D. Thierry, J. Serak, Corrosion mechanism of model zinc–magnesium alloys in atmospheric conditions, Corrosion Science, 50 (2008) 2216-2231. https://doi.org/10.1016/j.corsci.2008.06.008

[79] N. Murni, M. Dambatta, S. Yeap, G. Froemming, H. Hermawan, Cytotoxicity evaluation of biodegradable Zn–3Mg alloy toward normal human osteoblast cells, Materials Science and Engineering: C, 49 (2015) 560-566.

[80] J. Ma, N. Zhao, D. Zhu, Endothelial cellular responses to biodegradable metal zinc, ACS biomaterials science & engineering, 1 (2015) 1174-1182. https://doi.org/10.1021/acsbiomaterials.5b00319

[81] J. Ma, N. Zhao, D. Zhu, Bioabsorbable zinc ion induced biphasic cellular responses in vascular smooth muscle cells, Scientific reports, 6 (2016) 1-10. https://doi.org/10.1038/srep26661

[82] P.K. Bowen, J. Drelich, J. Goldman, Zinc exhibits ideal physiological corrosion behavior for bioabsorbable stents, Advanced materials, 25 (2013) 2577-2582. https://doi.org/10.1002/adma.201300226

[83] P.K. Bowen, R.J. Guillory II, E.R. Shearier, J.-M. Seitz, J. Drelich, M. Bocks, F. Zhao, J. Goldman, Metallic zinc exhibits optimal biocompatibility for bioabsorbable endovascular stents, Materials Science and Engineering: C, 56 (2015) 467-472. https://doi.org/10.1016/j.msec.2015.07.022

[84] B. Hennig, M. Toborek, C.J. McClain, Antiatherogenic properties of zinc: implications in endothelial cell metabolism, Nutrition, 12 (1996) 711-717. https://doi.org/10.1016/S0899-9007(96)00125-6

[85] M. Berger, E. Rubinraut, I. Barshack, A. Roth, G. Keren, J. George, Zinc reduces intimal hyperplasia in the rat carotid injury model, Atherosclerosis, 175 (2004) 229-234. https://doi.org/10.1016/j.atherosclerosis.2004.03.022

Nanohybrids
Materials Research Foundations **87** (2021) 156-201

Materials Research Forum LLC
https://doi.org/10.21741/9781644901076-7

Chapter 7

Application of Biopolymeric Electrospun Nanofibers in Biological Science

Mehdihasan I. Shekh[1,2]*, Jhaleh Amirian[1,2], Gisya Abdi[3], Dijit M. Patel[4], Bing Du[1]

[1]College of Materials Science and Engineering, Shenzhen Key Laboratory of Polymer Science and Technology, Guangdong Research Center for Interfacial Engineering of Functional Materials, Nanshan District Key Lab for Biopolymers and Safety Evaluation, Shenzhen University, Shenzhen 518055, PR China

[2]Key Laboratory of Optoelectronic Devices and Systems of Ministry of Education and Guangdong Province, College of Optoelectronic Engineering, Shenzhen University, Shenzhen 518060, PR China

[3]Center of new technologies, University of Warsaw, ul. Banacha 2c, 02-097 Warsaw, Poland

[4]Department of Advanced Organic Chemistry, P. D. Patel Institute of Applied Sciences, Charotar University of Science and Technology, Changa 388421, Gujarat, India

mehdi.shekh3@yahoo.com*

Abstract

Biopolymers are those class of macromolecules which are found in nature or extracted from the living organisms. Various structures and properties of the biopolymers-based materials are well researched till to date. These mainly includes hydrogels, bio glasses, bio inks, biocomposites, fibers and others. These biopolymers-based structures have some limitations. However, Biopolymers have some common advantages (i.e., non-toxicity, easy availability, monodispersity, degradability, and better solubility etc.) and disadvantages (i.e., poor thermal and chemical stabilities, brittleness etc.). To overcome these disadvantages, it is necessary to tailor these polymers by few emerging techniques like "Electrospinning". Electrospinning is one of the easiest techniques to prepare nanofibers from polymeric solutions by applying high voltage. Obtained nano/micro structural polymeric fibers have good properties like high surface area, porosity and low weights etc. The materials having high surface area and porosity can easily interact with cells and tissues, are better mobile vehicles for drugs, as well as possess good filtration and adsorption abilities. Thus, these one-dimensional structures of the biopolymers are very useful in various fields of biomedical especially water sanitation/desalination, tissue engineering, drug delivery and scaffolds. Various biopolymers like chitosan, chitin, sodium alginate, guar gum, polylactic acid and others are successfully fabricated as fibers and used in various fields of biomedical.

Keywords

Bioengineering Biopolymers, Electrospinning; Nanofibers, Scaffolds, Drug Delivery

Contents

1.   **Introduction**

Polymers are large molecules having a several repeating units (i.e., same or different units), thus they can act as the basic building block for fabrication of advanced materials having applications in various fields of science. Numerous classes of polymers are investigated and developed for fundamental applications. From past few decades, various synthetic polymers have taken the place of the biopolymers because of their extraordinary properties such as better stiffness, mechanical strength, less brittleness, environmental stability and other properties [1]. However, biopolymers have their own advantages over these synthetic polymers. The main advantages of biopolymers are non-toxicity, biodegradability, reusability and easy availability. In addition, these are also edible, biocompatible and digestible. As synthetic polymers are environmentally stable and thus, they are not easily degraded while biopolymers are easily degraded these limits their environmental usefulness. Moreover, biopolymers are well fits in environmental friendly materials and thus they are most precise polymers for biomedical applications [2].

All living organisms contain or made up of the biopolymers and these biopolymers are responsible for their structural aspects. There are three classes of biopolymers which are polynucleotides, polypeptides and polysaccharides. The polynucleotides are responsible for the heredity, immunological process and other important procedures (i.e., messengers, activation of enzymes etc.), which take place in living organisms. Polypeptides are long chains of the amino acids units which are part of protein synthesis. Polysaccharides are one of the most important classes of biopolymers which are useful for the preparation of various biomaterials (directly or indirectly). Polysaccharides contain long chains of carbohydrates. These various types of the biopolymers are successfully extracted or obtained from various living organisms. These includes polysaccharides (i.e., chitin, chitosan, cellulose, dextran and starch etc.), proteins (i.e. wheat protein, gelatine, albumin, collagen, and fibroin etc.) and other biopolymers (i.e., poly lactic acid, polyhydroxy butyrate etc.). These biopolymers are very useful as raw materials, coating agents, ingredients in industries like food, pharma, paint, cosmetics, polymer and others. Table 1 illustrates various types of biopolymers, their monomeric units and their sources.

Biopolymers can easily transformed into various structures which are very useful in biomedical applications, food industries, cosmetic industries, and pharmaceutical industries [3]. These structures are mainly hydrogels, bio glasses, bio inks, whiskers, nanoparticles, bio composites, micro/nanofibers. From past few decades, researchers are using nanotechnology to change the structural dimensions and properties of biopolymers (i.e., to take micro level structure to nanostructure). Thus, the obtained biomaterials have various applicable properties like high surface area, nano dimensions and light weight etc. Nano engineered materials are mainly nanoparticles, nanocomposites, nanofibers and nanotubes. These nanomaterials are one-, two- or three dimensional shapes and are used in energy harvesting, nanomedicine, as nanocarriers, nanoelectronics and other related fields [4]. In this chapter, we are going to describe the use of nanofibrous structures for active agent/drug carriers, antimicrobial agents and tissue engineering.

From past decades, various biopolymers based biomaterials like scaffolds, hydrogels, bio inks and other materials are used for biomedical applications but these have some structural limitations, as well as a poor adhesion properties and lower surface area. Some hydrogels and other biomaterials show limited porosity, low surface areas, and are difficult to sterilize for *in-vitro* experiments [5]. However, other nano-biomaterials possessed high porosity with high surface area, good mechanical properties, wide range of materials availability, biocompatibility, good adhesive properties, controlled releasing properties and are easy to handle during *in-vitro/vivo* experiments. Due to these advantageous properties of nano biomaterials, researchers have successfully fabricated the nano-dimensional structures which are very useful in the biomedical fields [6].

Materials Research Forum LLC
https://doi.org/10.21741/9781644901076-7

Among these nanostructures, the nanofibers are the most effective and compatible candidate for the fabrication of biomaterials. Various techniques were researched for fabrication of one-dimensional structures. These techniques are drawing [7], template [8], Self-assembly [9], phase inversion [10], and electrospinning [11] etc. Among all methods, electrospinning is one of the easiest and simple technique to obtain the nanostructured filaments. Several research groups have successfully fabricated the nanofibers or composite nanofibers using various synthetic polymers and biopolymers for their diverse applications in the fields of biomedical, coating, adsorbent, energy storage, sensors and other related fields [12-15].

Table 1: Types of Biopolymers, their structural units and Sources.

Type of Biopolymer	Examples	Structural Units	Source
Polysaccharides	Chitin	N-Acetylated Glucose	Shells of Crabs, Lobsters, Shrimps and cell walls of Fungi
	Chitosan	D-Glucosamine + *N*-acetyl-D-glucosamine	Deacetylation of Chitin
	Cellulose	D-Glucose	Cell wall of plants, Algae, Cotton fiber
	Alginates	D-mannuronic acid + L-guluronic acid	Cell wall of brown algae
	Guar Gum	Galactose + Mannose	Guar beans
	Dextran	D-glucopyranose	Bacteria (*Leuconostoc mesenteroides* and *Streptococcus mutans*), Dental Plaque,
Proteins	Silk Fibroin	Glycine, Serine, Alanine	Silk of spiders, larvae of *Bombyx mori*, other moth genera such as *Antheraea*, *Cricula*, *Samia* and *Gonometa*
	Casein	Phosphoproteins	mammalian milk
	Collagen	Glycine, Proline, Hydroxyproline, Arginine	Tendons, ligaments and skin of mammals
	Wheat Protein	Glutenin + Gliadin	Wheat
	Zein	Prolamine Protein	Maize (Corn)
	gelatine	Glycine, Proline, alanine, Hydroxiproline, Glutamic acid, Arginine, Aspartic acid, Tysine	Partial hydrolysis of Collagen

Nanohybrids   Materials Research Forum LLC
Materials Research Foundations **87** (2021) 156-201   https://doi.org/10.21741/9781644901076-7

Electrospinning of synthetic polymers are somehow easy because of their structural arrangements, vast range of molecular weights and solubility in various organic solvents. However, several disadvantages were encountered by the researchers during the electrospinning of the biopolymers which show poor solubility in organic solvents, easily forms hydrogels at higher concentrations, high level of crystallinity and poor mechanical strength [16]. These drawbacks limit the use of biopolymers in electrospinning. To overcome these limitations, it is necessary to blend the biopolymers with other synthetic or water-soluble polymers. Thus, the modification of biopolymers can lead to the obtention of fibers by electrospinning technique. In this chapter, we are going to explain various biopolymeric nanofibers composite blends for the tissue engineering, wound healing, carrier vehicles, bone regeneration and other biomedical fields.

2.   Electrospinning

Electrospinning is a technique in which the electrostatic force can be used for the fabrication of nanofibers using viscous polymer solutions. In other words, uniaxial stretching of the viscoelastic polymer solution by means of electrostatic force. Basically, electrospinning setup consists syringe pump, needle, high voltage DC supply and collector (Fig. 1). A polymer solution is placed into the syringe which is equipped with stainless steel needle. This syringe is then placed into the syringe pump. Syringe pump can permit the flow rate of polymer solution in a precise manner. High voltage DC supply is connected to the needle. When high voltage is applied on polymer solution, the drop of the polymer solution changes its structure to conical shape which is called "Taylor cone". This "Taylor cone" have very low stability due to various forces such as, gravitational force, interfacial force and charge repulsion force. When the charge repulsion forces overpass the interfacial force, "Taylor cone" can fly from the needle tip, and two processes occur simultaneously. First one, it is related to the solvent evaporation, and second one it is associated with the stretching of the polymer droplets. As the highly charged droplets have opposite charge than collector, this attraction force helps the polymer solution in stretching and collects on the collector.

There are various processing and solution parameters which have an effect on the fibre morphology. The processing parameters mainly include applied voltage, flow rate, distance between tip to collector, while solution parameters are related to viscosity of solution, type of solvent, and molecular weight of the polymers, among others. on the other hand, ambient parameters like temperature and humidity also determines fibre morphology to great extent. Furthermore, types of collector can also govern the fibre morphology.

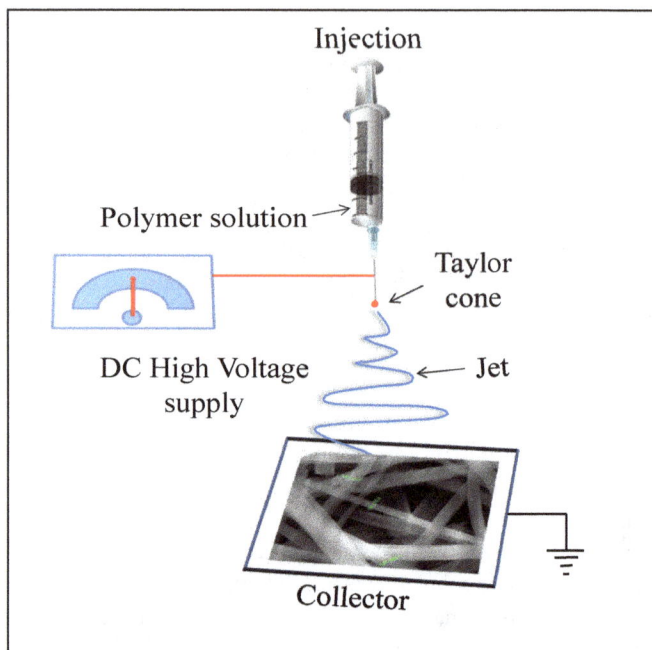

Figure 1: Schematic representation of electrospinning set up.

3. Electrospinning of the biopolymers and their biomedical applications

Polysaccharides are one class of the biopolymers in which are formed one type or different kinds of monosaccharaides (sugar units) as repeating units. Chitin, chitosan, alginates, and cellulose belongs to this class of biopolymers. These offer good antimicrobial properties, degradability and nontoxic properties which make them as excellent raw materials for fabrication of scaffolds in tissue engineering, food packaging, drug delivery and other biomedical applications [17]. In this chapter, selected and highly applicable biopolymers based nanofibrous materials are briefly illustrated.

3.1 Drug carrier system

Various materials are used as drug carriers. These materials are mainly hydrogels, emulsions, nanoparticles and nanofibers. Among them nanofibers prove to be great drug delivery carrier with controlled rate, due to their extraordinary one-dimensional structure, high surface area and porosity. Electrospun nanofibers can be used as carrier for numerous therapeutic agents such as drugs and other biologically active agents (i.e., proteins, nucleic acids). Various drug release mechanisms of nanofibers are known which are mainly desorption from the fibrous surface, diffusion through fibre surface, and fibre degradation [18]. Desorption is one of the simplest mechanisms for drug release. Different methods are employed to encapsulate the active molecules in nanofiber matrixes. These techniques are surface modification, direct blending, coaxial spinning and emulsion spinning. When drug loaded fibres comes in contact with by surrounding liquid, they burst thus desorption or diffusion of drug occurs from the surface of the fibres and also from inside of the fibres. Thus, this mechanism is not appropriate for controlled release of bioactive agents. The surface modification, coaxial spinning and emulsion spinning methods are favourable for preparation of drug loaded fibres. There are numerous types of the biodegradable synthetic or natural polymers favourable for the controlled drug release profiles. These polymers encapsulate the drug molecules in fibre matrix and release them in controlled manner. Thus, natural or biodegradable polymeric nanofibers are a good candidate for the encapsulation and controlled release of bioactive agents.

Zahedi and co-workers have successfully fabricated several composite nanofibers by varying the drug and polymers ratios [19]. Fabricated tetracycline hydrochloride (TC-HCl) loaded egg albumin (EA)/polyvinyl alcohol (PVA) composite nanofibers were studied for removal *S.aureus* and *E.coli* gram positive and gram negative bacteria, respectively. The results (Fig. 2) show that the order of the inhibition of the fibers is PVA/TC-HCl > PVA/EA/TC-HCl > PVA/EA ~PVA nanofibers. It is also shown from the results that drug loaded nanofibers effectively inhibit the growth of bacteria. They have also studied the kinetics of the drug release and noted that drug release profile is according to the Higuchi model which explain that drug follows diffusion mechanism.

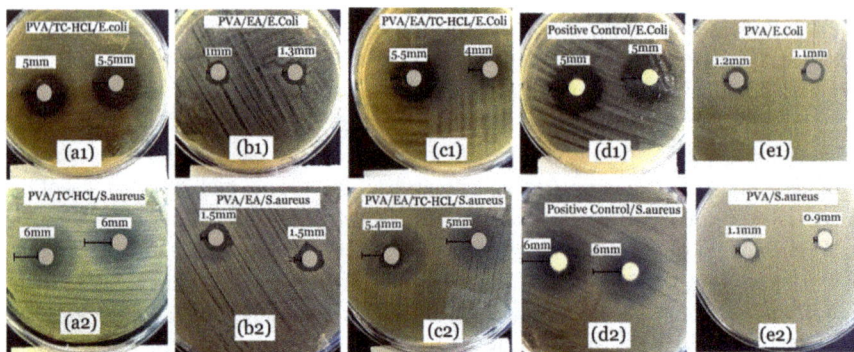

Figure 2: Antibacterial activities of the electrospun PVA nanofibrous samples containing TC-HCl in the exposure of E.coli (a1) and S.aureus (a2), the electrospun PVA nanofibrous samples containing EA in the exposure of E.coli (b1) and S.aureus (b2), the TC-HCl loaded electrospun PVA nanofibrous samples containing EA in the exposure of E.coli (c1) and S.aureus (c2), the TC-HCl solution with a concentration of 500 µg/ml (positive control) in the exposure of E.coli (d1) and S.aureus (d2), the neat electrospun PVA nanofibrous samples (negative control) in the exposure of E.coli (e1) and S.aureus (e2) after 24 hrs. [19] (Springer Nature, Fibers and Polymers, Zahedi et. al. © 2015)

In another study, ibuprofen (Ibu) and acetyl salicylic acid (ASA) loaded polyvinyl pyrolidone (PVP)/dextran (Dext.) composite nanofibers were spun by Maslakci and co-workers [20]. Two different types of dextran were used for this study, Dextran T10 (Dext T10) and Dextran T140 (Dext T40). The drug loaded composite nanofibers were shown in Figure 3. From Figure 3, it is confirmed that fabricated fibres are randomly oriented and some of them have beaded structure. The content of the drugs in the nanofibers were obtained from the chromatographic data while release profile is checked on four different bacteria namely, *P.aeruginosa*, *E.coli S.aureus* and *Bacillus subtilis*. Results (Figure 4) confirm that neat PVP and their composite nanofibers with dextran is unable to stop the growth of the microorganisms while their drug loaded nanofibers effectively reduce the growth of organisms. Overall, drug loaded nanofibers effectively reduce the gram positive bacteria, *S.aureus* and *Bacillus subtilis* while the growth of the gram negative bacteria is unaffected by drug loaded nanofibers. In Table 2, various drugs or bioactive molecules loaded different types of the natural and natural/synthetic composite nanofibers are illustrated.

Materials Research Forum LLC
https://doi.org/10.21741/9781644901076-7

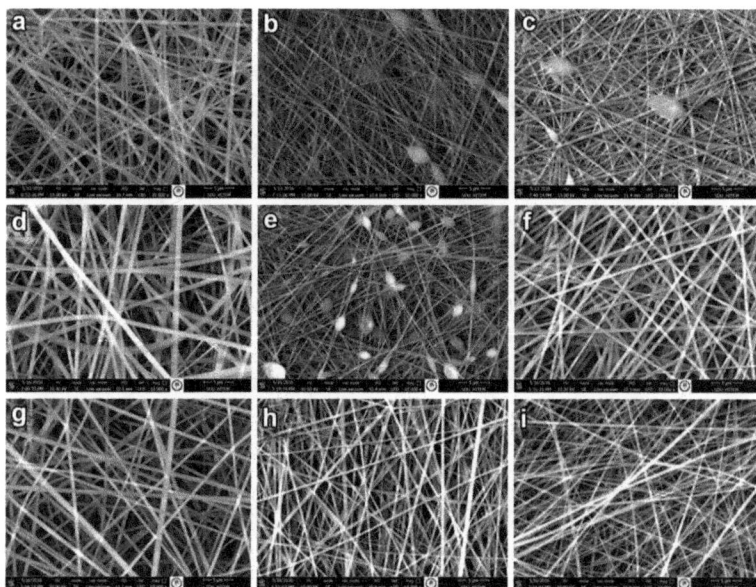

Figure 3: SEM images of the nanofibers prepared from aqueous solutions of PVP (a), PVP/Dext T10 (b), PVP/Dext T40 (c), PVP-Ibu (d), PVP-ASA (e), PVP/Dext T10-Ibu (f), PVP/Dext T10-ASA (g), PVP/ Dext T40-Ibu (h), PVP/Dext T40-ASA (i) [20] (Springer Nature, Polymer Bulletin, Maslakci et. al. © 2017)

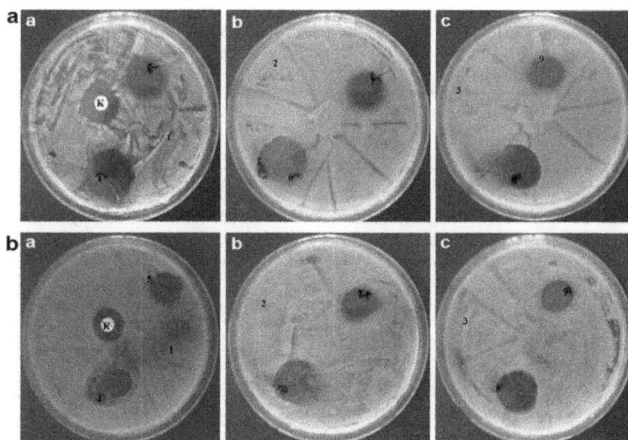

Figure 4: Inhibition zone diameter of the nanofiber samples against B. subtilis (a) and S. aureus (b) [20] (Springer Nature, Polymer Bulletin, Maslakci et. al. © 2017)

Table 2: Drugs or bioactive molecules loaded biopolymeric nanofibers.

No.	Biopolymeric nanofibers	Drugs/Bioactive agents	Reference
1	Halloysite nanotube (HNT)-reinforced alginate composite NFs	Cephalexin (CEF)	[21]
2	hordein/zein/surface-modified cellulose nanowhiskers (SCN)/fiber alignment composite NFs	Riboflavin	[22]
3	Polycaprolactone (PCL)/Gelatin/bioactive glass nanoparticles composite NFs	Dexamethasone (DEX)	[23]
4	soy-protein/ Poly ethylene terephthalate (PET) composite NFs	Riboflavin (Vitamine B$_2$)	[24]
5	Gelatin NFs	Cefradine (CE)	[25]
6	Polyvinyl alcohol (PVA)/ plasma modified Chitosan composite NFs	Ibuprofen	[26]
7	Polyvinyl alcohol (PVA) /Chitosan/ Cellulose acetate composite NFs		
8	Poly(lactide-co-caprolactone) (PLLC)/ collagen	Gentamicin	[27]
9	Chitosan/Polyethylene oxide (PEO) NFs	*5-Fluorouracil*	[28]
10	Wheat gluten (WG)/polyvinyl alcohol (PVA) NFs	Azathioprine	[29]
11	Chitosan/carboxymethyl-β-Cyclodextrin/Polyvinyl alcohol NFs	Salicyclic acid	[30]
12	cellulose nanocrystals/Polyethylene glycol (PEG)/ polylactic acid (PLA) composite NFs	Tetracycline hydrochloride	[31]
13	Starch/Polyvinyl alcohol (PVA) & Polyethylene Oxide (PEO) NFs	ampicillin	[32]
14	Amphiphilic alginate derivatives (AAD)/ Polyvinyl alcohol (PVA) NFs	λ-Cyhalothrin	[33]
15	Alginate/ polyethylene oxide (PEO) composite NFs	Ciprofloxacin hydrochloride	[34]

16	Xanthan Gum/Chitosan composite NFs	Curcumin	[35]
17	Gelatin NFs	Piperine	[36]
18	Polyvinyl alcohol/Dextran composite NFs	Ciprofloxacin	[37]
19	Cellulose acetate/Gliadin Protein composite NFs	Ferulic Acid	[38]
20	Polyvinyl alcohol (PVA)/Chitosan/3-aminopropyltriethoxysilane (APTES) modified Graphene oxide composite NFs	Curcumin	[39]
21	Gelatin NFs	Piperine	[40]
22	Gelatin/Sodium bicarbonate composite NFs	Ciprofloxacin	[41]
23	Cellolose nanocrystal/Polycaprolactone (PCL) composite NFs	Tetracycline	[42]
24	Collagen/Polyvinyl alcohol (PVA) composite NFs	Salycyclic acid	[43]
25	Polycaprolactone (PCL) / Polyethylene oxide (PEO)/ Polylactic acid (PLA)/ Polylactic-co-glycolic acid (PLGA) composite NFs	Quercetin	[44]
26	Glutinous rice starch/ Polyvinyl alcohol (PVA) composite NFs	Chlorpheniramine maleate (CPM	[45]
27	Gelatine NFs	Piperine	[46]
28	CuO NPs/Bacterial Cellulose/Chitosan	CuO Nps	[47]
29	Gelatein NFs	Bovine serum albumin (BSA)	[48]
30	Carboxymethyl cellulose (CMC)/ Polyvinyl alcohol (PVA) composite NFs	Diclofenac Sodium (DS)	[49]
31	Cellulose acetate phthalate	Curcumin	[50]

Nanohybrids   Materials Research Forum LLC
Materials Research Foundations **87** (2021) 156-201   https://doi.org/10.21741/9781644901076-7

3.2   Antimicrobial agents

The antimicrobial characteristics of these materials are very useful in the fields of medical science. These types of materials are not only used due to their antimicrobial characteristics and can be also employed host materials for drug deliveries, scaffolds, coatings and food packaging [51]. Some researchers have doped the inorganic nanoparticles such as Ag, metal oxides (i.e., ZnO, TiO_2, CuO, ZrO_2) and other biologically active compounds (i.e., curcumin, Phenolic acids, flavonoids, quercetin and other naturally occurred polyphenols or phenolic compounds etc.) for enhancing the antimicrobial characteristics [52-58]. Biopolymers have such antimicrobial characteristics, and thus, they are very important ingredients for the preparation of several biomedical devices.

Various research works have demonstrated that biopolymers can act as antimicrobial agents [56, 59-63]. This characteristic of the biopolymers is one of the most important factors which makes them more applicable in the field of biomedical. Due to their macro and complex structures with high functionality (i.e., -OH, $-NH_2$, -COOH, $-CONH_2$ groups) imparts the ability to attach with microorganisms, thus plays an important role in inhibition of microorganism growth. However, the action of biopolymers on microorganisms is still complicated and unclear. Few research reports confirm that biopolymers are attached with cell wall and can penetrate it. After penetration, they bind with nucleic acids and block the synthesis of the mRNA. These disturbs the production of essential proteins and enzymes [64-67]. Biopolymers have polyionic (i.e., cationic or anionic) structures. These functional groups are responsible of the interaction with the negatively or positively charged microorganisms, which enhance the contact with cell wall and these effects on the cell proliferation activity [68, 69].

Various types of biopolymers and their composites (with synthetic or degradable polymers or inorganic nanoparticles) nanofibers have been successfully investigated on different types of microorganisms which are illustrated in Table 3.

Table 3: Antimicrobial biopolymeric nanofibrous mats.

No.	Biopolymeric Nanofibers	Microorganisms	Reference
1	Ag NPs decorated hydroxyapatite/ Chitosan/Polyvinyl alcohol (PVA) composite NFs	*E.coli*	[70]
2	Chitosan/Gelatin composite NFs	*E.coli; S.aureus*	[71]
3	Ag NPs doped GO modified Poly(lactide-co-glycolide) (PLGA)/chitosan composite NFs	*P. aeruginosa; E.coli* and *S.aureus*	[72]
4	Ag NPs decorated Vitamin E/Polylactic acid composite NFs	*E.coli; L.monocytogenes* and *S. typhymurium*	[73]
5	α-Chitin NFs	*A.niger*	[74]
6	Polylactic acid (PLA)/ Chitosan Core-shell NFs	*E.coli*	[75]
7	Quaternized chitons/Organic rectorite (OREC)/Polyvinyl alcohol (PVA) composite NFs	*E.coli; S.aureus*	[76]
8	Ag NPs decorated Polyvinyl alcohol (PVA)/Carboxymethyl chitosan composite NFs	*E.coli*	[77]
9	Chitosan/Polyethylene oxide (PEO)/lauric alginate composite NFs	*E.coli; S.aureus*	[78]
10	Soy protein isolate blended Quaternized Chitosan/Cellulose acetate composite NFs	*E.coli; S.aureus*	[79]
11	Pectin/lysozyme bilayer cellulose composite NFs	*E.coli; S.aureus*	[80]
12	Chitosan-Epigallocatechin gallate bilayer cellulose composite NFs	*S.aureus*	[81]
13	Lysozyme-Chitosan-Organic rectorite blended cellulose actetate NFs	*E.coli; S.aureus*	[82]

14	Triclosan-cyclodextrin inclusion complex blended polylactic acid NFs	*E.coli; S.aureus*	[83]
15	Carboxymethyl Cellulose (CMC)/Chitin composite NFs	*E.coli; S.aureus*	[84]
16	Ag NPs loaded Chitosan NFs	*P.aeruginosa*	[85]
17	Ag NPs loaded Chitosan/polyvinyl alcohol (PVA) composite NFs	*E.coli*	[86]
18	Ag NPs loaded Silk Fibroin NFs	*S.aureus; P.aeruginosa*	[87]
19	Iminochitosan NFs	*E.coli; S.aureus; P.aeruginosa; B. subtilis*	[88]
20	Polyhexmethylene bigluanide (PHMB) blended cellulose acetate/polyester urethane (PEU) composite NFs	*E.coli*	[89]
21	Polyurethane/Hydrolyzed collagen or elastin or hyalauric acid composite NFs	*E.coli; S. typhymurium; L.monocytogenes*	[90]

3.3 Scaffolds for tissue engineering

The tissue engineering is the regeneration or healing of the tissues or parts of the bodies using biomimetic scaffolds. These biomimetic scaffolds have similar structures like natural extracellular matrixes (ECMs). ECMs are one type of proteins and poly saccharides mainly collagen, hyaluronic acid, proteoglycans, glycosaminoglycan, and elastin. Artificial extracellular matrixes (i.e., Scaffolds) have similar structures and have similar compatibility to regulate the regeneration of the tissues. ECMs play an important role like providing structural framework, promoting cells adhesions, cells migrations, cells differentiations, intracellular signaling and cell cytokine activities etc. During regeneration or healing, biological functions of the tissues are unaffected and artificial ECMs material degrades with time. Also newly regenerated cells or tissues will take place of the degraded materials. These two phenomena can simultaneously work. Various types of tissue engineered scaffolds are researched for bone regeneration, wound healing and tissue or muscle repairing.

For the fabrications of the scaffolds, nanofibers are one of the most favourable material because of its properties like high surface areas and low density. For tissue engineering,

Nanohybrids   Materials Research Forum LLC
Materials Research Foundations **87** (2021) 156-201   https://doi.org/10.21741/9781644901076-7

scaffolds must have indigenous characteristics like mechanical stability, sustain biological functions, biocompatible and should be biologically degradable. Biopolymers have similar structures as of the extracellular matrixes and thus are very suitable materials for the fabrication of the scaffolds for tissue engineering, wound healing and bone regenerations. However, biopolymeric nanofibers have poor mechanical strength and thus they are blended with various other biodegradable materials to obtain better scaffold. Various research groups have investigated the compatibility of the fibrous scaffolds on different tissues or organs [91-100].

Leszczak and co-workers have successfully fabricated the demineralized bone matrix (DBM) into electrospun nanofibers using 70:30 hexaflouro-2-propanol and trifluoroacetic acid. DBM fibers are crosslinked with various cross linking agents such as genipin, riboflavin, glutaraldehyde, 1- Ethyl-3-[3-(dimethylamino)propyl]carbodiimide hydrochloride (EDC) as well as dehydrothermal treatment and UV irradiation. They have examined the interaction of this crosslinked fibers with human dermal fibroblasts (HDF) and found that DBM fiber shows good cytocompatibility as well as they noted that the effect of residual electrospinning solvents and crosslinking agents on results are very less [101].Panda and co-workers have studied the hydroxyapatite blended Eri-Tasar silk fibroin nanofibrous scaffolds for ontogenesis of stem cells. In their study, they extracted the fibroin from Eri and Tasar, and then fabricates the fibers. On these fibers hydroxyapatite is loaded by surface deposition method. The obtained composite nanofibrous scaffolds are tested on cord blood human mesenchymal stem cells (CBhMSCs) for cellular attachment and viability studies. Results demonstrated that the fabricated fibrous scaffolds were favourable host material for cell proliferation and attachment. They have found that neat eri and tasar scaffolds have less attachment and proliferation capability compare to hydroxy apatite blended scaffolds. They also studied the cell differentiation of same stem cells by alkaline phosphate assay (ALP) and reverse-transcriptase PCR for determination of osteogenic primers (i.e., glyceraldehyde phosphate dehydrogenase [GAPDH], osteocalcin (OCN), osteopontin (OPN), osteonectin (ONN) and RUNX2). Overall, they have concluded that composite nanofirous scaffolds are favorable material for the osteogenic differentiation of CBhMSCs. And also noted that scaffolds shows enhanced differentiation characteristics in the absence of growth enhancing materials [102].

Two different types of silk fibroins are successfully blended with highly 70S bioactive glass ($70SiO_2.25CaO.5P_2O_5$) and thus fabricated fibers are used for repair of osteochondral defects (OCD). These two silk fibroins are namely, mulberry (*Bombyx mori*) and endemic northeast Indian non-mulberry (*Antheraea assama*). The compatibility of the fabricated composite nanofibers were confirmed from various biological studies

Materials Research Forum LLC

https://doi.org/10.21741/9781644901076-7

like gene expression studies (i.e., osteogenic genes namely bone sialoprotein (BSP), runt-related transcription factor 2 (runx2) and for chondrogenic genes namely aggrecan and sox-9), and *In Vivo* immune response assessments (i.e., murine macrophage Cells) etc. They have also studied the collagen leaching by human osteoblasts cell MG63 by growing a cell culture on fabricated composite mats. According to these studies, fabricated bioactive glass/silk fibroin composite nanofibers shows excellent osteochondral defects (OCD) repairs and managements [103]. In another study, silk fibroin and carboxymethyl cellulose (CMC) composite nanofibers mats were investigated for ability to nucleate the nano sized bioactive calcium phosphate by bio mineralization. For bone regeneration scaffolds must have this type of osteogenic properties. For cell viability, adhesion properties and osteogenic differentiation were investigated on human mesenchymal stem cells (hMSC). According to results, composite mats shown higher cell proliferation and attachments with cells were increased with increasing CMC content in mats. The osteogenic differentiation of HMSC cell were confirmed from the alkaline phosphate (ALP), RUNX2 transcription factor, osteocalcin and type-1 collagen expressions studies. All studies shows that biomineralization capabilities of the composite mats were excellent and scaffolds are capable for bone tissue engineering [104].

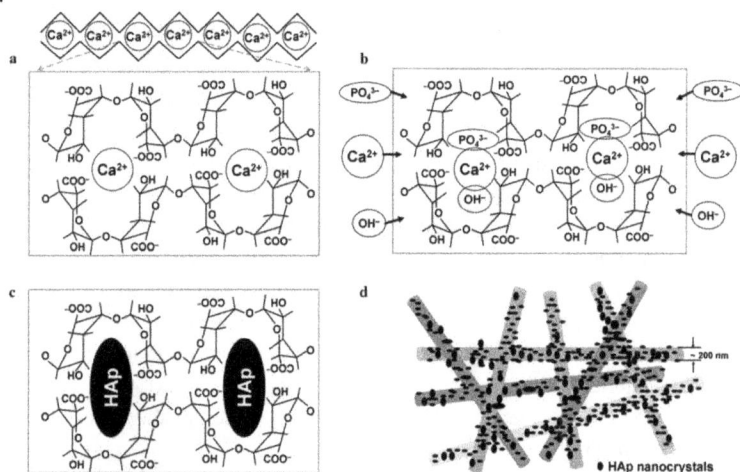

Figure 5. Chemical structure of (a) "egg box" model of calcium alginate, (b) "egg box" model of calcium alginate with precursor ions for HAp nucleation, and (c) mineralized "egg-box" structure with HAp and (d) illustration of cross-linked/in situ synthesized HAp/ alginate nanocomposite fibrous scaffold [105] (Springer Nature, Journal of Materials Science: Materials in Medicine, Chae et. al. © 2013)

Chae and coworkers have fabricated biomimetic scaffolds for bone tissue engineering [105]. The scaffolds were fabricated by synthesis of hydroxyl apatite (HAp) over alginate electrospun nanofibers. The Figure 5 represents, how Ca^{2+} ions forms chelating agents with COO⁻ group of alginates and start nucleation of the HAp. The Figure 5 (d) shows the nucleation of the HAp on alginate fibres. Fabricated Alginate/Ca^{2+} nanofibrous structure were shown in Figure 6. Scaffolds were assessed on rat calvarial osteoblasts cells for cell adhesion and viability studies. They have found that modified nanofiber scaffolds are effectively attached, and permits to grow the cells with spindle shape while neat alginate scaffolds permit comparatively poor attachment and cells are grown with round shape (Fig. 7). Overall, this scaffold is mimetic of mineralized collagen fibrils in bone tissue.

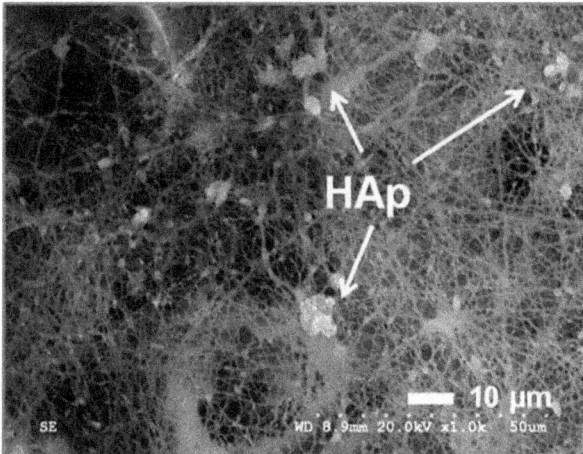

Figure 6: SEM micrograph of HAp/alginate scaffold fabricated using a mechanical blending/electrospinning method. The white arrows indicate the agglomerated HAp particles at micro-meter levels [105] (Springer Nature, Journal of Materials Science: Materials in Medicine, Chae et. al. © 2013)

Materials Research Forum LLC
https://doi.org/10.21741/9781644901076-7

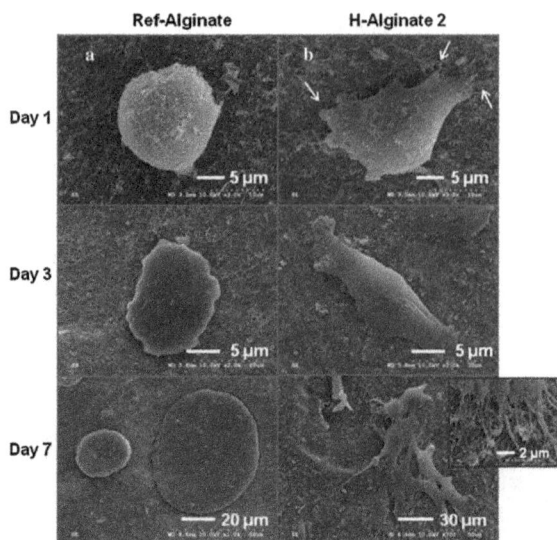

Figure 7: SEM micrographs of RCO cells cultured on (a) cross-linked Ref-Alginate and (b) cross-linked/in situ synthesized H-Alginate 2 scaffolds for 7 days. The white arrows indicate more closely and stably attached RCO cells on the HAp/alginate than the pure alginate scaffold at day 1 of post cell-seeding. The magnified inset SEM micrograph of the HAp/alginate scaffold shows the visible development of multiple filopodial connections to the surface at day 7 of post cell seeding [105] (Springer Nature, Journal of Materials Science: Materials in Medicine, Chae et. al. © 2013)

Donald and coworkers have developed the gelatin-arabinoxylan ferulate (AXF) composite fibers and studied on the diabetic chronic wound healing. The cell viability test was carried out on NIH3T3 mouse fibroblast cells and found that increasing the contents of the gelatin with respect to the drug, the proliferation activity of the scaffolds is decreased. They have noted that antioxidant properties of the arabinoxylan ferulate could promote the biocompatibility. These composite scaffold is impregnated with silver and tested on gram positive and gram negative bacteria and noted that the composite fibrous scaffold is very active against gram positive bacteria compared to gram negative bacteria [106]. In another study, silver nanoparticle (Ag NPs) doped chitosan (CS)/polyvinyl alcohol (PVA) composite nanofibrous scaffolds studied on aerobic bacteria. According to

Materials Research Forum LLC
https://doi.org/10.21741/9781644901076-7

this study Ag NPs loaded nanofibers effectively reduced the growth of the aerobic bacteria [86].

In other research study, researchers have fabricated bilayer electrospun nanofibers for wound healing. The first layer was made up of TiO_2 incorporated chitosan nanofibers and sub layer was human adipose derived extracellular matrix (ECM). The main objective of fabrication of such composite fibres is that TiO_2 doped nanofibers act as shield against bacteria and sponge like sub layer act as catalyst for fast tissue regeneration. The results show that bilayer composite nanofibrous scaffolds can help in cell regeneration with less cytotoxicity. The obtain results confirms the usability of the bilayer fibrous scaffolds for wound healing material [107]. Zhou and coworkers have successfully fabricated the composite nanofibers from the chitosan and sericin. Both chitosan and sericin are very useful natural biomaterials and both possess biocompatibility and better immunogenicity. Sericin consist of 18 kinds of amino acids and have antioxidant, antibacterial, moisture absorption and UV resistant properties. Obtain composite nanofibers were studied on L929 fibroblast cell using methyl thiazolyl tetrazolium (MTT) assay and antibacterial studies were carried out on *E.coli* and *B.subtilis* gram negative and gram positive bacteria respectively. MTT assay confirms the biocompatibilities of the nanofibers while antibacterial studies confirm the effectiveness of the fibrous mats towards gram negative and gram-positive bacteria [108].

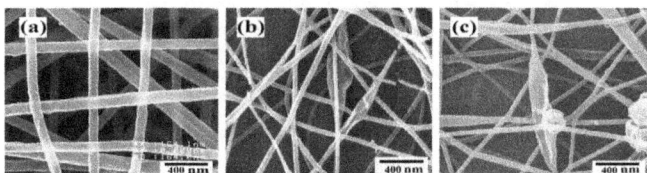

Figure 8: Structure of fibrous skin grafts prepared by electrospinning. a CS-based fibers. b and c contain 1% and 2% GO Nano sheets, respectively [109] (Springer Nature, Journal of Materials Science: Materials in Medicine, Mahmoudi et. al. © 2017).

Graphene oxide loaded chitosan-based nanofibers are used as temporary skin grafted biomaterials for effective wound healing [109]. For preparation of natural skin chitosan, graphene oxide and polyvinyl pyrolidone were spun into nanofibrous scaffolds (Fig. 8). Scaffolds were assessed on human skin fibroblast cells by dead assay and MTT assay, while in vivo studies were carried out on adult rat skin. Studies confirms the effectiveness of the grapheme oxide (GO) incorporation into the nanofibrous scaffolds. With increasing the GO content in scaffolds, the effectiveness of the scaffolds is increased. At 1.5 wt % of GO cell viability and proliferation activity is higher comparative to others (Fig.9).

Furthermore, GO promote the cell viability and heal the wound in less time. Overall, results confirm that fabricated scaffolds are used as temporary skin grafts (TSG).

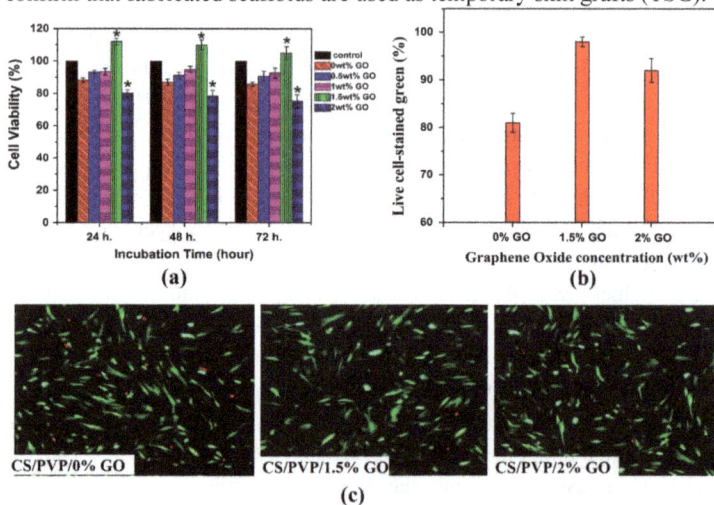

(a) (b)

(c)

*Figure 9: Cell viability of nanofibrous membranes containing different concentrations of GO nano sheets normalized to cell number determined by a MTT and b PrestoBlue® assay. c Live (stained green) and dead (stained red) staining of human fibroblast cells after 24 h incubation on chitosan-based nanofibers without GO addition and nanofibers containing 1.5 wt% and 2 wt% GO. For each experiment, three replicates were used per group. *Denotes significant difference between the control and the specimens (P < 0.05) [109] (Springer Nature, Journal of Materials Science: Materials in Medicine, Mahmoudi et. al. © 2017)*

Wang and coworkers have fabricated the 1 to 5 wt% heparin loaded gelatin composite nanofibrous mats for vascular tissue engineering [110]. Fabricated 5 wt% heparin loaded gelatin scaffolds were crosslinked for 7 days in various concentration of the glutaraldyhyde (Fig. 10). In Fig. 10 (A) it is clearly seen that fibers are fused with each other and compact structures are formed when 2.5 wt% (c) and 5 wt% (d) concentrated glutaraldehyde solution were used for crosslinking while Fig. 10 (B) demonstrate the crosslinked nanofibers obtained by dipping in 1 wt % of glutaraldehyde solution for different days and then dipped into water for 24 hours and dried in vacuum for 1 week.

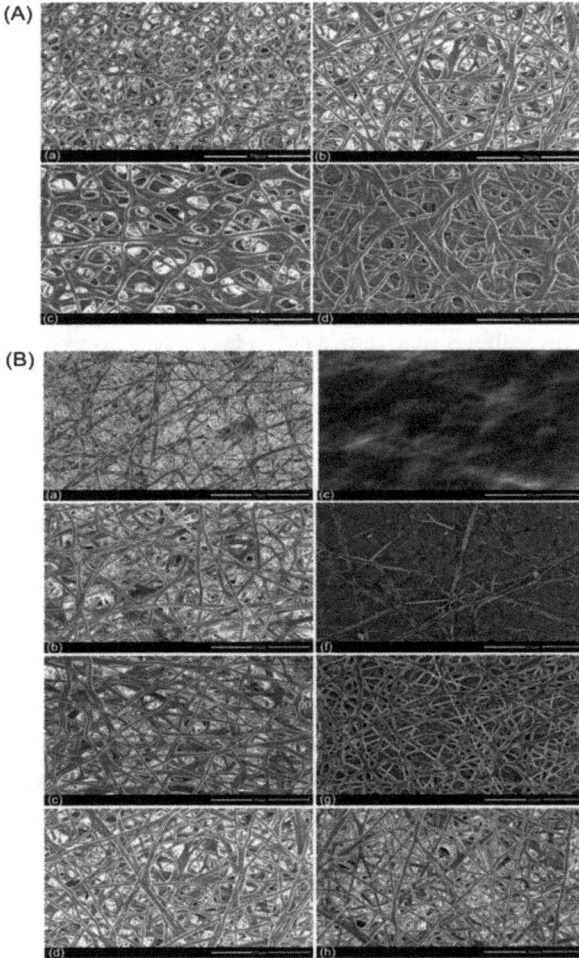

Figure 10: SEM images of gelatin-heparin (heparin content:5 wt%) fibrous scaffold (A) after different concentration of glutaraldehyde solutions crosslinking (crosslinking time: 7 days; the concentration of glutaraldehyde: (a) 0.5%, (b) 1%, (c) 2.5%, (d) 5%, w/w) and (B) through crosslinking of method glurataldehyde (glutaraldehyde concentration: 1%, w/w) during different time before (ad) and after (e-h) rising in water for 24 h after drying for 1 week in a vacuum oven ((a,e) 1 day, (b,f) 3 days, (c,g) 5 days, (d,h) 7 days). [110] (Springer Nature, Macromolecular Research, Wang et. al. © 2013)

Results confirms that after 1-day dipping in glutaraldehyde solution, fibres are completely destroyed when they are suspended into water while for 3 days and 5 days crosslinked nanofibrous scaffolds shows different morphologies than crosslinked fibrous mats. This suggests glutaraldehyde is dissolved in water during dipping in water. In case of 7 days, the negligible effects where observed on morphologies of the crosslinked fibres. Overall, minimum 7 days are required to obtain the proper crosslinked nanofibers by dipping the mats in 1 wt% glutaraldehyde solution. Due to the crosslinking the mechanical properties of the gelatin/heparin fibers are effectively improved. From Table 4 it is clearly seen that tensile strength and elasticity are increased with rising the heparin content in fibres while elongation to break was quite improved. Cytocompatibility of the crosslinked fibrous scaffolds were carried out on human umbilical vein endothelial cells (HUVECs) by growing the cells on fibrous mats. Overall results show (Figure 11), gelatin/heparin scaffolds have high biocompatibility and permits the cells to grow without any cytotoxic effects. It is noted that biocompatibility increased with rising heparin contents in fibrous mats.

Table 4. Mechanical Properties of Gelatin-Heparin Nanofibrous Scaffolds before and after Crosslinking [110] (Springer Nature, Macromolecular Research, Wang et. al. © 2013)

Heparin (wt%)	Tensile Strength (MPa)		Elongation at Break (%)		Elastic Modulus (MPa)	
	Before Crosslinking	After Crosslinking	Before Crosslinking	After Crosslinking	Before Crosslinking	After Crosslinking
0	1.83±0.21	13.7±1.9	6.47±2.30	4.21±2.1	180±28	1056±98
1	2.82±0.22	15.6±3.4	19±3.21	13.4±4.6	122±18	1090±110
3	3.24±0.58	19.8±4.4	11±5.42	8.2±2.3	154±31	1121±134
5	3.81±0.63	20.1±4.8	9.8±4.60	7.1±2.5	194±20	1197±167

Crosslinked collagen nanofibrous mats are very useful as extra cellular matrixes. Various methods are used to crosslink the collagen fibers such as milder enzymatic treatment procedure using transglutaminase, the use of N-[3-(dimethylamino)propyl]-N'-ethylcarbodiimide hydrochloride, N-hydroxysuccinimide, and genipin, as well as the use of a physical method based on exposure to ultraviolet light was carried out. Overall, crosslinked collagen mats based nanoscaffolds, effectively enhance the growth of the osteoblast cells and promote the cell proliferation [111]. Vatankhah and coworkers have studied the effects on contraction, ligand density and stiffness of the composite technophilic/gelatin nanofibrous mats. Results confirm that a higher amount of the gelatin gives higher ligand density while higher techophilic amount confirms the lower stiffness. Overall, contractibility of nanofibrous coated smooth muscle cells were maintained while the amount of the gelatin is higher in nanofibrous scaffolds [112]. Gu and coworkers have fabricated the scaffolds from the conductive polymer by means of increasing the cell proliferation through dielectric guidance [113]. Results reveal that conductive polymer-based scaffolds can easily receive the biochemical and electrical signals from the peripheral cells and promotes the cell growth and differentiation. Fabricated chitin/polyaniline composite nanofibrous scaffolds were assessed on human dermal fibroblast for cell viability. Results show that aligned nanofibrous scaffolds have high effectiveness and regeneration capabilities than random nanofibrous scaffolds. Few more nanofibrous scaffolds are listed in Table 5.

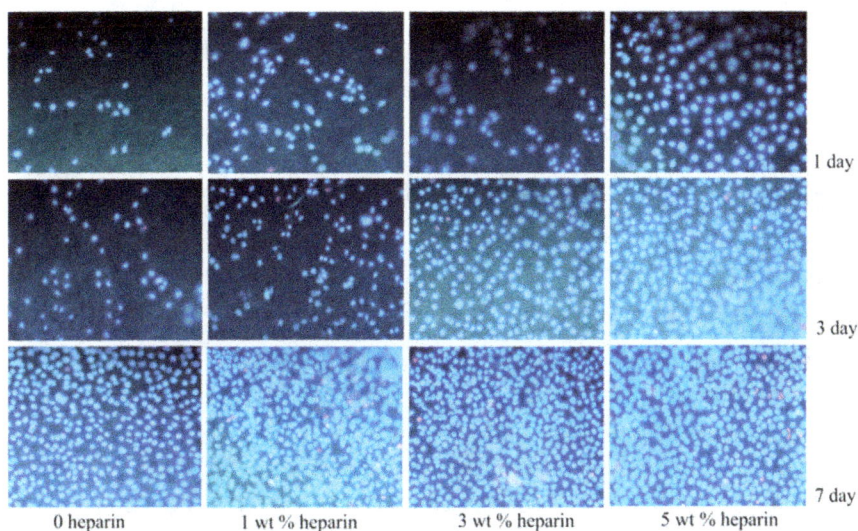

Figure 11. Live/dead cells staining fluorescent images of HUVECs cultured on the crosslinked gelatin fibrous scaffolds with different heparin content during 1, 3, and 7 culturing days. The live cells are stained blue and the dead cells red. The images were captured directly on an inverted fluorescent microscope [110]. (Springer Nature, Macromolecular Research, Wang et. al. © 2013)

Table 5: Various biopolymeric composite scaffolds for different applications.

Materials Research Forum LLC
https://doi.org/10.21741/9781644901076-7

No.	Biopolymeric scaffolds	Application	Reference
1	Poly(hydroxybutylate-co-hydroxy valerate) (PHBV)/m-keratin NFs	Wound dressing	[114]
2	PCL/Gelatin Cues NFs	Nerve outgrowth regeneration	[115]
3	Polycatecholamine coated collagen NFs	Burn wound healing	[116]
4	Hydroxy appetite (Hap)/Poly lactic acid (PLA) NFs	Wound dressing	[117]
5	Poly glycolic acid (PGA)/Chitin NFs	Tissue engineering scaffold	[118]
6	Polypropylene carbonate (PPC) /Gelatin NFs	Tissue engineering scaffold	[119]
7	Iminodiacetic acid (IDA) grafted Chitosan NFs	Hypertheramic tumour cell treatment	[120]
8	Collagen-Polyvinyl alcohol (PVA) NFs/collagen sponge	Surface cartilage repair scaffolds	[121]
9	Gelatin/Chitosan composite NFs	Tissue engineering scaffold	[122]
10	Polycaprolactone (PCL) macroparticles loaded Gelatin NFs	Tissue engineering scaffold	[123]
11	Gelatin NFs	Human mesenchymal stem cell regeneration	[124]
12	Porcine collagen/Polycaprolactone (PCL) NFs	Tissue engineering scaffold	[125]
13	Biotin and Galactose functionalized GEL NFs	Tissue engineering scaffold	[126]
14	Gelatin loaded Acrylic acid/Bacterial cellulose NFs	Tissue engineering scaffold	[127]
15	Cellulose acetate/Chitosan NFs	Scaffold	[128]

16	Silk Fibroin/Chitosan composite NFs	Scaffold	[129]
17	Gelatin NFs	Ophthalmic biomaterials	[130]
18	Polycaprolactone (PCL)/Chitosan composite NFs	Mesenchymal stem cells regeneration	[131]
19	Cellulose/Nylon 6 NFs	Membrane application	[132]
20	Carboxymethyl Chitin/Organic rectorite composite NFs	Food packaging, Tissue engineering, Wound healing	[133]
21	Chitosan/Silk fibroin composite NFs	Human mesenchymal stem cells regeneration	[134]
22	Chitosan/Pectin/Organic rectorite composite NFs	Food Packaging and Wound healing	[135]
23	Polyamide-6,6/Chitosan hybrid NFs	Tissue engineering scaffold	[[136]
24	Non-ionic cellulose ethers/polyvinyl alcohol (PVA) composite NFs	Tissue engineering scaffold	[137]
25	Curcumin-xanthan-Chitosan composite NFs	Carrier vehicle	[138]
26	Layer by Layer composite NFs of Chitosan/heparin/graphene oxide/poly lactic acid (PLA)	Tissue engineering scaffold	[139]
27	Gelatin/Aloe-vera/Polycaprolactone (PCL) hybrid NFs	Skin tissue engineering	[140]
28	Poly acrylic acid (PAA)/Chitosan (CS) and Poly acrylic acid (PAA)/Alginate NFs	For Bacteria bioreactor	[141]
29	Metronidazolee-Polycaprolactone (PCL)/Gelatin composite NFs	Tissue regeneration membranes	[142]
30	Carboxymethyl Chitin/polyvinyl	Tissue engineering scaffold	[143]

	alcohol (PVA) composite NFs		
31	Core-shell Sodium alginate (SA)/ Polyethylene oxide (PEO) NFs	Tissue engineering scaffold	[144]
32	Gelatin/Polyvinyl alcohol (PVA) NFs	Skin Tissue engineering	[145]
33	Modified Gelatin/Tyrosine NFs	Cartilage tissue engineering	[146]
34	Collagen NFs	Tissue engineering scaffold	[147]
35	Chitosan (CS)/Polyvinyl alcohol (PVA) NFs	Tissue engineering scaffold	[148]
36	Collagen Type-1 NFs	Cardiac Tissue engineering	[149]
37	Silk Fibroin (SF) NFs	Wound dressing	[150]
38	Collagen/Zein composite NFs	Wound dressing	[151]
39	Au NPs loaded PCL/Gelatin composite NFs	Wound healing	[152]
40	Chitosan (CS)/Polyvinyl alcohol (PVA) NFs	Wound healing	[153]
41	Chitosan (CS)/ Polycaprolactone (PCL) NFs	Dermal Wound dressing	[154]
42	Collagen loaded PHB/Gelatin NFs	Wound dressing	[155]
43	TGF-β1 inhibitor loaded Polycaprolactone (PCL)/ Gelatin composite NFs	For Clinical Hypertrophic scars prevention	[156]
44	Chitosan (CS)/Polyvinyl alcohol (PVA) NFs	Bone regeneration	[157]
45	Oxidized Starch-CS-calcium phosphate- Polycaprolactone (PCL) composite NFs	Bone Tissue engineering	[158]
46	Boron Nitride/Gelatine NFs	Bone Tissue engineering	[159]
47	Polycaprolactone (PCL)/	Stem cells based bone tissue	[160]

48	Gelatine composite NFs	engineering	
48	Gelatin/Genipin composite NFs	Tympanic membrane repair	[161]

Conclusion

Biopolymers are some of the most important raw ingredients for the preparation of the bio-based materials, because of their extraordinary properties like degradability, availability and non-toxicity. Above reported works described the electrospun nanofibers of biopolymers and their composites with synthetic or inorganic NPs. These biopolymers are mainly polysaccharides (i.e., chitin, chitosan, cellulose etc.), proteins (i.e., collagen, gelatin, silk fibroin etc.) and polynucleotides (i.e., DNA, RNA). Drug or bioactive agent loaded biopolymeric nanofibers can act as good drug carrier. Inorganic NPs or bioactive materials loaded fibres act as better antimicrobial agents.

Biopolymers are key raw materials in tissue engineering for preparation of biodegradable scaffolds. Additionally, a high functionality of some biopolymers (presence of polar groups-i.e., -ON, -NH$_2$, -CONH$_2$ and -COOH) enhances the interactions of the biopolymers with living organisms or cells, and act as extra cellular matrixes (ECM) which further enhance the cell attachment and cell viability. These high functionalities also improve the applicability of the biopolymers in the field associated with food packaging, pharmaceuticals and coatings. Finally, this mini-review could be insightful for the researchers to drawn their future research methodology. Overall, each class of the biopolymers have their own advantageous properties which are very useful to fabricate advanced materials for biomedical applications.

References

[1] B.-D. Gisela, H. Andrew, C. Jared, Electrospun nanofibers from biopolymers and their biomedical applications, in: J.V. Edwards (Ed.), Modified Fibers with Medical and Specialty Applications, Springer, Netherland, 2006, pp. 67-80.

[2] A. V. Vasenkov, Big data is the future of material science, Journal of Material Science & Engineering, S2(01) (2013). https://doi.org/10.4172/2169-0022.S1.010

[3] J.V. Edwards, B.-D. Gisela, S.C. Goheen, Modified Fibers with Medical and Specialty Applications, Springer, 2006. https://doi.org/10.1007/1-4020-3794-5

[4] S. Wang, Y. Zhao, M. Shen, X. Shi, Electrospun hybrid nanofibers doped with nanoparticles or nanotubes for biomedical applications, Therapeutic Delivery 3(10) (2012) 1155-1169. https://doi.org/10.4155/tde.12.103

[5] N. Chirani, L. Yahia, L. Gritsch, F.L. Motta, S. Chirani, S. Faré, History and Applications of Hydrogels, Journal of Biomedical Sciences 4(2) (2015) 13-23.

[6] B. M. Baker, A. M. Handorf, L. C. Lonescu, L. Wan-Ju, R. L. Mauck, New directions in nanofibrous scaffolds for soft tissue engineering and regeneration, Expert Rev. Med. Devices 6(5) (2009) 515-532. https://doi.org/10.1586/erd.09.39

[7] T. Ondaruchu, C. Jaochim, Drawing a single nanofibre over hundreds of microns, EUROPHYSICS LETTERS 42(2) (1998) 15-220. https://doi.org/10.1209/epl/i1998-00233-9

[8] M. R. Charles., Membrane-Based Synthesis of Nanomaterials, Chem. Mater. 8 (1996) 1739-1746. https://doi.org/10.1021/cm960166s

[9] P.X. Ma, R. Zhang, Synthetic nano-scale fibrous extracellular matrix, Journal of Biomedical Materials Research 46(1) (1999) 60-72. https://doi.org/10.1002/(SICI)1097-4636(199907)46:1<60::AID-JBM7>3.0.CO;2-H

[10] L. Guojun, D. Jianfu, Q. Lijie, G. Andrew, D.B. P., G.J. T., H. T., S. K., Polystyrene-block-poly(2-cinnamoylethyl methacrylate) Nanofibers: Preparation, Characterization, and Liquid Crystalline Properties, Chem. Eur. J 5(9) (1999) 2740-2749. https://doi.org/10.1002/(SICI)1521-3765(19990903)5:9<2740::AID-CHEM2740>3.0.CO;2-V

[11] J. M. Deitzal, J. D. Kleinmeyer, J. K. Hirvonen, N.C. Beck Tan, Controlled deposition of electrospun poly(ethylene oxide) fibers, Polymer 42 (2001) 8163-8170. https://doi.org/10.1016/S0032-3861(01)00336-6

[12] J. Li, S. Vadahanambi, C.D. Kee, I.K. Oh, Electrospun fullerenol-cellulose biocompatible actuators, Biomacromolecules 12(6) (2011) 2048-54. https://doi.org/10.1021/bm2004252

[13] S. An, A. Sankaran, A.L. Yarin, Natural Biopolymer-Based Triboelectric Nanogenerators via Fast, Facile, Scalable Solution Blowing, ACS Appl Mater Interfaces 10(43) (2018) 37749-37759. https://doi.org/10.1021/acsami.8b15597

[14] X. Huang, X. Li, Y. Li, X. Wang, Biopolymer as Stabilizer and Adhesive To in Situ Precipitate CuS Nanocrystals on Cellulose Nanofibers for Preparing Multifunctional Composite Papers, ACS OMEGA 3 (2018) 8083-8090. https://doi.org/10.1021/acsomega.8b01225

Nanohybrids   Materials Research Forum LLC
Materials Research Foundations **87** (2021) 156-201   https://doi.org/10.21741/9781644901076-7

[15]   M.I. Shekh, J. Amirian, F.J. Stadler, B. Du, Y. Zhu, Oxidized chitosan modified electrospun scaffolds for controllable release of acyclovir, Int J Biol Macromol 151 (2020) 787-796. https://doi.org/10.1016/j.ijbiomac.2020.02.230

[16]   S. Khansari, Ray Suman Sinha-, A.L. Yarin, B. Pourdeyhimi, Biopolymer-Based Nanofiber Mats and Their Mechanical Characterization, Ind. Eng. Chem. Res. 52 (2013) 15104−15113. https://doi.org/10.1021/ie402246x

[17]   J. D. Schiffman, C. L. Schauer, A Review: Electrospinning of Biopolymer Nanofibers and their Applications, Polymer Reviews 48(2) (2008) 317-352. https://doi.org/10.1080/15583720802022182

[18]   Q. Zhang, Y. Li, Z.Y.W. Lin, K.K.Y. Wong, M. Lin, L. Yildirimer, X. Zhao, Electrospun polymeric micro/nanofibrous scaffolds for long-term drug release and their biomedical applications, Drug discovery today 22(9) (2017) 1351-1366. https://doi.org/10.1016/j.drudis.2017.05.007

[19]   P. Zahedi , M. Fallah-Darrehchi, Electrospun Egg Albumin-PVA Nanofibers Containing Tetracycline Hydrochloride: Morphological, Drug Release, Antibacterial, Thermal and Mechanical Properties, Fibers and Polymers 16(10) (2015) 2184-2192. https://doi.org/10.1007/s12221-015-5457-9

[20]   N.N. Maslakci, S. Ulusoy, E. Uygun, H. Cevikbas, L. Oksuz, H.K. Can, A.U. Oksuz, Ibuprofen and acetylsalicylic acid loaded electrospun PVP-dextran nanofiber mats for biomedical applications, Polymer Bulletin 74 (2017) 3283–3299. https://doi.org/10.1007/s00289-016-1897-7

[21]   R.T. De Silva, R.K. Dissanayake, M. Mantilaka, W. Wijesinghe, S.S. Kaleel, T.N. Premachandra, L. Weerasinghe, G.A.J. Amaratunga, K.M.N. de Silva, Drug-Loaded Halloysite Nanotube-Reinforced Electrospun Alginate-Based Nanofibrous Scaffolds with Sustained Antimicrobial Protection, ACS Appl Mater Interfaces 10(40) (2018) 33913-33922. https://doi.org/10.1021/acsami.8b11013

[22]   Y. Wang, L. Chen, Cellulose nanowhiskers and fiber alignment greatly improve mechanical properties of electrospun prolamin protein fibers, ACS Appl Mater Interfaces 6(3) (2014) 1709-18. https://doi.org/10.1021/am404624z

[23] A. El-Fiqi, J.H. Kim, H.W. Kim, Osteoinductive fibrous scaffolds of biopolymer/mesoporous bioactive glass nanocarriers with excellent bioactivity and long-term delivery of osteogenic drug, ACS Appl Mater Interfaces 7(2) (2015) 1140-52. https://doi.org/10.1021/am5077759

[24] S. Khansari, S. Duzyer, S. Sinha-Ray, A. Hockenberger, A.L. Yarin, B. Pourdeyhimi, Two-stage desorption-controlled release of fluorescent dye and vitamin from solution-blown and electrospun nanofiber mats containing porogens, Molecular pharmaceutics 10(12) (2013) 4509-26. https://doi.org/10.1021/mp4003442

[25] Huarong Nie, S. Xu, J. Li, A. He, Qingsong Jiang, a.C.C. Han, Carrier System of Chemical Drugs and Isotope from Gelatin Electrospun Nanofibrous Membranes, Biomacromolecules 11 (2010) 2190–2194. https://doi.org/10.1021/bm100505j

[26] G. Celik, A.U. Oksuz, Controlled Release of Ibuprofen From Electrospun Biocompatible Nanofibers With In Situ QCM Measurements, Journal of Macromolecular Science, Part A Pure and Applied Chemistry 52 (2015) 76-83. https://doi.org/10.1080/10601325.2014.978200

[27] C.R. Reshmi, M. Tara, B. Anupama, M. Nandita, E.K. K., SujithcA., Poly(L-lactide-co-caprolactone)/collagen electrospun mat: Potential for wound dressing and controlled drug delivery, International Journal of Polymeric Materials and Polymeric Biomaterials 66(13) (2017) 645-657. https://doi.org/10.1080/00914037.2016.1252357

[28] Wei Li, T. Luo, Y. Shi, Y. Yang, X. Huang, K. Xing, L. Liu, M. Wang, Preparation, Characterization, and Property of Chitosan/Polyethylene Oxide Electrospun Nanofibrous Membrane for Controlled Drug Release, Integrated Ferroelectrics 151(1) (2014) 164-178. https://doi.org/10.1080/10584587.2014.901124

[29] A. Soroush, H. Leila, A. Elham, A.A. Hemati, Preparation of electrospun nanofibers based on wheat gluten containing azathioprine for biomedical application, International Journal of Polymeric Materials and Polymeric Biomaterials (2018).

[30] B. Maryam, N. Mahdi, M. Javad, Electrospinning of cyclodextrin functionalized chitosan/PVA nanofibers as a drug delivery system, Chinese Journal of Polymer Science 31(10) (2013) 1343 1351. https://doi.org/10.1007/s10118-013-1309-5

[31] Y. Hou-Yong, W. Chuang, S.Y.H. Abdalkarim, Cellulose nanocrystals/polyethylene glycol as bifunctional reinforcing/compatibilizing agents in poly(lactic acid) nanofibers for controlling long-term in vitro drug release, Cellulose 24 (2017) 4461–4477. https://doi.org/10.1007/s10570-017-1431-6

[32] S. Tang, Z. Zhao, G. Chen, Y. Su, L. Lu, B. Li, D. Liang, R. Jin, Fabrication of ampicillin/starch/polymer composite nanofibers with controlled drug release

properties by electrospinning, J Sol-Gel Sci Technol 77 (2016) 594–603.
https://doi.org/10.1007/s10971-015-3887-x

[33]   X. Chen, H. Yan, W. Sun, Y. Feng, J. Li, Q. Lin, Z. Shi, X. Wang, Synthesis of
amphiphilic alginate derivatives and electrospinning blend nanofibers: a novel
hydrophobic drug carrier, Polym. Bull. 72 (2017) 30973117.
https://doi.org/10.1007/s00289-015-1455-8

[34]   A. Kyziołq, J. Michna, I. Moreno, E. Gamezb, S. Irustab, Preparation and
characterization of electrospun alginate nanofibers loaded with ciprofloxacin
hydrochloride, European Polymer Journal 96 (2017) 350–360.
https://doi.org/10.1016/j.eurpolymj.2017.09.020

[35]   E. Shekarforoush, F. Ajalloueian, G. Zeng, A.C. Mendes, I.S. Chronakis,
Electrospun xanthan gum-chitosan nanofibers as delivery carrier of hydrophobic
bioactives, Materials Letters 228 (2018) 322–326.
https://doi.org/10.1016/j.matlet.2018.06.033

[36]   A. Laha, Y. Shital, M. Saptarshi, S.C. S., In-vitro release study of hydrophobic
drug using electrospun cross-linked gelatin nanofiber, Biochemical Engineering
Journal 105 (2016) 481–488. https://doi.org/10.1016/j.bej.2015.11.001

[37]   A.MeeraMoydeen, M.S.A. Padusha, E.F. Aboelfetoh, S. Al-Deyab3, M.H. El-
Newehy, Fabrication of electrospun poly(vinyl alcohol)/dextran nanofibers via
emulsion process as drugdelivery system: Kinetics and in vitro release study,
International Journal of Biological Macromolecules 116 (2018) 1250–1259.
https://doi.org/10.1016/j.ijbiomac.2018.05.130

[38]   L. Xinkuan, Y. Yaoyao, Y. Deng-Guang, Z. Ming-Jie, Z. Min, W.G. R., Tunable
zero-order drug delivery systems created by modified triaxial electrospinning,
Chemical Engineering Journal 356 (2019) 886–894.
https://doi.org/10.1016/j.cej.2018.09.096

[39]   R. Sedghi, A. Shaabani, Z. Mohammadi, F.Y. Samadi, E. Isaei, Biocompatible
electrospinning chitosan nanofibers: A novel delivery system with superior local
cancer therapy, Carbohydrate polymers 159 (2017) 1-10.
https://doi.org/10.1016/j.carbpol.2016.12.011

[40]   A. Laha, C.S. Sharma, S. Majumdar, Sustained drug release from multi-layered
sequentially crosslinked electrospun gelatin nanofiber mesh, Materials science &
engineering. C, Materials for biological applications 76 (2017) 782-786.
https://doi.org/10.1016/j.msec.2017.03.110

[41]   Q. Sang, G.R. Williams, H. Wu, K. Liu, H. Li, L.M. Zhu, Electrospun gelatin/sodium bicarbonate and poly(lactide-co-epsilon-caprolactone)/sodium bicarbonate nanofibers as drug delivery systems, Materials science & engineering. C, Materials for biological applications 81 (2017) 359-365. https://doi.org/10.1016/j.msec.2017.08.007

[42]   A. Hivechi, S.H. Bahrami, R.A. Siegel, Drug release and biodegradability of electrospun cellulose nanocrystal reinforced polycaprolactone, Materials science & engineering. C, Materials for biological applications 94 (2019) 929-937. https://doi.org/10.1016/j.msec.2018.10.037

[43]   X. Zhang, K. Tang, X. Zheng, Electrospinning and crosslinking of COL/PVA Nanofiber-microsphere Containing Salicylic Acid for Drug Delivery, Journal of Bionic Engineering 13(1) (2016) 143-149. https://doi.org/10.1016/S1672-6529(14)60168-2

[44]   Ş.M. Eskitoros-Togay, Y.E. Bulbul, N. Dilsiz, Quercetin-loaded and unloaded electrospun membranes: Synthesis, characterization and in vitro release study, Journal of Drug Delivery Science and Technology 47 (2018) 22-30. https://doi.org/10.1016/j.jddst.2018.06.017

[45]   P. Jaiturong, B. Sirithunyalug, S. Eitsayeam, C. Asawahame, P. Tipduangta, J. Sirithunyalug, Preparation of glutinous rice starch/polyvinyl alcohol copolymer electrospun fibers for using as a drug delivery carrier, Asian Journal of Pharmaceutical Sciences 13(3) (2018) 239-247. https://doi.org/10.1016/j.ajps.2017.08.008

[46]   A. Laha, C.S. Sharma, S. Majumdar, Electrospun gelatin nanofibers as drug carrier: effect of crosslinking on sustained release, Materials Today: Proceedings 3(10) (2016) 3484-3491. https://doi.org/10.1016/j.matpr.2016.10.031

[47]   H. Almasi, P. Jafarzadeh, L. Mehryar, Fabrication of novel nanohybrids by impregnation of CuO nanoparticles into bacterial cellulose and chitosan nanofibers: Characterization, antimicrobial and release properties, Carbohydrate polymers 186 (2018) 273-281. https://doi.org/10.1016/j.carbpol.2018.01.067

[48]   S. Liu, Y. Su, Y. Chen, Fabrication, surface properties and protein encapsulation/release studies of electrospun gelatin nanofibers, Journal of biomaterials science. Polymer edition 22(7) (2011) 945-55. https://doi.org/10.1163/092050610X496585

[49]   M.H. El-Newehy, M.E. El-Naggar, S. Alotaiby, H. El-Hamshary, M. Moydeen, S. Al-Deyab, Preparation of biocompatible system based on electrospun CMC/PVA nanofibers as controlled release carrier of diclofenac sodium, Journal of Macromolecular Science, Part A 53(9) (2016) 566-573. https://doi.org/10.1080/10601325.2016.1201752

[50]   R. Ravikumar, M. Ganesh, U. Ubaidulla, E. Young Choi, H. Tae Jang, Preparation, characterization, and in vitro diffusion study of nonwoven electrospun nanofiber of curcumin-loaded cellulose acetate phthalate polymer, Saudi pharmaceutical journal : SPJ : the official publication of the Saudi Pharmaceutical Society 25(6) (2017) 921-926. https://doi.org/10.1016/j.jsps.2017.02.004

[51]   J. Quirós, K. Boltes, R. Rosal, Bioactive Applications for Electrospun Fibers, Polymer Reviews 56(4) (2016) 631-667. https://doi.org/10.1080/15583724.2015.1136641

[52]   M.I. Shekh, N.N. Patel, K.P. Patel, R.M. Patel, A. Ray, Nano silver-embedded electrospun nanofiber of poly(4-chloro-3-methylphenyl methacrylate): use as water sanitizer, Environmental science and pollution research international 24(6) (2017) 5701-5716. https://doi.org/10.1007/s11356-016-8254-0

[53]   M.I. Shekh, D.M. Patel, K.P. Patel, R.M. Patel, Electrospun nanofibers of poly(NPEMA-co.-CMPMA): Used as Heavy metal ion remover and water sanitizer, Fibers and Polymers 17(3) (2016) 358-370. https://doi.org/10.1007/s12221-016-5861-9

[54]   M.I. Shekh, K.P. Patel, R.M. Patel, Electrospun ZnO Nanoparticles Doped Core–Sheath Nanofibers: Characterization and Antimicrobial Properties, Journal of Polymers and the Environment 26(12) (2018) 4376-4387. https://doi.org/10.1007/s10924-018-1310-8

[55] I. Esparza, N. Jimenez-Moreno, F. Bimbela, C. Ancin-Azpilicueta, L.M. Gandia, Fruit and vegetable waste management: Conventional and emerging approaches, J Environ Manage 265 (2020) 110510. https://doi.org/10.1016/j.jenvman.2020.110510

[56]   M. Thakur, G. Sharma, T. Ahamad, A.A. Ghfar, D. Pathania, M. Naushad, Efficient photocatalytic degradation of toxic dyes from aqueous environment using gelatin-Zr(IV) phosphate nanocomposite and its antimicrobial activity, Colloids Surf B Biointerfaces 157 (2017) 456-463. https://doi.org/10.1016/j.colsurfb.2017.06.018

[57]   G. Sharma, B. Thakur, M. Naushad, A. Kumar, F.J. Stadler, S.M. Alfadul, G.T. Mola, Applications of nanocomposite hydrogels for biomedical engineering and

environmental protection, Environmental Chemistry Letters 16(1) (2017) 113-146.
https://doi.org/10.1007/s10311-017-0671-x

[58] L. Rubio, M.J. Motilva, M.P. Romero, Recent advances in biologically active
compounds in herbs and spices: a review of the most effective antioxidant and anti-
inflammatory active principles, Crit Rev Food Sci Nutr 53(9) (2013) 943-53.
https://doi.org/10.1080/10408398.2011.574802

[59] D. Pathania, D. Gupta, N.C. Kothiyal, G. Sharma, G.E. Eldesoky, M. Naushad,
Preparation of a novel chitosan-g-poly(acrylamide)/Zn nanocomposite hydrogel and
its applications for controlled drug delivery of ofloxacin, Int J Biol Macromol 84
(2016) 340-8. https://doi.org/10.1016/j.ijbiomac.2015.12.041

[60] B.S. Rathore, G. Sharma, D. Pathania, V.K. Gupta, Synthesis, characterization and
antibacterial activity of cellulose acetate-tin (IV) phosphate nanocomposite,
Carbohydr Polym 103 (2014) 221-7. https://doi.org/10.1016/j.carbpol.2013.12.011

[61] D. Pathania, G. Sharma, R. Thakur, Pectin @ zirconium (IV) silicophosphate
nanocomposite ion exchanger: Photo catalysis, heavy metal separation and
antibacterial activity, Chemical Engineering Journal 267 (2015) 235-244.
https://doi.org/10.1016/j.cej.2015.01.004

[62] G. Sharma, S. Bhattacharya, V. Chauhan, A. Kumar, I. Inamuddin, A.M. Asiri,
K.A. Alamry, Chemical modification of raw Quercus leucotricophora wood strips
and studies of its physicochemical properties and antifungal behavior, Desalination
and Water Treatment 150 (2019) 252-262. https://doi.org/10.5004/dwt.2019.23696

[63] V.K. Gupta, S. Agarwal, I. Tyagi, D. Pathania, B.S. Rathore, G. Sharma,
Synthesis, characterization and analytical application of cellulose acetate-tin (IV)
molybdate nanocomposite ion exchanger: binary separation of heavy metal ions and
antimicrobial activity, Ionics 21(7) (2015) 2069-2078.
https://doi.org/10.1007/s11581-015-1368-4

[64] S. Xin, X. Li, Y. Zhu, T. Zhang, Z. Lei, W. Li, X. Zhou, H. Deng, Nanofibrous
mats coated by homocharged biopolymer-layered silicate nanoparticles and their
antitumor activity, Colloids and surfaces. B, Biointerfaces 105 (2013) 137-43.
https://doi.org/10.1016/j.colsurfb.2012.12.010

[65] D. Kai, S.S. Liow, X.J. Loh, Biodegradable polymers for electrospinning: towards
biomedical applications, Materials science & engineering. C, Materials for biological
applications 45 (2014) 659-70. https://doi.org/10.1016/j.msec.2014.04.051

Materials Research Forum LLC
https://doi.org/10.21741/9781644901076-7

[66]   A. Anitha, S. Sowmya, P.T.S. Kumar, S. Deepthi, K.P. Chennazhi, H. Ehrlich, M. Tsurkan, R. Jayakumar, Chitin and chitosan in selected biomedical applications, Progress in Polymer Science 39(9) (2014) 1644-1667. https://doi.org/10.1016/j.progpolymsci.2014.02.008

[67]   B.S. de Farias, T.R. Sant'Anna Cadaval Junior, L.A. de Almeida Pinto, Chitosan-functionalized nanofibers: A comprehensive review on challenges and prospects for food applications, Int J Biol Macromol 123 (2019) 210-220. https://doi.org/10.1016/j.ijbiomac.2018.11.042

[68]   R. Jayakumar, M. Prabaharan, S.V. Nair, H. Tamura, Novel chitin and chitosan nanofibers in biomedical applications, Biotechnology advances 28(1) (2010) 142-50. https://doi.org/10.1016/j.biotechadv.2009.11.001

[69]   K. Kalantari, A.M. Afifi, H. Jahangirian, T.J. Webster, Biomedical applications of chitosan electrospun nanofibers as a green polymer - Review, Carbohydr Polym 207 (2019) 588-600. https://doi.org/10.1016/j.carbpol.2018.12.011

[70]   H. Celebi, M. Gurbuz, S. Koparal, A. Dogan, Development of antibacterial electrospun chitosan/poly(vinyl alcohol) nanofibers containing silver ion-incorporated HAP nanoparticles, Composite Interfaces 20(9) (2013) 799-812. https://doi.org/10.1080/15685543.2013.819700

[71] S. Habibi, K. Hajinasrollah, Electrospinning of Nanofibers Based on Chitosan/Gelatin Blend for Antibacterial Uses, Russian Journal of Applied Chemistry 91(5) (2018) 877-881. https://doi.org/10.1134/S1070427218050191

[72]   A.F. de Faria, F. Perreault, E. Shaulsky, L.H. Arias Chavez, M. Elimelech, Antimicrobial Electrospun Biopolymer Nanofiber Mats Functionalized with Graphene Oxide-Silver Nanocomposites, ACS Appl Mater Interfaces 7(23) (2015) 12751-9. https://doi.org/10.1021/acsami.5b01639

[73]   B.S. Munteanu, Z. Aytac, G.M. Pricope, T. Uyar, C. Vasile, Polylactic acid (PLA)/Silver-NP/VitaminE bionanocomposite electrospun nanofibers with antibacterial and antioxidant activity, Journal of Nanoparticle Research 16(10) (2014) 2643-2655. https://doi.org/10.1007/s11051-014-2643-4

[74] A.M. Salaberria, S.C. Fernandes, R.H. Diaz, J. Labidi, Processing of alpha-chitin nanofibers by dynamic high pressure homogenization: characterization and antifungal activity against A. niger, Carbohydrate polymers 116 (2015) 286-91. https://doi.org/10.1016/j.carbpol.2014.04.047

[75]   T.T.T. Nguyen, O.H. Chung, J.S. Park, Coaxial electrospun poly(lactic acid)/chitosan (core/shell) composite nanofibers and their antibacterial activity, Carbohydrate polymers 86(4) (2011) 1799-1806. https://doi.org/10.1016/j.carbpol.2011.07.014

[76]   H. Deng, P. Lin, S. Xin, R. Huang, W. Li, Y. Du, X. Zhou, J. Yang, Quaternized chitosan-layered silicate intercalated composites based nanofibrous mats and their antibacterial activity, Carbohydrate polymers 89(2) (2012) 307-13. https://doi.org/10.1016/j.carbpol.2012.02.009

[77]   Y. Zhao, Y. Zhou, X. Wu, L. Wang, L. Xu, S. Wei, A facile method for electrospinning of Ag nanoparticles/poly (vinyl alcohol)/carboxymethyl-chitosan nanofibers, Applied Surface Science 258(22) (2012) 8867-8873. https://doi.org/10.1016/j.apsusc.2012.05.106

[78]   L. Deng, M. Taxipalati, A. Zhang, F. Que, H. Wei, F. Feng, H. Zhang, Electrospun Chitosan/Poly(ethylene oxide)/Lauric Arginate Nanofibrous Film with Enhanced Antimicrobial Activity, Journal of agricultural and food chemistry 66(24) (2018) 6219-6226. https://doi.org/10.1021/acs.jafc.8b01493

[79]   Y. Pan, X. Huang, X. Shi, Y. Zhan, G. Fan, S. Pan, J. Tian, H. Deng, Y. Du, Antimicrobial application of nanofibrous mats self-assembled with quaternized chitosan and soy protein isolate, Carbohydrate polymers 133 (2015) 229-35. https://doi.org/10.1016/j.carbpol.2015.07.019

[80]   T. Zhang, P. Zhou, Y. Zhan, X. Shi, J. Lin, Y. Du, X. Li, H. Deng, Pectin/lysozyme bilayers layer-by-layer deposited cellulose nanofibrous mats for antibacterial application, Carbohydrate polymers 117 (2015) 687-93. https://doi.org/10.1016/j.carbpol.2014.10.064

[81]   J. Tian, H. Tu, X. Shi, X. Wang, H. Deng, B. Li, Y. Du, Antimicrobial application of nanofibrous mats self-assembled with chitosan and epigallocatechin gallate, Colloids and surfaces. B, Biointerfaces 145 (2016) 643-652. https://doi.org/10.1016/j.colsurfb.2016.05.008

[82]   W. Huang, H. Xu, Y. Xue, R. Huang, H. Deng, S. Pan, Layer-by-layer immobilization of lysozyme–chitosan–organic rectorite composites on electrospun nanofibrous mats for pork preservation, Food Research International 48(2) (2012) 784-791. https://doi.org/10.1016/j.foodres.2012.06.026

[83]   F. Kayaci, O.C. Umu, T. Tekinay, T. Uyar, Antibacterial electrospun poly(lactic acid) (PLA) nanofibrous webs incorporating triclosan/cyclodextrin inclusion

complexes, Journal of agricultural and food chemistry 61(16) (2013) 3901-8.
https://doi.org/10.1021/jf400440b

[84] M.C. Li, Q. Wu, K. Song, H.N. Cheng, S. Suzuki, T. Lei, Chitin Nanofibers as
Reinforcing and Antimicrobial Agents in Carboxymethyl Cellulose Films: Influence
of Partial Deacetylation, ACS Sustainable Chemisty & Engineering 4(8) (2016)
4385-4395. https://doi.org/10.1021/acssuschemeng.6b00981

[85] S.J. Lee, D.N. Heo, J.H. Moon, W.K. Ko, J.B. Lee, M.S. Bae, S.W. Park, J.E.
Kim, D.H. Lee, E.C. Kim, C.H. Lee, I.K. Kwon, Electrospun chitosan nanofibers
with controlled levels of silver nanoparticles. Preparation, characterization and
antibacterial activity, Carbohydrate polymers 111 (2014) 530-7.
https://doi.org/10.1016/j.carbpol.2014.04.026

[86] A.M. Abdelgawad, S.M. Hudson, O.J. Rojas, Antimicrobial wound dressing
nanofiber mats from multicomponent (chitosan/silver-NPs/polyvinyl alcohol)
systems, Carbohydrate polymers 100 (2014) 166-78.
https://doi.org/10.1016/j.carbpol.2012.12.043

[87] P. Uttayarat, S. Jetawattana, P. Suwanmala, J. Eamsiri, T. Tangthong, S. Pongpat,
Antimicrobial electrospun silk fibroin mats with silver nanoparticles for wound
dressing application, Fibers and Polymers 13(8) (2012) 999-1006.
https://doi.org/10.1007/s12221-012-0999-6

[88] Rupesh Gajanan Nawalakhe, Samuel M. Hudson, A.-F.M. Seyam, Ahmed I. Waly,
Nabil Y. Abou-Zeid, Hassan M. Ibrahim, Development of Electrospun Iminochitosan
for Improved Wound Healing Application, Journal of Engineered Fibers and Fabrics,
7(2) (2012) 47-55. https://doi.org/10.1177/155892501200700208

[89] X. Liu, T. Lin, Y. Gao, Z. Xu, C. Huang, G. Yao, L. Jiang, Y. Tang, X. Wang,
Antimicrobial electrospun nanofibers of cellulose acetate and polyester urethane
composite for wound dressing, Journal of biomedical materials research. Part B,
Applied biomaterials 100(6) (2012) 1556-65. https://doi.org/10.1002/jbm.b.32724

[90] D. Macocinschi, D. Filip, E. Paslaru, B.S. Munteanu, R.P. Dumitriu, G.M.
Pricope, M. Aflori, M. Dobromir, V. Nica, C. Vasile, Polyurethane–extracellular
matrix/silver bionanocomposites for urinary catheters, Journal of Bioactive and
Compatible Polymers 30(1) (2014) 99-113.
https://doi.org/10.1177/0883911514560661

[91] F. Kalhori, E. Arkan, F. Dabirian, G. Abdi, P. Moradipour, Controlled Preparation and Characterization of Nigella Sativa Electrospun Pad for Controlled Release, 11(2) Silicon (2018), 593-601. https://doi.org/10.1007/s12633-018-9931-z

[92] E. Yeniay, L. Öcal, E. Altun, B. Giray, F. Nuzhet Oktar, A. Talat Inan, N. Ekren, O. Kilic, O. Gunduz, Nanofibrous wound dressing material by electrospinning method, International Journal of Polymeric Materials and Polymeric Biomaterials (2018) 1-8. https://doi.org/10.1080/00914037.2018.1525718

[93] G. Avsar, D. Agirbasli, M.A. Agirbasli, O. Gunduz, E.T. Oner, Levan based fibrous scaffolds electrospun via co-axial and single-needle techniques for tissue engineering applications, Carbohydrate polymers 193 (2018) 316-325. https://doi.org/10.1016/j.carbpol.2018.03.075

[94] T. Sultana, J. Amirian, C. Park, S.J. Lee, B.T. Lee, Preparation and characterization of polycaprolactone-polyethylene glycol methyl ether and polycaprolactone-chitosan electrospun mats potential for vascular tissue engineering, Journal of biomaterials applications 32(5) (2017) 648-662. https://doi.org/10.1177/0885328217733849

[95] J. Amirian, S.-Y. Lee, B.-T. Lee, Designing of Combined Nano and Microfiber Network by Immobilization of Oxidized Cellulose Nanofiber on Polycaprolactone Fibrous Scaffold, Journal of Biomedical Nanotechnology 12(10) (2016) 1864-1875. https://doi.org/10.1166/jbn.2016.2308

[96] S. Agarwal, J.H. Wendorff, A. Greiner, Use of electrospinning technique for biomedical applications, Polymer 49(26) (2008) 5603-5621. https://doi.org/10.1016/j.polymer.2008.09.014

[97] F. Mohammadian, A. Eatemadi, Drug loading and delivery using nanofibers scaffolds, Artificial cells, nanomedicine, and biotechnology 45(5) (2017) 881-888. https://doi.org/10.1080/21691401.2016.1185726

[98] V. J, R. S, Applications of Polymer Nanofibers in Biomedicine and Biotechnology, Applied Biochemistry and Biotechnology 125 (2005) 147-157. https://doi.org/10.1385/ABAB:125:3:147

[99] H. Cao, T. Liu, S.Y. Chew, The application of nanofibrous scaffolds in neural tissue engineering, Advanced drug delivery reviews 61(12) (2009) 1055-64. https://doi.org/10.1016/j.addr.2009.07.009

[100] F. Naghizadeh, A. Solouk, S.B. Khoulenjani, Osteochondral scaffolds based on electrospinning method: General review on new and emerging approaches,

International Journal of Polymeric Materials and Polymeric Biomaterials 67(15) (2017) 913-924. https://doi.org/10.1080/00914037.2017.1393682

[101] V. Leszczak, L.W. Place, N. Franz, K.C. Popat, M.J. Kipper, Nanostructured biomaterials from electrospun demineralized bone matrix: a survey of processing and crosslinking strategies, ACS Appl Mater Interfaces 6(12) (2014) 9328-37. https://doi.org/10.1021/am501700e

[102] N. Panda, A. Bissoyi, K. Pramanik, A. Biswas, Directing osteogenesis of stem cells with hydroxyapatite precipitated electrospun eri-tasar silk fibroin nanofibrous scaffold, Journal of biomaterials science. Polymer edition 25(13) (2014) 1440-57. https://doi.org/10.1080/09205063.2014.943548

[103] J.C. M, P.J. Reardon, R. Konwarh, J.C. Knowles, B.B. Mandal, Mimicking Hierarchical Complexity of the Osteochondral Interface Using Electrospun Silk-Bioactive Glass Composites, ACS Appl Mater Interfaces 9(9) (2017) 8000-8013. https://doi.org/10.1021/acsami.6b16590

[104] B.N. Singh, N.N. Panda, R. Mund, K. Pramanik, Carboxymethyl cellulose enables silk fibroin nanofibrous scaffold with enhanced biomimetic potential for bone tissue engineering application, Carbohydrate polymers 151 (2016) 335-347. https://doi.org/10.1016/j.carbpol.2016.05.088

[105] T. Chae, H. Yang, V. Leung, F. Ko, T. Troczynski, Novel biomimetic hydroxyapatite/alginate nanocomposite fibrous scaffolds for bone tissue regeneration, Journal of materials science. Materials in medicine 24(8) (2013) 1885-94. https://doi.org/10.1007/s10856-013-4957-7

[106] D.C. Aduba, S.-S. An, G.S. Selders, W.A. Yeudall, G.L. Bowlin, T. Kitten, H. Yang, Electrospun gelatin–arabinoxylan ferulate composite fibers for diabetic chronic wound dressing application, International Journal of Polymeric Materials and Polymeric Biomaterials (2018) 1-9. https://doi.org/10.1080/00914037.2018.1482466

[107] C.H. Woo, Y.C. Choi, J.S. Choi, H.Y. Lee, Y.W. Cho, A bilayer composite composed of TiO2-incorporated electrospun chitosan membrane and human extracellular matrix sheet as a wound dressing, Journal of biomaterials science. Polymer edition 26(13) (2015) 841-54. https://doi.org/10.1080/09205063.2015.1061349

[108] R. Zhao, X. Li, B. Sun, Y. Zhang, D. Zhang, Z. Tang, X. Chen, C. Wang, Electrospun chitosan/sericin composite nanofibers with antibacterial property as

potential wound dressings, Int J Biol Macromol 68 (2014) 92-7.
https://doi.org/10.1016/j.ijbiomac.2014.04.029

[109] N. Mahmoudi, N. Eslahi, A. Mehdipour, M. Mohammadi, M. Akbari, A. Samadikuchaksaraei, A. Simchi, Temporary skin grafts based on hybrid graphene oxide-natural biopolymer nanofibers as effective wound healing substitutes: pre-clinical and pathological studies in animal models, Journal of materials science. Materials in medicine 28(5) (2017) 73. https://doi.org/10.1007/s10856-017-5874-y

[110] H. Wang, Y. Feng, Z. Fang, R. Xiao, W. Yuan, M. Khan, Fabrication and characterization of electrospun gelatin-heparin nanofibers as vascular tissue engineering, Macromolecular Research 21(8) (2013) 860-869. https://doi.org/10.1007/s13233-013-1105-7

[111] S. Torres-Giner, J.V. Gimeno-Alcaniz, M.J. Ocio, J.M. Lagaron, Comparative performance of electrospun collagen nanofibers cross-linked by means of different methods, ACS Appl Mater Interfaces 1(1) (2009) 218-23. https://doi.org/10.1021/am800063x

[112] E. Vatankhah, M.P. Prabhakaran, D. Semnani, S. Razavi, M. Zamani, S. Ramakrishna, Phenotypic modulation of smooth muscle cells by chemical and mechanical cues of electrospun tecophilic/gelatin nanofibers, ACS Appl Mater Interfaces 6(6) (2014) 4089-101. https://doi.org/10.1021/am405673h

[113] B.K. Gu, S.J. Park, C.H. Kim, Beneficial effect of aligned nanofiber scaffolds with electrical conductivity for the directional guide of cells, Journal of biomaterials science. Polymer edition 29(7-9) (2018) 1053-1065. https://doi.org/10.1080/09205063.2017.1364097

[114] Jiang Yuan, Zhi-Cai Xing, Suk-Woo Park, Jia Geng, I.-K. Kang, J. Yuan, J. Shen, W. Meng, K.-J. Shim, I.-S. Han, J.-C. Kim, Fabrication of PHBV/Keratin Composite Nanofibrous Mats for Biomedical Applications, Macromolecular Research 17(11) (2009) 850-855. https://doi.org/10.1007/BF03218625

[115] Alvarez-Perez Marco Antonio , G. Vincenzo, C. Valentina, A. Luigi, Influence of Gelatin Cues in PCL Electrospun Membranes on Nerve Outgrowth, Biomacromolecules 11 (2010) 2238–2246. https://doi.org/10.1021/bm100221h

[116] C. Dhand, V.A. Barathi, S.T. Ong, M. Venkatesh, S. Harini, N. Dwivedi, E.T. Goh, M. Nandhakumar, J.R. Venugopal, S.M. Diaz, M.H. Fazil, X.J. Loh, L.S. Ping, R.W. Beuerman, N.K. Verma, S. Ramakrishna, R. Lakshminarayanan, Latent Oxidative Polymerization of Catecholamines as Potential Cross-linkers for

Biocompatible and Multifunctional Biopolymer Scaffolds, ACS Appl Mater Interfaces 8(47) (2016) 32266-32281. https://doi.org/10.1021/acsami.6b12544

[117] F. Peng, M.T. Shaw, J.R. Olson, M. Wei, Hydroxyapatite Needle-Shaped Particles/Poly(l-lactic acid) Electrospun Scaffolds with Perfect Particle-along-Nanofiber Orientation and Significantly Enhanced Mechanical Properties, The Journal of Physical Chemistry C 115(32) (2011) 15743-15751. https://doi.org/10.1021/jp201384q

[118] P.K. Eun, K.H. Ki, L.S. Jin, M. Byung-Moo, P.W. Ho, Biomimetic Nanofibrous Scaffolds: Preparation and Characterization of PGA/Chitin Blend Nanofibers, Biomacromolecules 7 (2006) 635-643. https://doi.org/10.1021/bm0509265

[119] X. Jing, M.R. Salick, T. Cordie, H.-Y. Mi, X.-F. Peng, L.-S. Turng, Electrospinning Homogeneous Nanofibrous Poly(propylene carbonate)/Gelatin Composite Scaffolds for Tissue Engineering, Industrial & Engineering Chemistry Research 53(22) (2014) 9391-9400. https://doi.org/10.1021/ie500762z

[120] T.C. Lin, F.H. Lin, J.C. Lin, In vitro characterization of magnetic electrospun IDA-grafted chitosan nanofiber composite for hyperthermic tumor cell treatment, Journal of biomaterials science. Polymer edition 24(9) (2013) 1152-63. https://doi.org/10.1080/09205063.2012.743061

[121] H.Y. Lin, W.C. Tsai, S.H. Chang, Collagen-PVA aligned nanofiber on collagen sponge as bi-layered scaffold for surface cartilage repair, Journal of biomaterials science. Polymer edition 28(7) (2017) 664-678. https://doi.org/10.1080/09205063.2017.1295507

[122] Y.F. Qian, K.H. Zhang, F. Chen, Q.F. Ke, X.M. Mo, Cross-linking of gelatin and chitosan complex nanofibers for tissue-engineering scaffolds, Journal of biomaterials science. Polymer edition 22(8) (2011) 1099-113. https://doi.org/10.1163/092050610X499447

[123] Y.Y. Huang, D.Y. Wang, L.L. Chang, Y.C. Yang, Fabricating microparticles/nanofibers composite and nanofiber scaffold with controllable pore size by rotating multichannel electrospinning, Journal of biomaterials science. Polymer edition 21(11) (2010) 1503-14. https://doi.org/10.1163/092050609X12519805625997

[124] X. Gui, J. Hu, Y. Han, Random and aligned electrospun gelatin nanofiber mats for human mesenchymal stem cells, Materials Research Innovations (2018) 1-8. https://doi.org/10.1080/14328917.2018.1428073

[125] Z.C. Chen, A.K. Ekaputra, K. Gauthaman, P.G. Adaikan, H. Yu, D.W. Hutmacher, In vitro and in vivo analysis of co-electrospun scaffolds made of medical grade poly(epsilon-caprolactone) and porcine collagen, Journal of biomaterials science. Polymer edition 19(5) (2008) 693-707. https://doi.org/10.1163/156856208784089580

[126] R. Selvakumar, S.N. Mohaideen, S. Aravindh, C. Sabarinath, M. Ananthasubramanian, Effect of biotin and galactose functionalized gelatin nanofiber membrane on HEp-2 cell attachment and cytotoxicity, The Journal of membrane biology 247(1) (2014) 35-43. https://doi.org/10.1007/s00232-013-9608-x

[127] Y.-M. Lim, S.I. Jeong, Y.M. Shin, J.-S. Park, H.-J. Gwon, Y.-C. Nho, S.-J. An, J.-B. Choi, J.-O. Jeong, J.-W. Choi, Physicochemical characterization of gelatin-immobilized, acrylic acid-bacterial cellulose nanofibers as cell scaffolds using gamma-irradiation, Biotechnology and Bioprocess Engineering 20(5) (2015) 942-947. https://doi.org/10.1007/s12257-015-0175-0

[128] W. Li, X. Li, W. Li, T. Wang, X. Li, S. Pan, H. Deng, Nanofibrous mats layer-by-layer assembled via electrospun cellulose acetate and electrosprayed chitosan for cell culture, European Polymer Journal 48(11) (2012) 1846-1853. https://doi.org/10.1016/j.eurpolymj.2012.08.001

[129] C. Jyh-Ping, C. Shih-Hsien, L. Guo-Jyun, Preparation and characterization of biomimetic silk fibroin/chitosan composite nanofibers by electrospinning for osteoblasts culture, Nanoscale Research Letters 7 (2012) 170-180. https://doi.org/10.1186/1556-276X-7-170

[130] S.F. Chou, L.J. Luo, J.Y. Lai, D.H. Ma, Role of solvent-mediated carbodiimide cross-linking in fabrication of electrospun gelatin nanofibrous membranes as ophthalmic biomaterials, Materials science & engineering. C, Materials for biological applications 71 (2017) 1145-1155. https://doi.org/10.1016/j.msec.2016.11.105

[131] S. Hong, G. Kim, Fabrication of electrospun polycaprolactone biocomposites reinforced with chitosan for the proliferation of mesenchymal stem cells, Carbohydrate polymers 83(2) (2011) 940-946. https://doi.org/10.1016/j.carbpol.2010.09.002

[132] M.K. Joshi, A.P. Tiwari, B. Maharjan, K.S. Won, H.J. Kim, C.H. Park, C.S. Kim, Cellulose reinforced nylon-6 nanofibrous membrane: Fabrication strategies, physicochemical characterizations, wicking properties and biomimetic mineralization, Carbohydrate polymers 147 (2016) 104-113. https://doi.org/10.1016/j.carbpol.2016.02.056

[133] S. Xin, Y. Li, W. Li, J. Du, R. Huang, Y. Du, H. Deng, Carboxymethyl chitin/organic rectorite composites based nanofibrous mats and their cell compatibility, Carbohydrate polymers 90(2) (2012) 1069-74. https://doi.org/10.1016/j.carbpol.2012.06.045

[134] G.J. Lai, K.T. Shalumon, S.H. Chen, J.P. Chen, Composite chitosan/silk fibroin nanofibers for modulation of osteogenic differentiation and proliferation of human mesenchymal stem cells, Carbohydrate polymers 111 (2014) 288-97. https://doi.org/10.1016/j.carbpol.2014.04.094

[135] S. Xin, X. Li, Z. Ma, Z. Lei, J. Zhao, S. Pan, X. Zhou, H. Deng, Cytotoxicity and antibacterial ability of scaffolds immobilized by polysaccharide/layered silicate composites, Carbohydrate polymers 92(2) (2013) 1880-6. https://doi.org/10.1016/j.carbpol.2012.11.040

[136] B.K. Shrestha, H.M. Mousa, A.P. Tiwari, S.W. Ko, C.H. Park, C.S. Kim, Development of polyamide-6,6/chitosan electrospun hybrid nanofibrous scaffolds for tissue engineering application, Carbohydrate polymers 148 (2016) 107-14. https://doi.org/10.1016/j.carbpol.2016.03.094

[137] A. Wali, Y. Zhang, P. Sengupta, Y. Higaki, A. Takahara, M.V. Badiger, Electrospinning of non-ionic cellulose ethers/polyvinyl alcohol nanofibers: Characterization and applications, Carbohydrate polymers 181 (2018) 175-182. https://doi.org/10.1016/j.carbpol.2017.10.070

[138] A. Faralli, E. Shekarforoush, F. Ajalloueian, A.C. Mendes, I.S. Chronakis, In vitro permeability enhancement of curcumin across Caco-2 cells monolayers using electrospun xanthan-chitosan nanofibers, Carbohydrate polymers 206 (2019) 38-47. https://doi.org/10.1016/j.carbpol.2018.10.073

[139] K. Wu, X. Zhang, W. Yang, X. Liu, Y. Jiao, C. Zhou, Influence of layer-by-layer assembled electrospun poly (l -lactic acid) nanofiber mats on the bioactivity of endothelial cells, Applied Surface Science 390 (2016) 838-846. https://doi.org/10.1016/j.apsusc.2016.08.178

[140] S. Baghersad, S. Hajir Bahrami, M.R. Mohammadi, M.R.M. Mojtahedi, P.B. Milan, Development of biodegradable electrospun gelatin/aloe-vera/poly(epsiloncaprolactone) hybrid nanofibrous scaffold for application as skin substitutes, Materials science & engineering. C, Materials for biological applications 93 (2018) 367-379. https://doi.org/10.1016/j.msec.2018.08.020

[141] R.D. Velasco-Barraza, R. Vera-Graziano, E.A. López-Maldonado, M.T. Oropeza-Guzmán, S.G. Dastager, A. Álvarez-Andrade, A.L. Iglesias, L.J. Villarreal-Gómez, Study of nanofiber scaffolds of PAA, PAA/CS, and PAA/ALG for its potential use in biotechnological applications, International Journal of Polymeric Materials and Polymeric Biomaterials 67(13) (2017) 800-807. https://doi.org/10.1080/00914037.2017.1378887

[142] J. Xue, M. He, H. Liu, Y. Niu, A. Crawford, P.D. Coates, D. Chen, R. Shi, L. Zhang, Drug loaded homogeneous electrospun PCL/gelatin hybrid nanofiber structures for anti-infective tissue regeneration membranes, Biomaterials 35(34) (2014) 9395-405. https://doi.org/10.1016/j.biomaterials.2014.07.060

[143] K.T. Shalumon, N.S. Binulal, N. Selvamurugan, S.V. Nair, D. Menon, T. Furuike, H. Tamura, R. Jayakumar, Electrospinning of carboxymethyl chitin/poly(vinyl alcohol) nanofibrous scaffolds for tissue engineering applications, Carbohydrate polymers 77(4) (2009) 863-869. https://doi.org/10.1016/j.carbpol.2009.03.009

[144] G. Ma, D. Fang, Y. Liu, X. Zhu, J. Nie, Electrospun sodium alginate/poly(ethylene oxide) core–shell nanofibers scaffolds potential for tissue engineering applications, Carbohydrate polymers 87(1) (2012) 737-743. https://doi.org/10.1016/j.carbpol.2011.08.055

[145] P. Vashisth, K. Nikhil, P. Roy, P.A. Pruthi, R.P. Singh, V. Pruthi, A novel gellan-PVA nanofibrous scaffold for skin tissue regeneration: Fabrication and characterization, Carbohydrate polymers 136 (2016) 851-9. https://doi.org/10.1016/j.carbpol.2015.09.113

[146] M. Agheb, M. Dinari, M. Rafienia, H. Salehi, Novel electrospun nanofibers of modified gelatin-tyrosine in cartilage tissue engineering, Materials science & engineering. C, Materials for biological applications 71 (2017) 240-251. https://doi.org/10.1016/j.msec.2016.10.003

[147] I.J. Hall Barrientos, E. Paladino, P. Szabo, S. Brozio, P.J. Hall, C.I. Oseghale, M.K. Passarelli, S.J. Moug, R.A. Black, C.G. Wilson, R. Zelko, D.A. Lamprou, Electrospun collagen-based nanofibres: A sustainable material for improved antibiotic utilisation in tissue engineering applications, International journal of pharmaceutics 531(1) (2017) 67-79. https://doi.org/10.1016/j.ijpharm.2017.08.071

[148] P. Agrawal, K. Pramanik, Chitosan-poly(vinyl alcohol) nanofibers by free surface electrospinning for tissue engineering applications, Tissue engineering and regenerative medicine 13(5) (2016) 485-497. https://doi.org/10.1007/s13770-016-9092-3

Nanohybrids   Materials Research Forum LLC
Materials Research Foundations **87** (2021) 156-201   https://doi.org/10.21741/9781644901076-7

[149] A. Elamparithi, A.M. Punnoose, S. Kuruvilla, Electrospun type 1 collagen matrices preserving native ultrastructure using benign binary solvent for cardiac tissue engineering, Artificial cells, nanomedicine, and biotechnology 44(5) (2016) 1318-25. https://doi.org/10.3109/21691401.2015.1029629

[150] S.P. Miguel, D. Simoes, A.F. Moreira, R.S. Sequeira, I.J. Correia, Production and characterization of electrospun silk fibroin based asymmetric membranes for wound dressing applications, Int J Biol Macromol 121 (2019) 524-535. https://doi.org/10.1016/j.ijbiomac.2018.10.041

[151] J. Lin, C. Li, Y. Zhao, J. Hu, L.M. Zhang, Co-electrospun nanofibrous membranes of collagen and zein for wound healing, ACS Appl Mater Interfaces 4(2) (2012) 1050-7. https://doi.org/10.1021/am201669z

[152] X. Yang, J. Yang, L. Wang, B. Ran, Y. Jia, L. Zhang, G. Yang, H. Shao, X. Jiang, Pharmaceutical Intermediate-Modified Gold Nanoparticles: Against Multidrug-Resistant Bacteria and Wound-Healing Application via an Electrospun Scaffold, ACS nano 11(6) (2017) 5737-5745. https://doi.org/10.1021/acsnano.7b01240

[153] A.C. Alavarse, F.W. de Oliveira Silva, J.T. Colque, V.M. da Silva, T. Prieto, E.C. Venancio, J.J. Bonvent, Tetracycline hydrochloride-loaded electrospun nanofibers mats based on PVA and chitosan for wound dressing, Materials science & engineering. C, Materials for biological applications 77 (2017) 271-281. https://doi.org/10.1016/j.msec.2017.03.199

[154] P. Guha Ray, P. Pal, P.K. Srivas, P. Basak, S. Roy, S. Dhara, Surface Modification of Eggshell Membrane with Electrospun Chitosan/Polycaprolactone Nanofibers for Enhanced Dermal Wound Healing, ACS Applied Bio Materials 1(4) (2018) 985-998. https://doi.org/10.1021/acsabm.8b00169

[155] S. Kandhasamy, S. Perumal, B. Madhan, N. Umamaheswari, J.A. Banday, P.T. Perumal, V.P. Santhanakrishnan, Synthesis and Fabrication of Collagen-Coated Ostholamide Electrospun Nanofiber Scaffold for Wound Healing, ACS Appl Mater Interfaces 9(10) (2017) 8556-8568. https://doi.org/10.1021/acsami.6b16488

[156] L. Wang, J. Yang, B. Ran, X. Yang, W. Zheng, Y. Long, X. Jiang, Small Molecular TGF-beta1-Inhibitor-Loaded Electrospun Fibrous Scaffolds for Preventing Hypertrophic Scars, ACS Appl Mater Interfaces 9(38) (2017) 32545-32553. https://doi.org/10.1021/acsami.7b09796

[157] A. Islam, T. Yasin, M.A. Rafiq, T.H. Shah, A. Sabir, S.M. Khan, T. Jamil, In-situ crosslinked nanofiber mats of chitosan/poly(vinyl alcohol) blend: Fabrication,

characterization and MTT assay with cancerous bone cells, Fibers and Polymers 16(9) (2015) 1853-1860. https://doi.org/10.1007/s12221-015-5353-3

[158] J. Nourmohammadi, A. Ghaee, S.H. Liavali, Preparation and characterization of bioactive composite scaffolds from polycaprolactone nanofibers-chitosan-oxidized starch for bone regeneration, Carbohydrate polymers 138 (2016) 172-9. https://doi.org/10.1016/j.carbpol.2015.11.055

[159] S. Nagarajan, H. Belaid, C. Pochat-Bohatier, C. Teyssier, I. Iatsunskyi, E. Coy, S. Balme, D. Cornu, P. Miele, N.S. Kalkura, V. Cavailles, M. Bechelany, Design of Boron Nitride/Gelatin Electrospun Nanofibers for Bone Tissue Engineering, ACS Appl Mater Interfaces 9(39) (2017) 33695-33706. https://doi.org/10.1021/acsami.7b13199

[160] N.S. Binulal, A. Natarajan, D. Menon, V.K. Bhaskaran, U. Mony, S.V. Nair, PCL-gelatin composite nanofibers electrospun using diluted acetic acid-ethyl acetate solvent system for stem cell-based bone tissue engineering, Journal of biomaterials science. Polymer edition 25(4) (2014) 325-40. https://doi.org/10.1080/09205063.2013.859872

[161] L. Li, W. Zhang, M. Huang, J. Li, J. Chen, M. Zhou, J. He, Preparation of gelatin/genipin nanofibrous membrane for tympanic member repair, Journal of biomaterials science. Polymer edition 29(17) (2018) 2154-2167. https://doi.org/10.1080/09205063.2018.1528519

Nanohybrids
Materials Research Foundations **87** (2021) 202-229

Materials Research Forum LLC
https://doi.org/10.21741/9781644901076-8

Chapter 8

Development of Hybrid Materials Based on Polymers for Biomedical Applications: A Short Introduction

A. García-Peñas[1,*], S.C. Cifuentes[2], Y. Wang[3], V. San-Miguel[1,*]

[1]Departamento de Ciencia e Ingeniería de Materiales e Ingeniería Química (IAAB), Universidad Carlos III de Madrid, 28911 Leganés, Madrid, Spain

[2]Área de Ciencia e Ingeniería de Materiales, ESCET, Universidad Rey Juan Carlos, C/Tulipán s/n, 28933 Móstoles, Madrid, Spain

[3]College of Materials Science and Engineering, Shenzhen Key Laboratory of Polymer Science and Technology, Guangdong Research Center for Interfacial Engineering of Functional Materials, Nanshan District Key Laboratory for Biopolymers and Safety Evaluation, Shenzhen University, Shenzhen, 518055, P. R. China

alberto.garcia.penas@uc3m.es*, veronica.sanmiguel@uc3m.es*

Abstract

This chapter is focused on some of the most important polymers used for preparing hybrid materials applied in biomedicine. The work is divided into two parts: Non-degradable polymers used for hybrid materials in implants and degradable polymers employed to fabricate biomedical implants and devices. Each part describes the main characteristics of these structures, followed by a list of the most significant polymers and derivatives. This brief introduction could be useful for industry, students, or people interested in the recent advances of biomedical applications where polymers play an important role.

Keywords

Composites, Non-Degradable Polymers, Degradable Materials, Biomedical Applications

Contents

1.   Non-degradable polymers used for hybrid materials in implants

1.1   Introduction

The properties of polymers or hybrid materials used for biomedical applications must respond to necessities of the patient and provide specific functionalities. In general, biocompatibility is essential, but the implant or device will combine other characteristics associated with the final use. For instance, drug carriers need small size of particles to avoid thrombosis, or some scaffolds act as reinforcement and matrix for cell growth, and consequently the morphology of the surface and the biodegradability will play an important role.

There are different types of applications, and some of them are focused on the replacement of parts of the body (bones), the confinement or protection of organs and other permanent devices, where the lifetime will be another important parameter [1]. These structures should avoid degradable materials, and can be classified as permanent, prolonged, and limited implants. The lifetime of limited and prolonged implants will be lower than permanent devices [2]. Then, the main parameters associated with these non-degradable devices will be biocompatibility and durability, and will be required for selecting suitable materials and avoiding secondary reactions [3].

The use of polymers regarding other materials as ceramics or metals, can be understood from their ease to be processed and their wide spectra of properties which can be easily modulated through the molecular features [4].

The interactions between implants and the body will be defined by intrinsic and external parameters. The intrinsic parameters are defined by the implant, and its shape, size, composition, morphology, porosity, and surface. The extrinsic factors are associated with the function of the implant or device, and the genetic or state of the patient [5]. Numerous diseases can derive from interactions between implants and body fluids or cells. Thus, materials for implants must be non-toxic, avoiding thrombosis, allergies, inflammations, immune response, or cancer [6].

The combination of these factors decreases the spectrum of polymeric structures suitable for permanent, limited, and prolonged implants. Generally, polymeric structures offer many possibilities because they can be easily adapted (shape and properties) to different necessities. Nevertheless, some shortcomings associated with biocompatibility can take place, and they may be addressed by modifying the surface, which could provide other benefits, as preserving the original mechanical properties of the pure polymer [7].

Hybrid materials or composites can deliver multiple properties assuring a good biocompatibility. Unfortunately, the hybrid materials require a long procedure of the biocompatibility analysis using numerous steps and testing methods, given that a standard test has not been reported yet. The composition, heterogeneity, and the complex parameters of these structures make difficult to study the biological and chemical interactions with the body [8].

There are many kinds of implants and devices prepared with hybrid materials based on polymers as dental post, brackets, spine cage, plates, rods, screws, discs, hips, knees, tendons, ligaments, intramedullary nails, abdominal prosthesis, vascular grafts, or finger joints [7]. The list of non-degradable hybrid polymers for implants can be prepared by diverse criteria, according to the application, filler, or type of material, among others. For that reason, this chapter listed some of the most important polymers used as components of hybrid materials.

1.2 General features and properties of implants

The implants should combine biocompatibility and durability, but also other features will be required. The contribution of each characteristic or property depends on the final application [8].

The next list shows some of the most important necessary factors or parameters to take into account:

- Biocompatibility

- Durability and stability: The degradation must be avoided as it can lead to undesirable interactions with the body. The degradation of the polymeric matrix may be associated with different effects as physical, chemical, and/or thermal degradation. Moreover, polymeric structures need to keep their properties and degradation could destroy their purpose.

- Good mechanical properties: Implants should be strong enough, in terms of mechanical properties, to prevent the wear or break. A good ratio between stiffness, elasticity, and hardness needs to be defined depending on the requirements of implant.

- Sterility: The implants require a complete sterilization before use.

- Processability: The shape needs to be tailored to the final application, and the easy processability could raise the production and reduce the costs.

- Easy to manufacture.

- Low cost

1.3 Main polymers and derivatives used in hybrid materials for implants

This is the list conformed by some of the most important polymers used for preparation of non-degradable hybrid polymers. In this context, some derivatives and composites are detailed here given the major interest on hybrid materials applied for this purpose.

1.3.1 Polyethylene (PE)

There are diverse types of polyethylene as high density polyethylene (HDPE), low density polyethylene (LDPE), linear low density polyethylene (LLDPE), and ultra-high molecular-weight polyethylene (UHMWPE), which are widely used for biomedical applications.

Polyethylene or derivatives can offer good characteristics for biomedical applications, in terms of mechanical properties and high porosity. In addition, it is an inert and non-degradable material used for many permanent implants.

It is used for joint replacements, orthopedics, tubing, catheters, and non-woven textiles. Nevertheless, the use of crossed-linked polyethylene (XLPE) is growing as polymer for hip implants thanks to its better behavior in comparison to other type of polyethylene based materials [4,8].

In general, the surface modification coating provides a better bioactivity in HDPE, which improves the stability of implants [9,10]. On the other hand, composites and derivatives of polyethylene provide other features and functionalities. The following list includes some hybrid materials based on polyethylene:

- UHMWPE/Ti composites (hip joint) [11].

- Al-Cu-Fe quasicrystal/UHMWPE composites [12-14].

- HDPE/hydroxyapatite (HA) composite is commercialized as HAPEX, and it is indicated for several implants [15].

- HDPE/graphene nanoplatelets (GNP) composites [16-19].

- HDPE/multiwall carbon nanotube (MWCNT) composites [20].

- Reinforced HDPE hybrid composite (10% titanium oxide and 20% alumina) [21].

- HMWPE reinforced with quartz powder [22].

- Glass-ceramic A-W-polyethylene composites [23].

- UHMWPE/polypropylene (PP) blends [24].

- Polyamide-6(PA-6)/UHMWPE blends [15].

- HDPE/tricalcium phosphate/UHMWPE nanocomposites [15].

1.3.2 Polypropylene (PP)

Polypropylene exhibits a wide range of properties, which can be easily modulated by molecular weight, molar mass distribution, tacticity, and composition (in case of copolymers or terpolymers). In addition, its versatility plays an important role regarding other polymers because of its low density and low cost of production.

A good balance of properties can be defined in structures based on polypropylene, which can combine high strength, great rigidity, and chemical resistance [8]. The tacticity can define different kinds of polypropylene based materials which are known as isotactic, syndiotactic, and atactic polypropylene, and whereby the properties considerably change.

Polypropylene is involved as component of surgical meshes (hernia repairs), catheters, and sutures, among others.

Hybrid materials based on polypropylene enhance its lifetime and mechanical properties. Some of the most important composites and recent advances are listed below:

- PP/silicone rubber composites (finger joint replacements) [25].

- PP/ fumarate and calcium phosphate composites (bone cements) [26].

- PP/ ultra-short CNT nanocomposites (bone tissue engineering) [27].

- PP/ HA composites [28].

- PP/hexagonal boron nitride [28].

- Polypropylene carbonates/poly(lactic acid) (PLA) composites nanofibers [29].

- PP/carbon nanofiber (CNF) composites [30].

1.3.3 Poly(methyl methacrylate) (PMMA)

Poly(methyl methacrylate) (PMMA) is used in many biomedical applications due to its high biocompatibility, consistency, and low toxicity. In addition, the shape can be tailored to many implants due to its easy processability [31].

It is applied as bone cement, and many applications are focused on dental implants and artificial joints (orthopedic surgery). For instance, PMMA is incorporated as filler for vertebrae stabilization of osteoporotic patients. Additionally, PMMA is a specific material for intraocular lens (corneal prosthesis).

Some composites and derivatives improve the properties of the pure polymer, and are listed below:

- PMMA/HDPE composites.

- PMMA/HA composites [32].

- PMMA/brushite composites [32].

- Composites of PMMA and TiO_2 nanoparticles [33,34].

- PMMA/tricalcium phosphate (TCP) composites [35].

There are other hybrid materials, as PMMA/Polystyrene/Silica nanocomposites, that can incorporate specific functionalities [36].

1.3.4 Polyvinyl chloride (PVC)

Polyvinyl chloride (PVC) is usually modified by some plasticizers to meet the degree of flexibility required by several applications. The normal plasticizers used for this purpose are based on phthalates of low molecular weight. However, these plasticizers can be degraded or released in contact with body fluids, and consequently infections or diseases could take place.

Its good mechanical properties and resistance are ideal for many devices. In case of biomedical applications, PVC plays an important role due to its resilience and also is easily sterilized.

PVC can be incorporated to catheters, lung bypass, endotracheal tubes, hearing aid devices, IV tubing, or shrink tubing.

Some of the most important hybrid materials of PVC are composed by polycaprolactone (PCL) and/or polycarbonates (PC) [37].

1.3.5 Fluorinated polymers (PTFE)

The family of fluorinated polymers (PTFE) is mainly composed by polytetrafluoroethylene (PTFE), perfluoroalkoxy alkane (PFA), polychlorotrifluoroethylene (PCTFE), and polyvinylidene fluoride (PVDF) [39]. These can be involved in numerous implants and medical devices as matrices [38].

These polymers are defined by a good chemical resistance and are insoluble in most chemicals. It is well known the commercial *Teflon*, which exhibits a high thermal stability. Some shortcomings of PTFE are associated with its difficult processing which requires thermomechanical treatments as injection and extrusion molding given its high viscosity.

These polymers can be involved in artificial joints and several vascular devices.

Some composites based on PVDF are used for tissue engineering, and incorporate titanyl phthalocyanine (TiOPc) and indium-tin-oxide (ITO). In addition, tissue regeneration is using PVDF porous membranes with β-cyclodextrin (β-CD) because it allows the incoporation of diverse drugs [40].

1.3.6 Polyamides

Polyamides offer a wide range of properties, and their production is inexpensive. Therefore, the use of polyamides is preferred over other polymers for several applications because similar properties are achieved at a reasonably reduced cost [41, 42].

They are used for scaffolds due to their good mechanical properties and easy processing, which is especially useful for the production of fibers. For instance, polyamide 6 (PA6) is used for getting 3D textile scaffolds [41, 42]. The presence of polyamides is identified in catheters, kidney dialysis, or nylon spike.

The properties can be varied through the preparation of composites with other polymers, glass fibers, ceramics, or hydroxyapatite. The aliphatic polyamides, denominated as nylon, are used for the production of films and fibers. The typical composites based on Nylon are:

- Nylon-6/gelatin composites.

- CuS-nylon composites.

- Plasma-modified Nylon meshes.

- Collagen/Nylon composites.

1.3.7 Polyesters

Polyethylene terephthalate (PET or PETE) is known by its commercial name *Dacron*. This polymer provides excellent mechanical properties, in terms of hardness and stiffness. Furthermore, PET combines a good chemical resistance and dimensional stability [4].

It is used in sutures, vascular grafts, taxillofemoral bypass, and ligaments [43]. The incorporation of coatings spread its range of features and applications to surgical meshes, vascular grafts, woven vascular prostheses, valves, scaffolds, sutures, and catheters. For instance, the use of silver can provide antibacterial properties.

On the other hand, the functionalization of polyesters can lead to better interactions between the polymer and body. There are many kinds of composites, and these are some of them [44]:

- Vascular grafts based on PET incorporate collagen, heparin, and albumin proteins.

- PET/polyacrylamide (PAAN) composites.

- PET/polyethylenglicol (PEG) composites.

1.3.8 Polysulfones

Polysulfones (PSF) and polyphenylsulfones show good mechanical properties and dimensional stability, and are chemical inert, resistant to temperature, and biocompatible [45-47].

Materials Research Forum LLC
https://doi.org/10.21741/9781644901076-8

These materials are used in handles for dental instruments, ophthalmic scopes, and lenses. There are diverse types of composites based on polysulfones such as:

- PSF/HA composites (Tissue engineering) [1, 48].
- Polyether sulfone/Sulfonated polyether sulfone nanofibers (Bone regeneration) [49].
- Sulfonated polyether sulfone/Ca [50].

1.3.9 Polyurethanes (PU)

Polyurethane exhibits good mechanical properties, durability, compliance, resistance, and elasticity. Thus, this polymer provides numerous advantages in comparison with other polymers [4], and are specially indicated for bone regeneration thanks to its calcifying properties [1].

It can be involved in catheters, ligament replacements, heart valves, vascular grafts, connectors, feeding tubes, pacemaker leads, breast prosthesis, and tubing.

The composites improve its characteristics, as conductivity, among others. Some examples of composites are:

- Flexible conductive graphene/polyurethane composites [51].
- Phosphates/polyurethane composites [52].
- Polyurethane/Biocellulose nanocomposites [53].
- Polyuretane-ceramic matrices [54].

1.3.10 Polydimethylsiloxane (PDMS)

Polydimethylsiloxane (PDMS) is an elastomeric polymer, non-toxic, biocompatible, and transparent. In addition, this polymer combines elasticity and durability. Nevertheless, the hydrophobic behavior observed in this polymer affects the cells adhesion, and consequently, some treatments are required before its use. One of the most effective methods to improve the cell adhesion is the use of a coating based on fibronectin, laminin, or collagen [4, 55].

PDMS is used for artificial skin, joint replacement, artificial hearts, breast implants, and catheters.

The preparation of some composites enhances its properties, and also provides some functionalities:

- PDMS/HA composites [56].
- PDMS/Ag composites [57].
- PDMS/Zr composites [1].

1.3.11 Polyketones

The production of polyketones, polyether ether ketone (PEEK), and polyaryletherketone is really expensive in comparison with other polymers. Nevertheless, its excellent wear, heat, electrical, and chemical resistance allow getting exceptional properties for specific cardiovascular implants, dental implants, and tubing [58].

There are not many composites associated with polyketones due to its hard processing. The carbon fiber reinforced PEEK composites were recently studied.

1.3.12 Other possibilities

Other polymers can be used for implants as acrylics, silicones, ABS copolymers, acetals, or polycarbonates, among others. Nevertheless, this chapter tries to enumerate some of the most relevant polymers and derivatives used in hybrid materials for implants.

2.   Degradable hybrid polymers for implants

2.1   Introduction

The development of biocompatible and biodegradable polymeric materials for biomedical applications has attracted a significant attention in the last years [59, 60]. A biodegradable biomaterial requires exceptional biocompatibility – the capacity of a material to perform with an adequate host response in a specific situation –over time. Biocompatibility of materials may be influenced, for instance, by molecular weight, hydrophilicity/hydrophobicity, solubility, degradation and/or erosion mechanism, surface energy, lubricity, and shape and structure of the implant [61]. An ideal biodegradable polymeric material should lead to nontoxic and easily metabolized degradation products from the body. Furthermore, other important properties should be considered when a biodegradable material is selected. On one hand, time degradation of the biomaterial should occur simultaneously with the regeneration and/or healing process to ensure an adequate adaptation of the tissue. On the other hand, biomaterial should preserve proper permeability and processability for its function and initial mechanical properties should be maintained for the entire duration that the object remains in the body.

Medical devices and implants, along with the material, need to comply certain requirements and properties for long-term use in the human body. Material requirements for implants and devices include, among others, proper interfacial material between the body and the device, small size of the object, mechanical resistance to withstand stresses and shocks, and thermal resistance to perform its function for the required time [6]. From a biological point of view, device and implant materials should not directly or indirectly provoke unfavourable local or system level results, or have adverse development and reproductive effects. Williams *et al.* proposed a new definition of biocompatibility focused on the ability of the biomaterial to perform the proper function respect to a medical therapy, without producing adverse effects in the body, but performing the most adequate beneficial cellular or tissue response in that specific situation [62].

2.2   Materials

The materials used for implants and devices are essential for its survivability within the human body environment. In general, these materials include metals, ceramic, polymers, and polymer composites [7]. Regarding polymers, they are available in wide variety of properties, compositions, and forms and they can also be fabricated into complex structures and shapes. As undesirable behaviours, polymers may absorb fluids, leach inappropriate products, and swell up. However, if polymers are used as part of a composite, additional advantages, such us high strength and low elastic modulus with a greater potential for structural biocompatibility, would be presented.

A broad diversity of biodegradable and biocompatible polymers can be used in biomedical implants and devices. They can be classified into two categories: natural and synthetic polymeric biomaterials. Natural biodegradable polymers (such as collagen, fibrin, cellulose, silk, starch, chitosan, hyaluronic acid derivatives, etc) have given rise to a greater attention for their potential use as nano-carriers in drug delivery and for their ability to generate efficient nanocomposite adhesives [63, 64]. Main properties and biomedical applications of some natural biodegradable polymers are set forth in Table 1. Despite natural biodegradable polymers have proven to present a potential advantage of supporting cell function and adhesion, there are certain limitations with regard to their use. For instance, it is difficult to control their mechanical properties and degradation rates [65]. Conversely, synthetic biodegradable polymers, mainly aliphatic polyesters (such as poly(glycolic acid) (PGA), poly(lactic acid) (PLA), poly(lactic-*co*-glycolic acid) (PLGA), and poly(ε-caprolactone) (PCL)), have shown to improve certain mechanical and chemical properties compared to natural polymeric materials. Synthetic polymers can be structurally controllable and thus exhibit reproducible and predictable physical and mechanical properties, among them elastic modulus, tensile strength, and degradation rate [66]. A further advantage of pure synthetic polymeric materials is the control of impurities which minimizes the risks with regard to infections, toxicity, and immunogenicity. This chapter is focused on the most typical biodegradable synthetic polymers.

Table 1. Chemical structure, main properties, and biomedical applications of some natural biodegradable polymers.

Polymer	Chemical structure	Natural Source	Properties	Biomedical Applications	Ref.
Alginate		Anionic polysaccharide derived from the cell walls of brown algae and extracellularly in some bacteria.	- Non-toxic - Non-inflammatory - Absorbs water quickly - Chemical and physical cross-linking abilities - Non-thrombogenic	Alginate hydrogels are used in wound healing, and delivery of bioactive agents.	[67, 68]
Chitosan		Derivative of chitin, the second most abundant natural polymer in nature obtained from exoskeletons of crustacean and insects.	- Low toxicity - Good adsorption properties - Hydrophilic - Anti-ulcer, anti-acid, and anti-tumor - Hemostatic properties	Chitosan is presented as an excellent material in drug delivery systems, among them, in cancer therapy, or in treatment of infectious diseases. They are also widely used in wound healing and tissue engineering applications.	[69, 70]
Hyaluro-nic acid (HA)		Essential component of the extracellular matrix (ECM).	- Forms hydrated network of tissues - Antioxidant effect - Acts as space filler, lubricant, and osmotic buffer in ECM - High absorption and water retention capacity	HA plays an important role in wound healing to direct tissue repair and regeneration. In tissue engineering the use of PHAs has been mainly focused on poly(hydroxybutyrate) and the copolymer poly(hydroxybutyrate-*co*-valerate.	[71, 72]

Starch		Polysaccharide produced by most plants as energy storage.	- High Young's modulus with low levels of elongation at break - Non-cytotoxic - Excellent substrate for cell adhesion	Starch is widely applied in tissue-engineering applications, particularly in scaffolds for bone regeneration. It is also used in drug delivery systms.	[73, 74]
Cellulose		The most predominant protein in the human body.	- High tensile strength - Non-toxic - Enzymatic degradability - High thrombogenicity - Slight immunogenicity	Cellulose plays an important role in tissue regeneration applications. It is a key initiator of the coagulation and a notable material for the delivery of drugs, genes, proteins, and plasmids.	[75-77]

2.2.1 Poly(glycolic acid), PGA

Poly(glycolic acid) with very simple structure (Fig. 1) is a rigid thermoplastic material with a high crystallinity (around 45-55%), hence insoluble in water. Its glass transition temperature is in range of 35-40 °C and its melting point is lying between 225 and 230 °C. PGA has notable biodegradable and biocompatible properties, mechanical stability [78], and non-toxic degradation products [79]. Poly(glycolic acid) also presents a relatively fast degradation rate. In aqueous solutions or in vivo, it can degrade rapidly and lose its mechanical properties within only 1-2 months, depending on the degradation conditions, the degree of crystallinity, and the molecular weight [80]. The poor hydrolytic stability of poly(glycolic acid) is owing to the ester linkages in its backbone. The degradation process of PGA takes place in two steps. Firstly, water infiltration is produced through the amorphous areas of polymer matrix, which then suffers hydrolysis of the ester groups within the chain. Secondly, the crystalline areas of PGA are hydrolysed after degradation of the amorphous part. The by-product from degradation, glycolic acid, is non-toxic but possesses acid character. Tissue may be damaged by the absorption of the degradation product at high concentration, and thus it can prejudice its use in some biomedical applications. This limitation can be resolved by blending PGA with other polymers, such as poly(lactic acid) or poly(caprolactone) [81].

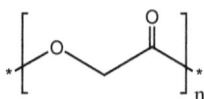

Figure 1. Structure of poly(glycolic acid) (PGA).

Poly(glycolic acid) polymer has great versatility to modify material properties and physical parameters such as tortuosity and pore size, both relevant for developing tissue engineering scaffolds [82]. Hence, PGA can be applied not only in tissue engineering, but also in other biomedical applications such as biological adhesives or drug delivery [83-86].

The biodegradability and thermoplasticity of a poly(glycolic acid) scaffold may enhance cell proliferation. Besides, it should be taken into account that scaffolds must have porous structures to get host cells migrating into inside the tissue. An alternative is to prepare spongiform scaffolds in order to recruit host cells quickly. Nanocomposites nanofibers of PGA and collagen were previously reported to accomplish the recruitment of peripheral blood vessels and host cells without growth factors matter [87]. Kim *et al.* reported the fabrication of silk fibroin (SF) nanofiber membranes and PGA hybrid scaffolds by electrospinning and 3D printing for their use as guided bone tissue regeneration. SF-PGA scaffolds exhibited superior bone volume regeneration compared to scaffolds without nanofiber membranes [88].

2.2.2 Poly(lactic acid), PLA

Poly(lactic acid) is a biodegradable polymer which can be produced from renewable resources. It is linear with one pendent methyl group (Fig. 2), which makes greater hydrophobic and amorphous behaviour than poly(glycolic acid). PLA has reasonably good optical, mechanical, physical, and barrier properties. Figure 2 shows the structure of the three isomers of PLA (L-PLA, D-PLA, and a racemic mixture of D,L-PLA).

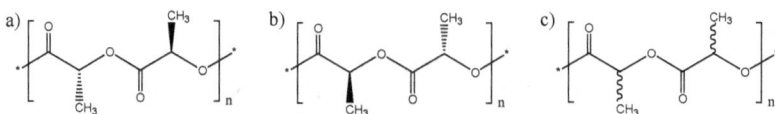

Figure 2. Structure of poly(lactic acid) (PLA) isomers: a) L-PLA, b) D-PLA, and c) D,L-PLA.

PLA is considered as an outstanding material for biomedical applications due to its properties, such as its biodegradability, biocompatibility, processability, and mechanical strength. It has been employed to fabricate tissue-engineering scaffolds, delivery systems, bio-adsorbable implants, covering membranes, and sutures in cosmetics and dermatology [89-91]. Blending poly(lactic acid) with other polymers overcomes certain limitations of PLA related to its slow degradation rate, low impact toughness for its use, and hydrophobicity. PLA blends have been explored for biomedical applications such as implants, drug delivery, sutures, and tissue engineering [92]. Regeneration of annulus fibrosus (AF) tissue to alleviate chronic lower back pain was carried out by using a bilayer scaffold prepared from synthetic biopolymer blend of poly(ε-caprolactone) and poly(L-lactic acid). These electrospun fiber scaffolds mimicked the AF tissue and supported cell viability and orientation, secretion of collagen, and desirable tensile properties [93]. Nanaki *et al.* reported the fabrication of risperidone controlled release microspheres as long-acting injectable formulations. The pharmaceutical microspheres were based on biodegradable and biocompatible poly(lactic acid) and poly(propylene adipate) (PLA/PPAd) polymer blends [94].

2.2.3 Poly(lactic-co-glycolic acid), PLGA

The copolymerization of lactide with other polyglycolides can improve crystallinity, solubility, and melting point of poly(glycolic acid). Thus, the copolymer poly(lactic-*co*-glycolic acid) (PLGA) (Fig. 3) presents better properties than PGA and PLA. Furthermore, PLGA can be applied not only in tissue engineering, but also in food products, wastewater treatment, and other biomedical applications such as biological adhesives or drug delivery [95, 96].

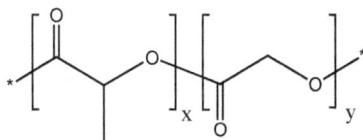

Figure 3. Structure of poly(lactic-co-glycolic acid) (PLGA). x = number of units of lactic acid and y = number of units of glycolic acid.

PLGA has shown to be biocompatible, non-toxic, and non-inflammatory when applied in bone repair although owing to its hydrophobicity it cannot completely promote cell adhesion for improving the proliferation and the bone ingrowth [97]. Furthermore, its suboptimal mechanical properties and low osteoinductivity hinder the clinical application of pure PLGA for bone regeneration. Thus, PLGA is mainly used combined with other materials, such as ceramics, and it is modified in order to improve mechanical properties and support bone repair [98]. Kankala *et al.* reported the fabrication of 3D porous Gelatin/Hydroxyapatite/PLGA scaffolds using a 3D printing method for bone tissue engineering. Scaffolds presented exceptional mechanical properties, biodegradation, and biocompatibility which enabled osteoblasts adhesion and efficiently promoted their growth and differentiation [99]. Recently, PLGA was directly grafted to hydroxyapatite (HA) surface in order to enhance the biocompatibility and bone formation. The composite material presented increased tensile strengths by more than doubled comparted with a PLGA/HA blended composite [100].

2.2.4 Poly(ε-caprolactone), (PCL)

Poly(ε-caprolactone) is a semicrystalline polyester with a low glass transition temperature (~ −54 °C) and a melting temperature of 55-60 °C. It can be degraded by microorganisms, enzymatic, hydrolytic, or intracellular mechanisms under physiological conditions. However, its slow degradation rate along with its hydrophobicity restrict its use for bone tissue engineering applications and make it more attractive for drug-delivery systems and long-term implants [101].

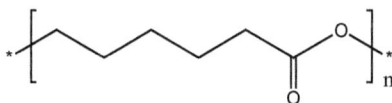

Figure 4. Structure of poly(ε-caprolactone) (PCL).

Degradation rate of PCL can be modified by adding graphene oxide (GO), hydrophilic nHA, or hybrid nHA/GO nanofillers. Recently, Li et al. fabricated electrospun PCL/nHA mats to promote bone cell infiltration. The nHA additions increased the tensile elongation and tensile strength of PCL as well as cell viability and infiltration demonstrated that it is a potential candidate for bone tissue regeneration [102]. Porous scaffolds with an interconnected network were fabricated by an extrusion-based 3D printer from a blend of poly(ε-caprolactone) and graphene oxide. Scaffolds presented good mechanical properties and GO enhanced the cellular response. Furthermore, PCL showed enhanced cell proliferation and differentiation, suggesting that 3D printed PCL/GO blends are an advantageous system for bone tissue engineering applications Zhou et al. [103] reported the synthesis of PCL/nHA-GO hybrid nanocomposites through hydrothermal process of the HA/GO hybrid followed by incorporation into PCL using solvent casting. Hybrid nanocomposites showed an enhanced mechanical properties and better protein adsorption resistance than the pure PCL and PCL/GO nanocomposites which indicated that hybrid nanocomposites are promising materials for biomedical applications [104].

References

[1]   S.V. Gohil, S. Suhail, J. Rose, T. Vella, L.S. Nair, Chapter 8 - Polymers and Composites for Orthopedic Applications, in: S. Bose, A. Bandyopadhyay (Eds.) Materials for Bone Disorders, Academic Press, (2017) 349-403. https://doi.org/10.1016/B978-0-12-802792-9.00008-2

[2]   A. Bigi, S. Fare, P. Petrini, N. Roveri, M.C. Tanzi, Biointegrable 3D Polyurethane/a-TCP Composites for Bone Reconstruction, (2002).

[3]   S. Lerouge, A. Simmons, Sterilisation of Biomaterials and Medical Devices, Elsevier Science, (2012). https://doi.org/10.1533/9780857096265

[4]   A. Subramaniam, S. Sethuraman, Chapter 18 - Biomedical Applications of Nondegradable Polymers, in: S.G. Kumbar, C.T. Laurencin, M. Deng (Eds.) Natural and Synthetic Biomedical Polymers, Elsevier, Oxford, (2014) 301-308. https://doi.org/10.1016/B978-0-12-396983-5.00019-3

[5]   A. Gopanna, K.P. Rajan, S.P. Thomas, M. Chavali, Chapter 6 - Polyethylene and polypropylene matrix composites for biomedical applications, in: V. Grumezescu, A.M. Grumezescu (Eds.) Materials for Biomedical Engineering, Elsevier, (2019) 175-216. https://doi.org/10.1016/B978-0-12-816874-5.00006-2

[6]   A.J.T. Teo, A. Mishra, I. Park, Y.-J. Kim, W.-T. Park, Y.-J. Yoon, Polymeric Biomaterials for Medical Implants and Devices, ACS Biomaterials Science & Engineering, 2 (2016) 454-472. https://doi.org/10.1021/acsbiomaterials.5b00429

[7]   S. Ramakrishna, J. Mayer, E. Wintermantel, K.W. Leong, Biomedical applications of polymer-composite materials: a review, Composites science and technology, 61 (2001) 1189-1224. https://doi.org/10.1016/S0266-3538(00)00241-4

[8]   T. Hutley, M. Ouederni, Polyolefin Compounds and Materials—Fundamentals and Industrial Applications; AlMa'adeed, MAA, Krupa, I., Eds, in, Springer: Heidelberg, Germany, (2016).

[9]   M.J. Dalby, L. Di Silvio, E.J. Harper, W. Bonfield, Increasing hydroxyapatite incorporation into poly(methylmethacrylate) cement increases osteoblast adhesion and response, Biomaterials, 23 (2002) 569-576. https://doi.org/10.1016/S0142-9612(01)00139-9

Nanohybrids   Materials Research Forum LLC
Materials Research Foundations **87** (2021) 202-229   https://doi.org/10.21741/9781644901076-8

[10] M.J. Dalby, S.J. Yarwood, M.O. Riehle, H.J.H. Johnstone, S. Affrossman, A.S.G. Curtis, Increasing Fibroblast Response to Materials Using Nanotopography: Morphological and Genetic Measurements of Cell Response to 13-nm-High Polymer Demixed Islands, Experimental Cell Research, 276 (2002) 1-9. https://doi.org/10.1006/excr.2002.5498

[11] Q. Wang, D. Zhang, S. Ge, Biotribological behaviour of ultra-high molecular weight polyethylene composites containing Ti in a hip joint simulator, Proceedings of the Institution of Mechanical Engineers, Part J: Journal of Engineering Tribology, 221 (2007) 307-313. https://doi.org/10.1243/13506501JET232

[12] B.C. Anderson, P.D. Bloom, K.G. Baikerikar, V.V. Sheares, S.K. Mallapragada, Al–Cu–Fe quasicrystal/ultra-high molecular weight polyethylene composites as biomaterials for acetabular cup prosthetics, Biomaterials, 23 (2002) 1761-1768. https://doi.org/10.1016/S0142-9612(01)00301-5

[13] X.L. Xie, C.Y. Tang, K.Y.Y. Chan, X.C. Wu, C.P. Tsui, C.Y. Cheung, Wear performance of ultrahigh molecular weight polyethylene/quartz composites, Biomaterials, 24 (2003) 1889-1896. https://doi.org/10.1016/S0142-9612(02)00610-5

[14] Y. Huang, W. Wang, C. Liu, A.J. Rosakis, Analysis of intersonic crack growth in unidirectional fiber-reinforced composites, Journal of the Mechanics and Physics of Solids, 47 (1999) 1893-1916. https://doi.org/10.1016/S0022-5096(98)00124-0

[15] V. Grumezescu, A. Grumezescu, Materials for Biomedical Engineering: Thermoset and Thermoplastic Polymers, Elsevier, (2019).

[16] X. Jiang, L.T. Drzal, Properties of injection molded high density polyethylene nanocomposites filled with exfoliated graphene nanoplatelets, Some critical issues for injection molding, (2012) 251-270. https://doi.org/10.5772/35328

[17] C. Liu, Y.-X. Guo, S. Xiao, Capacitively loaded circularly polarized implantable patch antenna for ISM band biomedical applications, IEEE transactions on antennas and propagation, 62 (2014) 2407-2417. https://doi.org/10.1109/TAP.2014.2307341

[18] W. Bonfield, M. Grynpas, A. Tully, J. Bowman, J. Abram, Hydroxyapatite reinforced polyethylene--a mechanically compatible implant material for bone replacement, Biomaterials, 2 (1981) 185-186. https://doi.org/10.1016/0142-9612(81)90050-8

[19] M. Wang, D. Porter, W. Bonfield, Processing, characterisation, and evaluation of hydroxyapatite reinforced polyethylene, Br. Ceram. Trans, 93 (1994) 91-95.

[20] S. Kanagaraj, F.R. Varanda, T.V. Zhil'tsova, M.S. Oliveira, J.A. Simões, Mechanical properties of high density polyethylene/carbon nanotube composites, Composites Science and Technology, 67 (2007) 3071-3077. https://doi.org/10.1016/j.compscitech.2007.04.024

[21] M. Haneef, J.F. Rahman, M. Yunus, S. Zameer, S. Patil, T. Yezdani, Hybrid polymer matrix composites for biomedical applications, Int. J. Modern. Eng. Res, 3 (2013) 970-979. ISSN: 2249-6645

[22] C. Liu, L. Ren, R. Arnell, J. Tong, Abrasive wear behavior of particle reinforced ultrahigh molecular weight polyethylene composites, Wear, 225 (1999) 199-204. https://doi.org/10.1016/S0043-1648(99)00011-3

[23] J. Juhasz, S. Best, R. Brooks, M. Kawashita, N. Miyata, T. Kokubo, T. Nakamura, W. Bonfield, Mechanical properties of glass-ceramic A–W-polyethylene composites: effect of filler content and particle size, Biomaterials, 25 (2004) 949-955. https://doi.org/10.1016/j.biomaterials.2003.07.005

[24] S. Hashmi, S. Neogi, A. Pandey, N. Chand, Sliding wear of PP/UHMWPE blends: effect of blend composition, Wear, 247 (2001) 9-14. https://doi.org/10.1016/S0043-1648(00)00513-5

[25] S. Ziraki, S.M. Zebarjad, M.J. Hadianfard, A study on the tensile properties of silicone rubber/polypropylene fibers/silica hybrid nanocomposites, Journal of the Mechanical Behavior of Biomedical Materials, 57 (2016) 289-296. https://doi.org/10.1016/j.jmbbm.2016.01.019

[26] S.J. Peter, P. Kim, A.W. Yasko, M.J. Yaszemski, A.G. Mikos, Crosslinking characteristics of an injectable poly (propylene fumarate)/β-tricalcium phosphate paste and mechanical properties of the crosslinked composite for use as a biodegradable bone cement, Journal of Biomedical Materials Research: An Official Journal of The Society for Biomaterials, The Japanese Society for Biomaterials, and The Australian Society for Biomaterials, 44 (1999) 314-321. https://doi.org/10.1002/(SICI)1097-4636(19990305)44:3<314::AID-JBM10>3.0.CO;2-W

[27] X. Shi, B. Sitharaman, Q.P. Pham, F. Liang, K. Wu, W.E. Billups, L.J. Wilson, A.G. Mikos, Fabrication of porous ultra-short single-walled carbon nanotube nanocomposite scaffolds for bone tissue engineering, Biomaterials, 28 (2007) 4078-4090. https://doi.org/10.1016/j.biomaterials.2007.05.033

[28] K.W. Chan, H.M. Wong, K.W.K. Yeung, S.C. Tjong, Polypropylene biocomposites with boron nitride and nanohydroxyapatite reinforcements, Materials, 8 (2015) 992-1008. https://doi.org/10.3390/ma8030992

[29] W. Ding, D. Jahani, E. Chang, A. Alemdar, C.B. Park, M. Sain, Development of PLA/cellulosic fiber composite foams using injection molding: Crystallization and foaming behaviors, Composites Part A: Applied Science and Manufacturing, 83 (2016) 130-139. https://doi.org/10.1016/j.compositesa.2015.10.003

[30] C.Z. Liao, H.M. Wong, K.W.K. Yeung, S.C. Tjong, The development, fabrication, and material characterization of polypropylene composites reinforced with carbon nanofiber and hydroxyapatite nanorod hybrid fillers, International journal of nanomedicine, 9 (2014) 1299-1310. https://doi.org/10.2147/IJN.S58332

[31] L.C. Jones, L.T. Topoleski, A. Tsao, Biomaterials in orthopaedic implants, in: Mechanical Testing of Orthopaedic Implants, Elsevier, (2017) 17-32. https://doi.org/10.1016/B978-0-08-100286-5.00002-0

[32] S. Aghyarian, X. Hu, I.H. Lieberman, V. Kosmopoulos, H.K. Kim, D.C. Rodrigues, Two novel high performing composite PMMA-CaP cements for vertebroplasty: An ex vivo animal study, Journal of the mechanical behavior of biomedical materials, 50 (2015) 290-298. https://doi.org/10.1016/j.jmbbm.2015.06.022

[33] C. Fukuda, K. Goto, M. Imamura, M. Neo, T. Nakamura, Bone bonding ability and handling properties of a titania–polymethylmethacrylate (PMMA) composite bioactive bone cement modified with a unique PMMA powder, Acta biomaterialia, 7 (2011) 3595-3600. https://doi.org/10.1016/j.actbio.2011.06.006

[34] S. Khaled, P.A. Charpentier, A.S. Rizkalla, Physical and mechanical properties of PMMA bone cement reinforced with nano-sized titania fibers, Journal of biomaterials applications, 25 (2011) 515-537. https://doi.org/10.1177/0885328209356944

[35] M. Fini, G. Giavaresi, N.N. Aldini, P. Torricelli, R. Botter, D. Beruto, R. Giardino, A bone substitute composed of polymethylmethacrylate and α-tricalcium phosphate: results in terms of osteoblast function and bone tissue formation, Biomaterials, 23 (2002) 4523-4531. https://doi.org/10.1016/S0142-9612(02)00196-5

[36] H.S. Costa, M.F. Rocha, G.I. Andrade, E.F. Barbosa-Stancioli, M.M. Pereira, R.L. Orefice, W.L. Vasconcelos, H.S. Mansur, Sol–gel derived composite from bioactive glass–polyvinyl alcohol, Journal of Materials Science, 43 (2008) 494-502. https://doi.org/10.1007/s10853-007-1875-4

221

[37] M. Hakkarainen, New PVC materials for medical applications—the release profile of PVC/polycaprolactone–polycarbonate aged in aqueous environments, Polymer Degradation and Stability, 80 (2003) 451-458. https://doi.org/10.1016/S0141-3910(03)00029-6

[38] V.F. Cardoso, D.M. Correia, C. Ribeiro, M.M. Fernandes, S. Lanceros-Méndez, Fluorinated polymers as smart materials for advanced biomedical applications, Polymers, 10 (2018) 161-187. https://doi.org/10.3390/polym10020161

[39] B. Ameduri, Chlorotrifluoroethylene Copolymers for Energy-applied Materials, in: Fluorinated Polymers, (2016) 265-300. https://doi.org/10.1039/9781782629368-00265

[40] F. Boschin, N. Blanchemain, M. Bria, E. Delcourt-Debruyne, M. Morcellet, H. Hildebrand, B. Martel, Improved drug delivery properties of PVDF membranes functionalized with β-cyclodextrin—Application to guided tissue regeneration in periodontology, Journal of Biomedical Materials Research Part A: An Official Journal of The Society for Biomaterials, The Japanese Society for Biomaterials, and The Australian Society for Biomaterials and the Korean Society for Biomaterials, 79 (2006) 78-85. https://doi.org/10.1002/jbm.a.30769

[41] A. Abdal-Hay, K.A. Khalil, F.F. Al-Jassir, A.M. Gamal-Eldeen, Biocompatibility properties of polyamide 6/PCL blends composite textile scaffold using EA. hy926 human endothelial cells, Biomedical Materials, 12 (2017) 035002. https://doi.org/10.1088/1748-605X/aa6306

[42] G. Kubyshkina, B. Zupančič, M. Štukelj, D. Grošelj, L. Marion, I. Emri, Sterilization effect on structure, thermal and time-dependent properties of polyamides, in: Mechanics of Time-Dependent Materials and Processes in Conventional and Multifunctional Materials, 3(2011) 11-19. https://doi.org/10.1007/978-1-4614-0213-8_3

[43] S. Samavedi, L.K. Poindexter, M. Van Dyke, A.S. Goldstein, Synthetic biomaterials for regenerative medicine applications, in: Regenerative Medicine Applications in Organ Transplantation, Elsevier, (2014) 81-99. https://doi.org/10.1016/B978-0-12-398523-1.00007-0

[44] S. Swar, V. Zajícová, M. Rysová, I. Lovětinská-Šlamborová, L. Voleský, I. Stibor, Biocompatible surface modification of poly (ethylene terephthalate) focused on pathogenic bacteria: Promising prospects in biomedical applications, Journal of Applied Polymer Science, 134 (2017) 44990. https://doi.org/10.1002/app.44990

[45] S. Teoh, Z. Tang, G.W. Hastings, Thermoplastic polymers in biomedical applications: structures, properties and processing, in: Handbook of biomaterial properties, Springer, (1998) 270-301. https://doi.org/10.1007/978-1-4615-5801-9_19

[46] L. Rubin, Polyethylene as a bone and cartilage substitute: a 32-year retrospective, Biomaterials in Plastic Surgery. St Louis, MO: CV Mosby, (1983) 477-493.

[47] H. Toiserkani, G. Yilmaz, Y. Yagci, L. Torun, Functionalization of polysulfones by click chemistry, Macromolecular Chemistry and Physics, 211 (2010) 2389-2395. https://doi.org/10.1002/macp.201000245

[48] H. Wang, Y. Li, Y. Zuo, J. Li, S. Ma, L. Cheng, Biocompatibility and osteogenesis of biomimetic nano-hydroxyapatite/polyamide composite scaffolds for bone tissue engineering, Biomaterials, 28 (2007) 3338-3348. https://doi.org/10.1016/j.biomaterials.2007.04.014

[49] I. Shabani, V. Haddadi-Asl, M. Soleimani, E. Seyedjafari, S.M. Hashemi, Ion-Exchange Polymer Nanofibers for Enhanced Osteogenic Differentiation of Stem Cells and Ectopic Bone Formation, ACS Applied Materials & Interfaces, 6 (2014) 72-82. https://doi.org/10.1021/am404500c

[50] A. Akinci, Mechanical and structural properties of polypropylene composites filled with graphite flakes, Archives of Materials Science and Engineering, 35 (2009) 91-94.

[51] G. Kaur, R. Adhikari, P. Cass, M. Bown, A.V. Vashi, P. Gunatillake, Flexible conductive graphene/polyurethane composite films for biomedical applications, Frontiers in Bioengineering and Biotechnology, (2016).

[52] T. Yoshii, J.E. Dumas, A. Okawa, D.M. Spengler, S.A. Guelcher, Synthesis, characterization of calcium phosphates/polyurethane composites for weight-bearing implants, J Biomed Mater Res B Appl Biomater, 100 (2012) 32-40. https://doi.org/10.1002/jbm.b.31917

[53] F. Khan, Y. Dahman, A novel approach for the utilization of biocellulose nanofibres in polyurethane nanocomposites for potential applications in bone tissue implants, Designed Monomers and Polymers, 15 (2012) 1-29. https://doi.org/10.1163/156855511X606119

[54] S. Sahan, P. Hosseinian, D. Ozdil, M. Turk, H.M. Aydin, Polyurethane–Ceramic matrices as orbital implants, International Journal of Polymeric Materials and Polymeric Biomaterials, 67 (2018) 487-493. https://doi.org/10.1080/00914037.2017.1354194

[55] L. Wang, B. Sun, K.S. Ziemer, G.A. Barabino, R.L. Carrier, Chemical and physical modifications to poly (dimethylsiloxane) surfaces affect adhesion of Caco-2 cells, Journal of Biomedical Materials Research Part A: An Official Journal of The Society for Biomaterials, The Japanese Society for Biomaterials, and The Australian Society for Biomaterials and the Korean Society for Biomaterials, 93 (2010) 1260-1271. https://doi.org/10.1002/jbm.a.32621

[56] N. Ignjatović, J. Jovanović, E. Suljovrujić, D. Uskoković, Injectable polydimethylsiloxane–hydroxyapatite composite cement, Bio-medical materials and engineering, 13 (2003) 401-410.

[57] A. Larmagnac, S. Eggenberger, H. Janossy, J. Vörös, Stretchable electronics based on Ag-PDMS composites, Scientific Reports, 4 (2014) 7254. https://doi.org/10.1038/srep07254

[58] F. Rahmitasari, Y. Ishida, K. Kurahashi, T. Matsuda, M. Watanabe, T. Ichikawa, PEEK with Reinforced Materials and Modifications for Dental Implant Applications, Dentistry journal, 5 (2017). https://doi.org/10.3390/dj5040035

[59] S.A. Stewart, J. Domínguez-Robles, R.F. Donnelly, E. Larrañeta, Implantable polymeric drug delivery devices: Classification, manufacture, materials, and clinical applications, Polymers, 10 (2018) 1379. https://doi.org/10.3390/polym10121379

[60] J. Karlsson, H.J. Vaughan, J.J. Green, Biodegradable polymeric nanoparticles for therapeutic cancer treatments, Annual review of chemical and biomolecular engineering, 9 (2018) 105-127. https://doi.org/10.1146/annurev-chembioeng-060817-084055

[61] D.S. Kohane, R. Langer, Polymeric biomaterials in tissue engineering, Pediatric research, 63 (2008) 487-491. https://doi.org/10.1203/01.pdr.0000305937.26105.e7

[62] D.F. Williams, On the mechanisms of biocompatibility, Biomaterials, 29 (2008) 2941-2953. https://doi.org/10.1016/j.biomaterials.2008.04.023

[63] S. Tabasum, A. Noreen, M.F. Maqsood, H. Umar, N. Akram, S.A.S. Chatha, K.M. Zia, A review on versatile applications of blends and composites of pullulan with natural and synthetic polymers, International journal of biological macromolecules, 120 (2018) 603-632. https://doi.org/10.1016/j.ijbiomac.2018.07.154

[64] C. Sharma, N.K. Bhardwaj, Bacterial nanocellulose: Present status, biomedical applications and future perspectives, Materials Science and Engineering: C, (2019) 109963. https://doi.org/10.1016/j.msec.2019.109963

[65] C.H. Lee, A. Singla, Y. Lee, Biomedical applications of collagen, International journal of pharmaceutics, 221 (2001) 1-22. https://doi.org/10.1016/S0378-5173(01)00691-3

[66] B. Seal, T. Otero, A. Panitch, Polymeric biomaterials for tissue and organ regeneration, Materials Science and Engineering: R: Reports, 34 (2001) 147-230. https://doi.org/10.1016/S0927-796X(01)00035-3

[67] M. Zare-Gachi, H. Daemi, J. Mohammadi, P. Baei, F. Bazgir, S. Hosseini-Salekdeh, H. Baharvand, Improving anti-hemolytic, antibacterial and wound healing properties of alginate fibrous wound dressings by exchanging counter-cation for infected full-thickness skin wounds, Materials Science & Engineering C-Materials for Biological Applications, 107 (2020) 1-42. https://doi.org/10.1016/j.msec.2019.110321

[68] A.C. Hernandez-Gonzalez, L. Tellez-Jurado, L.M. Rodriguez-Lorenzo, Alginate hydrogels for bone tissue engineering, from injectables to bioprinting: A review, Carbohydrate Polymers, 229 (2020) 1-52. https://doi.org/10.1016/j.carbpol.2019.115514

[69] R.K. Eid, D.S. Ashour, E.A. Essa, G.M. El Maghraby, M.F. Arafa, Chitosan coated nanostructured lipid carriers for enhanced in vivo efficacy of albendazole against Trichinella spiralis, Carbohydrate Polymers, 232 (2020) 1-13. https://doi.org/10.1016/j.carbpol.2019.115826

[70] Y. Wang, C. He, Y. Feng, Y. Yang, Z. Wei, W. Zhao, C. Zhao, A chitosan modified asymmetric small-diameter vascular graft with anti-thrombotic and anti-bacterial functions for vascular tissue engineering, Journal Of Materials Chemistry B, 8 (2020) 568-577. https://doi.org/10.1039/C9TB01755K

[71] W. Chen, Y. Zhu, Z. Zhang, Y. Gao, W. Liu, Q. Borjihan, H. Qu, Y. Zhang, Y. Zhang, Y.-J. Wang, L. Zhang, A. Dong, Engineering a multifunctional N-halamine-based antibacterial hydrogel using a super-convenient strategy for infected skin defect therapy, Chemical Engineering Journal, 379 (2020) 1-13. https://doi.org/10.1016/j.cej.2019.122238

[72] N. Goonoo, A. Bhaw-Luximon, P. Passanha, S.R. Esteves, D. Jhurry, Third generation poly(hydroxyacid) composite scaffolds for tissue engineering, Journal Of Biomedical Materials Research Part B-Applied Biomaterials, 105 (2017) 1667-1684. https://doi.org/10.1002/jbm.b.33674

[73] I. Manavitehrani, A. Fathi, Y. Wang, P.K. Maitz, F. Mirmohseni, T.L. Cheng, L. Peacock, D.G. Little, A. Schindeler, F. Dehghani, Fabrication of a Biodegradable

Implant with Tunable Characteristics for Bone Implant Applications, Biomacromolecules, 18 (2017) 1736-1746. https://doi.org/10.1021/acs.biomac.7b00078

[74] L. Wang, X. Zhao, F. Yang, W. Wu, M. Wu, Y. Li, X. Zhang, Loading paclitaxel into porous starch in the form of nanoparticles to improve its dissolution and bioavailability, International Journal Of Biological Macromolecules, 138 (2019) 207-214. https://doi.org/10.1016/j.ijbiomac.2019.07.083

[75] H. Luo, R. Cha, J. Li, W. Hao, Y. Zhang, F. Zhou, Advances in tissue engineering of nanocellulose-based scaffolds: A review, Carbohydrate Polymers, 224 (2019) 1-15. https://doi.org/10.1016/j.carbpol.2019.115144

[76] I. Khan, M. Apostolou, R. Bnyan, C. Houacine, A. Elhissi, S.S. Yousaf, Paclitaxel-loaded micro or nano transfersome formulation into novel tablets for pulmonary drug delivery via nebulization, International Journal Of Pharmaceutics, 575 (2020) 1-15. https://doi.org/10.1016/j.ijpharm.2019.118919

[77] Y. Poetzinger, L. Rahnfeld, D. Kralisch, D. Fischer, Immobilization of plasmids in bacterial nanocellulose as gene activated matrix, Carbohydrate Polymers, 209 (2019) 62-73. https://doi.org/10.1016/j.carbpol.2019.01.009

[78] S.I. Moon, K. Deguchi, M. Miyamoto, Y. Kimura, Synthesis of polyglactin by melt/solid polycondensation of glycolic/L-lactic acids, Polymer international, 53 (2004) 254-258. https://doi.org/10.1002/pi.1335

[79] E. Marin, M.I. Briceño, C. Caballero-George, Critical evaluation of biodegradable polymers used in nanodrugs, International journal of nanomedicine, 8 (2013) 3071-3091. https://doi.org/10.2147/IJN.S47186

[80] D.J. Cameron, M.P. Shaver, Aliphatic polyester polymer stars: synthesis, properties and applications in biomedicine and nanotechnology, Chemical Society Reviews, 40 (2011) 1761-1776. https://doi.org/10.1039/C0CS00091D

[81] J.B. Jonnalagadda, I.V. Rivero, J. Warzywoda, In-vitro degradation characteristics of poly (e-caprolactone)/poly (glycolic acid) scaffolds fabricated via solid-state cryomilling, Journal of biomaterials applications, 30 (2015) 472-483. https://doi.org/10.1177/0885328215592853

[82] H. Seyednejad, A.H. Ghassemi, C.F. van Nostrum, T. Vermonden, W.E. Hennink, Functional aliphatic polyesters for biomedical and pharmaceutical applications, Journal of Controlled release, 152 (2011) 168-176. https://doi.org/10.1016/j.jconrel.2010.12.016

[83] R. Gaudin, C. Knipfer, A. Henningsen, R. Smeets, M. Heiland, T. Hadlock, Approaches to peripheral nerve repair: generations of biomaterial conduits yielding to replacing autologous nerve grafts in craniomaxillofacial surgery, BioMed research international, 2016 (2016) 1-18. https://doi.org/10.1155/2016/3856262

[84] M. Okamoto, B. John, Synthetic biopolymer nanocomposites for tissue engineering scaffolds, Progress in Polymer Science, 38 (2013) 1487-1503. https://doi.org/10.1016/j.progpolymsci.2013.06.001

[85] V. Singh, M. Tiwari, Structure-processing-property relationship of poly (Glycolic Acid) for drug delivery systems 1: Synthesis and catalysis, International Journal of Polymer Science, 2010 (2010) 1-23. https://doi.org/10.1155/2010/652719

[86] K. Okuyama, S. Yanamoto, T. Naruse, Y. Sakamoto, S. Rokutanda, S. Ohba, I. Asahina, M. Umeda, Clinical complications in the application of polyglycolic acid sheets with fibrin glue after resection of mucosal lesions in oral cavity, Oral surgery, oral medicine, oral pathology and oral radiology, 125 (2018) 541-546. https://doi.org/10.1016/j.oooo.2017.12.013

[87] H. Kobayashi, D. Terada, Y. Yokoyama, D.W. Moon, Y. Yasuda, H. Koyama, T. Takato, Vascular-inducing poly (glycolic acid)-collagen nanocomposite-fiber scaffold, Journal of biomedical nanotechnology, 9 (2013) 1318-1326. https://doi.org/10.1166/jbn.2013.1638

[88] B.N. Kim, Y.-G. Ko, T. Yeo, E.J. Kim, O.K. Kwon, O.H. Kwon, Guided Regeneration of Rabbit Calvarial Defects Using Silk Fibroin Nanofiber–Poly (glycolic acid) Hybrid Scaffolds, ACS Biomaterials Science & Engineering, 5 (2019) 5266-5272. https://doi.org/10.1021/acsbiomaterials.9b00678

[89] M. Porgham Daryasari, M. Dusti Telgerd, M. Hossein Karami, A. Zandi-Karimi, H. Akbarijavar, M. Khoobi, E. Seyedjafari, G. Birhanu, P. Khosravian, F. SadatMahdavi, Poly-l-lactic acid scaffold incorporated chitosan-coated mesoporous silica nanoparticles as pH-sensitive composite for enhanced osteogenic differentiation of human adipose tissue stem cells by dexamethasone delivery, Artificial cells, nanomedicine, and biotechnology, 47 (2019) 4020-4029. https://doi.org/10.1080/21691401.2019.1658594

[90] Y. Wang, L. Sun, Z. Mei, F. Zhang, M. He, C. Fletcher, F. Wang, J. Yang, D. Bi, Y. Jiang, 3D printed biodegradable implants as an individualized drug delivery system for local chemotherapy of osteosarcoma, Materials & Design, 186 (2020) 108336. https://doi.org/10.1016/j.matdes.2019.108336

Nanohybrids   Materials Research Forum LLC
Materials Research Foundations **87** (2021) 202-229   https://doi.org/10.21741/9781644901076-8

[91] R. Zhang, X. Wang, J. Wang, M. Cheng, Synthesis and characterization of konjac glucomannan/carrageenan/nano-silica films for the preservation of postharvest white mushrooms, Polymers, 11 (2019) 6. https://doi.org/10.3390/polym11010006

[92] P. Saini, M. Arora, M.R. Kumar, Poly (lactic acid) blends in biomedical applications, Advanced Drug Delivery Reviews, 107 (2016) 47-59. https://doi.org/10.1016/j.addr.2016.06.014

[93] A.H. Shamsah, S.H. Cartmell, S.M. Richardson, L.A. Bosworth, Tissue Engineering the Annulus Fibrosus Using 3D Rings of Electrospun PCL: PLLA Angle-Ply Nanofiber Sheets, Frontiers in Bioengineering and Biotechnology, 7 (2019) 1-13. https://doi.org/10.3389/fbioe.2019.00437

[94] S. Nanaki, P. Barmpalexis, A. Iatrou, E. Christodoulou, M. Kostoglou, D.N. Bikiaris, Risperidone Controlled Release Microspheres Based on Poly (Lactic Acid)-Poly (Propylene Adipate) Novel Polymer Blends Appropriate for Long Acting Injectable Formulations, Pharmaceutics, 10 (2018) 130 1-21. https://doi.org/10.3390/pharmaceutics10030130

[95] X. Si, X. Quan, Top capping of nanosilver-loaded titania nanotubes with norspermidine-incorporated polymer for sustained anti-biofilm effects, International Biodeterioration & Biodegradation, 123 (2017) 228-235. https://doi.org/10.1016/j.ibiod.2017.07.003

[96] C. Liu, S. Zhang, D.J. McClements, D. Wang, Y. Xu, Design of astaxanthin-loaded core–shell nanoparticles consisting of chitosan oligosaccharides and poly (lactic-co-glycolic acid): Enhancement of water solubility, stability, and bioavailability, Journal of agricultural and food chemistry, 67 (2019) 5113-5121. https://doi.org/10.1021/acs.jafc.8b06963

[97] S.J. Lee, G. Khang, Y.M. Lee, H.B. Lee, Interaction of human chondrocytes and NIH/3T3 fibroblasts on chloric acid-treated biodegradable polymer surfaces, Journal of Biomaterials Science, Polymer Edition, 13 (2002) 197-212. https://doi.org/10.1163/156856202317414375

[98] Z. Pan, J. Ding, Poly (lactide-co-glycolide) porous scaffolds for tissue engineering and regenerative medicine. Interface Focus. 2 (2012) 366–377. https://doi.org/10.1098/rsfs.2011.0123

[99] R.K. Kankala, X.-M. Xu, C.-G. Liu, A.-Z. Chen, S.-B. Wang, 3D-printing of microfibrous porous scaffolds based on hybrid approaches for bone tissue engineering, Polymers, 10 (2018) 807. https://doi.org/10.3390/polym10070807

[100] J.-W. Park, J.-U. Hwang, J.-H. Back, S.-W. Jang, H.-J. Kim, P.-S. Kim, S. Shin, T. Kim, High strength PLGA/Hydroxyapatite composites with tunable surface structure using PLGA direct grafting method for orthopedic implants, Composites Part B: Engineering, 178 (2019) 107449. https://doi.org/10.1016/j.compositesb.2019.107449

[101] R. Boia, P.A. Dias, J.M. Martins, C. Galindo-Romero, I.D. Aires, M. Vidal-Sanz, M. Agudo-Barriuso, H.C. de Sousa, A.F. Ambrósio, M.E. Braga, Porous poly (ε-caprolactone) implants: A novel strategy for efficient intraocular drug delivery, Journal of Controlled Release, 316 (2019) 331-348. https://doi.org/10.1016/j.jconrel.2019.09.023

[102] H. Li, C. Huang, X. Jin, Q. Ke, An electrospun poly (ε-caprolactone) nanocomposite fibrous mat with a high content of hydroxyapatite to promote cell infiltration, RSC advances, 8 (2018) 25228-25235. https://doi.org/10.1039/C8RA02059K

[103] J.M. Unagolla, A.C. Jayasuriya, Enhanced cell functions on graphene oxide incorporated 3D printed polycaprolactone scaffolds, Materials Science and Engineering: C, 102 (2019) 1-11. https://doi.org/10.1016/j.msec.2019.04.026

[104] K. Zhou, R. Gao, S. Jiang, Morphology, thermal and mechanical properties of poly (ε-caprolactone) biocomposites reinforced with nano-hydroxyapatite decorated graphene, Journal of colloid and interface science, 496 (2017) 334-342. https://doi.org/10.1016/j.jcis.2017.02.038

Nanohybrids
Materials Research Foundations **87** (2021) 230-243

Materials Research Forum LLC
https://doi.org/10.21741/9781644901076-9

Chapter 9

Role of Liposomes Composite as Drug Transport Mechanism

P. Senthil Kumar[1, 2]*, P.R. Yaashikaa[1]

[1]Department of Chemical Engineering, SSN College of Engineering, Chennai 603110, India

[2]SSN-Centre for Radiation, Environmental Science and Technology (SSN-CREST), SSN College of Engineering, Chennai 603110, India

senthilkumarp@ssn.edu.in*

Abstract

Liposomes are spherical shaped vesicles comprising of at least one phospholipid bilayer that serve as a novel drug delivery framework. They are microscopic structures in which a fluid system is totally encased by a film made out of lipid bilayers. It varies in size, conformation, charge and drug transporter stacked with assortment of particles, for example, small molecules of drug, plasmids, nucleotides or proteins and so on. Ongoing advances in nanotherapeutics have brought about engineered liposomes rising in nanomedicine, giving better restorative control of diseased states. This has made ready for the improvement of second-stage liposomes for increased efficiency and could at last lead to a change in perspective from the regular drug delivery methods.

Keywords

Drug delivery, Liposomes, Methods and Mechanism, Target, Therapeutics

Contents

1.   Introduction

Liposomes have gotten broad consideration as a transporter framework for restoratively dynamic mixes, because of their remarkable qualities, for example, ability to join hydrophilic and hydrophobic medications, great biocompatibility, low poisonous quality, absence of safe framework actuation, and focused on conveyance of bioactive mixes to the site of activity. A liposome is a circular vesicle with a layer made out of a phospholipid bilayer used to convey drugs or hereditary material into a cell. Liposomes can be made out of normally determined phospholipids with blended lipid chain. Figure 1 shows the structure of liposome.

Furthermore, a few accomplishments since the disclosure of liposomes are controlled size from microscale to nanoscale and surface-built polymer conjugates functionalized with peptide, protein, and immune response. In spite of the fact that liposomes have been widely read as promising bearers for remedially dynamic intensifies, a portion of the significant disadvantage for liposomes utilized in pharmaceutics are the fast debasement due to the Reticulo Endothelial System (RES) and failure to accomplish supported medication conveyance over a delayed timeframe [1].

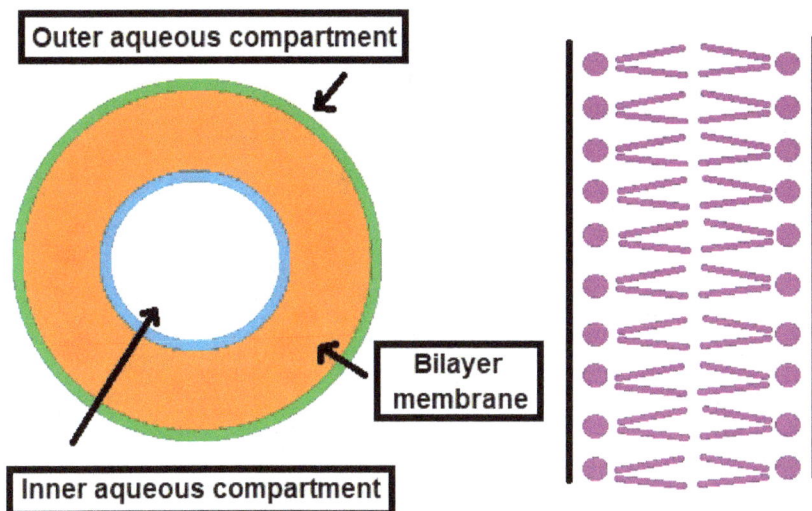

Figure 1. Structure of liposome.

The acknowledgment of the biological cells misusing surface-dynamic properties of lipids to define anatomical layers prompted the advancement of model frameworks dependent on the direction of lipids at the interfaces for the examination of transport capacities and instruments, penetration properties, just as bond and combination energy. Model lipid layers that copy numerous parts of cell membranes have been exceptionally valuable in helping examinations to recognize the component of communication between bioactive molecules and lipids. Artificial membrane frameworks give a reasonable stage to examination of bio-physical communications of medications and medication conveyance frameworks bypassing the complexities of cell film structures and related powerful nature of lipid–lipid and lipid–protein connections in cell layers and furthermore permit experimentation to be performed under conditions that living cells will most likely be unable to withstand and stay suitable [2,3].

The likeness of their lamellar structure with common layers, the capability to segregate particles, and weakness to adjustment by bioactive atoms like organic layers have rendered liposomes a versatile apparatus in the field of science, natural chemistry, and prescription. With the acknowledgment of the biocompatibility, biodegradability, low danger and immunogenicity, and the ability to ensnare particles, liposomes have moved far from being simply one more intriguing object of biophysical research to turning into a pharmaceutical bearer of decision for various handy applications [4]. The advantages of a composite framework, be that as it may, incorporate improvement of liposome solidness, the capacity of the liposome to control tranquilize discharge over a drawn out timeframe, and conservation of the bio activeness of the medications in polymeric-based innovation. Also, expanded adequacy might be accomplished from this coordinated conveyance framework when contrasted with that of simply polymeric-based or liposome-based frameworks [5]. The point of this article accordingly, is to survey the present liposome-based and polymeric-based advances, just as the coordination of liposome-based innovation inside impermanent terminal polymeric-based innovation for supported medication discharge [6]. The conversation will concentrate on various kinds of liposome-based innovation and stop polymeric framework advances, different strategies for installing drug-stacked liposomes inside a terminal, and different methodologies answered to control the pace of continued medication discharge inside warehouse frameworks over a delayed timeframe [7,8].

2.   Types of liposome

In general, there are four key sorts of liposomal conveyance frameworks—regular liposomes, sterically-settled liposomes, ligand-focused on liposomes. Regular liposomal plans diminished the lethality of mixes in vivo, through adjusting pharmacokinetics and biodistribution to upgrade sedate conveyance to unhealthy tissue in correlation with free medication. A wide range of liposomes can actuate the supplement framework. Liposome size, morphology, charge, lipid arrangement, bilayer bundling, surface attributes, and regulated lipid portion all manage supplement actuation. Explicit liposomal qualities that improve the affinity for supplement enactment incorporate a positive or negative surface charge, expanding size, absence of liposomal homogeneity, endotoxin pollution, nearness of totals, nearness of drugs that can tie to and total liposomes [9,10].

Contingent on the structure there are two kinds of liposome.

2.1   Unilamellar liposomes

Unilamellar vesicle has a solitary phospholipid bilayer circle encasing fluid arrangement.

2.2   Multilamellar liposomes

Multilamellar vesicles have onion structure. Ordinarily, a few unilamellar vesicles will frame one inside the other in reducing size, making a multilamellar structure of concentric phosphlipid circles isolated by layers of water.

Nanohybrids   Materials Research Forum LLC
Materials Research Foundations **87** (2021) 230-243   https://doi.org/10.21741/9781644901076-9

2.3   Customary liposomes

Customary liposome-based innovation is the original of liposome to be utilized in pharmaceutical applications. Ordinary liposome definitions are for the most part involved regular phospholipids, sphingomyelin, egg phosphatidylcholine, and monosialoganglioside. Since this definition is comprised of phospholipids just, liposomal details have experienced numerous difficulties; one of the significant ones being the shakiness in plasma, which brings about short blood dissemination half-life. Liposomes that are contrarily or decidedly charged have been accounted for to have shorter half-lives, are dangerous, and quickly expelled from the course. A few different endeavours to beat these difficulties have been made, explicitly in the control of the lipid film. One of the endeavors concentrated on the control of cholesterol. Expansion of cholesterol to customary definitions diminishes fast arrival of the typified bioactive compound into the plasma.

2.4   Stealth liposomes

Stealth liposome innovation is one of the frequently utilized liposome-based frameworks for conveyance of dynamic particles. This procedure was created to conquer the majority of the difficulties experienced by traditional liposome innovation, for example, the powerlessness to sidestep interference by the insusceptible framework, poisonous quality because of charged liposomes, low blood course half-life, and steric dependability. Stealth liposome system was accomplished just by changing the outside of the liposome layer, a procedure that was accomplished by building hydrophilic polymer conjugates. The utilized hydrophilic polymers were either common or manufactured polymers such polyethylene glycol (PEG), chitosan, silk-fibroin, and polyvinyl liquor. A few properties that would add favourable circumstances to polymeric conjugate were viewed as, for example, high biocompatibility, nontoxicity, low immunogenicity, and antigenicity. In spite of the fact that most of hydrophilic polymers meet the above criteria, PEG remains the most generally utilized polymer conjugate. It is explicitly utilized to build the hydrophilicity of the liposome surface by means of a cross-connected lipid.

2.5   Different types of liposomes

2.5.1 Virosomes and stimuli-responsive liposomes

Liposomal advances, for example, regular, stealth, and focused on liposomes have just gotten clinical endorsement. New age sorts of liposome have been created to increment bioactive particle conveyance to the cytoplasm by escape endosome. New methodologies that utilize liposomes as pharmaceutical bearers are virosomes and boosts type liposomes.

The animating specialists right now pH, light, attraction, temperature, and ultrasonic waves. A virosome is another kind of liposome plan. It includes noncovalent coupling of a liposome and a fusogenic viral wrap. A boosts delicate liposome is a kind of liposome that for the most part relies upon various natural factors so as to trigger medication, protein, and quality conveyance [11-13].

2.5.2 Gene based liposomes

The portrayal of human genome combined with recombinant DNA innovation has made open doors for quality treatment that never existed. Applicant ailments for such innovation incorporate malignant growth, arteriosclerosis, cystic fibrosis, hemophilia, sickle cell frailty, and other hereditary sicknesses. In a perfect world, the organization of the quality of intrigue should bring about the outflow of the helpful protein. In any case, the conveyance of the huge anionic bioactive DNA across cell has been one of the most troublesome undertakings [14]. DNA is effectively debased by circling and intracellular deoxyribonucleases. In any case, it should likewise be conveyed flawless over the cell and nucleolar layers to the core. Liposomes have along these lines demonstrated to accomplish proficient intracellular conveyance of DNA. Such liposomes are set up from phospholipids with an amine hydrophilic head gathering. The amines might be either quaternary ammonium, tertiary, auxiliary, or essential, and the liposomes arranged right now usually alluded to as cationic liposomes, since they have a positive surface charge at physiological pH.

Despite the fact that the trial information have shown that cationic liposomes can encourage the exchange of DNA into live mammalian cells, there are as yet serious issues that should be defeated so as to adequately accomplish the objective. These remember a decrease for the quick leeway of cationic liposomes and the creation of productively focused on liposomes. At the cell level, the issues might be overwhelmed by improving receptor intervened take-up utilizing proper ligands. The blessing of liposomes with endosomal get away from instruments, combined with progressively proficient translocation of DNA to the core and the effective separation of the liposome complex just before the passage of free DNA into the core may give an ideal foundation answer for the issue.

2.6   Compositions and qualities of liposomes

Normally liposomes made out of cholesterol and phospholipids, synthesis and extent being basically equivalent to in the host cell layers. The phospholipids have a hydrophobic tail structure and a hydrophilic head part and compose in the accompanying way when disintegrated in water: the hydrophobic tails commonly draw in, while the hydrophilic heads contact with the fluid medium outside and inward to the liposome surface. Right now, lipid layers are shaped which close to frame little vesicles like the body cells and their organelles. These circles or liposomes establish little stores that can be made to contain an antigen, an anti-infection, an allergen, a medication substance or a quality. The liposomes can thus be presented in the body without activating insusceptible dismissal responses. Phospholipid Bilayers are the center structure of liposome and cell film developments.

2.7   Instrument of transportation through liposome

Liposome can cooperate with cells by four distinct systems

Endocytosis by phagocytic cells of the reticulo endothelial framework, for example, macrophages and neutrophils.

➤ Adsorption to the cell surface either by vague powerless hydrophobic or electrostatic powers or by explicit collaborations with cell-surface segments.

➤ Fusion with the plasma cell film by addition of the lipid bilayer of the liposome into the plasma layer, with synchronous arrival of liposomal content into the cytoplasm.

➤ Transfer of liposomal lipids to cell or subcellular films, or the other way around, with no relationship of the liposome substance.

It regularly is hard to figure out what instrument is usable and more than one may work simultaneously.

3.   Liposome-based technology

A liposome is a small vesicle comprising of a fluid centre ensnared within one or more regular phospholipids shaping shut bilayered structures. Liposomes have been widely utilized as potential conveyance frameworks for an assortment of mixes basically because of their high level of biocompatibility and the colossal decent variety of structures and organizations [4]. The lipid segments of liposomes are transcendently phosphatidylcholine gotten from egg or soybean lecithins. Liposomes are biphasic an element that renders them the capacity to go about as bearers for both lipophilic and hydrophilic medications. It has been seen that medication particles are found contrastingly in the liposomal condition and relying on their dissolvability and parceling qualities, they show distinctive entanglement and discharge properties. Lipophilic medications are for the most part entangled totally in the lipid bilayers of liposomes and since they are inadequately water dissolvable, issues like loss of a captured medication on capacity are once in a while experienced. Hydrophilic medications may either be entangled inside the watery centres of liposomes or be situated in the outer water stage.

Directed liposome based framework was proposed after ordinary stealth liposome neglected to dodge take-up of dynamic particles by touchy typical cells or vague focuses in vivo. In contrast to stealth liposome, site-explicit focusing on liposome has been built or functionalized with various sorts of focusing on moieties such antibodies, peptide, glycoprotein, oligopeptide, polysaccharide, development factors, folic corrosive, starch, and receptors. Moreover, directed ligand can additionally expand the pace of liposomal medicate collection in the perfect tissues/cells by means of overexpressed receptors, antigen, and unregulated selectin. Peptides, protein, and antibodies have been most broadly read as a ligand for coordinating medication stacked liposomes into locales of activity, because of their particle structures, which are basically made out of known amino corrosive arrangements. Besides, it has been hypothesized that ligands can be conjugated onto pegylated liposomes through various sorts of coupling strategies, for example, covalent and noncovalent official. Covalent coupling happens when novel ligands are by implication designed on the outside of liposome through a hydrophobic stay by means of thioether, hydrazone bonds, avidin–biotin communication, cross-connecting between carboxylic acids and additionally amines. Noncovalent coupling is seen when novel ligands are legitimately added to the blend of phospholipids during the liposomal plan.

This compound depicts its anticancer action by initiating irreversible vascular shutdown in strong tumors. Regardless of its anticancer potential, the medication has appeared to have a few unfortunate symptoms to the basic ordinary tissues. These issues might be lightened by focusing on the medication explicitly to the strong tumor vasculature. These surface markers segregate tumor endothelial cells from the typical endothelial cells and can be utilized as an objective for antivascular tranquilize conveyance.

3.1   Liposomal drug delivery

Passive and active targeting in request to evade the issues related with customary medication conveyance, which incorporates inefficient biodistribution all through the body and absence of specific conveyance and latently focusing on tissues and organs that have broken endothelium (e.g., liver, spleen, and bone marrow), liposomal definitions of a few key dynamic atoms were created. The clamorous tumor vessel design portraying any strong tumor encourages inactive focusing on. Defective vasculature and broken lymphatic seepage describe any strong tumor. The fenestrations of the broken vasculature permit the liposomes to escape into the tumor tissue. Extravasation is progressively trailed by expanded maintenance of the medication stacked nanocarrier in the tumor

Nanohybrids   Materials Research Forum LLC
Materials Research Foundations **87** (2021) 230-243   https://doi.org/10.21741/9781644901076-9

tissue. Not at all like low-atomic weight medicates, the development of high-sub-atomic weight anticancer d floor coverings/nanoparticles takes place by convection as opposed to by dissemination from the circulatory framework through the interstitial space [15]. A significant dynamic focusing on technique includes the utilization of ligands with affinity for the receptors communicated on the plasma film of malignant growth cells or tumor neovasculature, to expand the collection of anticancer medications even with high IFP condition of the tumor tissues and improve the remedial adequacy. In conjugation with nanotherapeutics, helpful intercessions intended to diminish IFP could be applied to increase the treatment of malignant growth. Figure 2 shows that targeted drug delivery system.

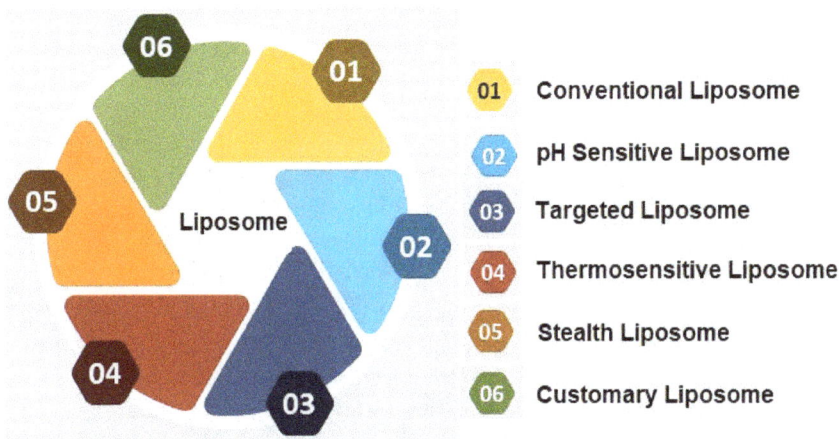

01   Conventional Liposome

02   pH Sensitive Liposome

03   Targeted Liposome

04   Thermosensitive Liposome

05   Stealth Liposome

06   Customary Liposome

Figure 2. Targeted Drug Delivery System.

3.2   Liposome based therapeutics

Liposomes have been generally answered to be a remedial instrument of decision since they have various points of interest as pharmaceutical transporters. Be that as it may, the major related restriction of traditional liposomes for helpful use lies in its quick end from the blood and acknowledgment by RES. The efficient take-up of liposomes by macrophages and ensuing expulsion from foundational flow upon intravenous organization are, nonetheless, seriously influenced when the objective site is past the mononuclear phagocyte framework. The customary liposomes were additionally seen to show significant precariousness in plasma, which brought about the quick arrival of the embodied payload attributable to their communications with both high-and low-thickness

Nanohybrids   Materials Research Forum LLC
Materials Research Foundations **87** (2021) 230-243   https://doi.org/10.21741/9781644901076-9

lipoproteins. So as to sidestep the low-fundamental course time of traditional liposomes, amalgamation of long circling liposomes (Stealth liposomes) has been endeavoured by covering the liposome surface with polymers.

This brought about altogether expanded liposome solidness, which delayed, by a few sets of greatness, their blood dissemination times after foundational organization and at last prompted the advancement of customized liposomal definition with expanded dependability both in vitro and in vivo, improved biodistribution, and enhanced habitation time in fundamental course [16]. Late advances report the rise of another class of liposomes for malignancy specific treatment, which effectively defeats the restrictions of ordinary liposomes. As opposed to customary liposomes, upgrades responsive vesicles experience moderately enormous and sudden physical and concoction changes in sharp reaction to applied boosts. This is the fate specifically noteworthy when the boosts to which these vesicles respond are infection or foundational organic chemistry specific, (for example, pH). Strong tumors are portrayed by poor vasculature, which causes commonness of anaerobic conditions, and the extracellular pH is additionally altogether acidic (~6–7) than foundational pH (7.4). The pH move of the specific tissues can go about as inner improvements of substance and bio-synthetic starting point that trigger medication discharge from the upgrades responsive nanocarriers. Outside physical boosts activating arrival of epitomized load incorporate warmth, light, and attractive field.

Liposomal antibody detailing bearing antigenic peptide got from choriomen-ingitis infection and safe stimulatory oligonucleotides has been accounted for to evoke antiviral and antitumor resistance [17]. One of the latest increments to the collection of liposomes is the multifunctional theranostic liposomes, which can be considered as a key progression in nanomedicine and has opened up a plenty of potential outcomes for concurrent malignant growth treatment and analysis. The flow focal point of medication conveyance look into in clinical trials has been on dynamic focused on sedate conveyance. By following the development of liposomes as intense pharmaceutical bearers for anticancer medications, one can absorb that liposomes have gone far and at present various alluring and diversified systems are as a rule effectively applied preclinically or clinically for improved and viable conveyance of medications.

3.3 Liposomes in clinical use

Liposomes act as supplies typifying the medication, shielding it from debasement and diminishing the unintended reactions, for example, cardiotoxicity, nephrotoxicity, neurotoxicity, or dermal lethality [18]. Various liposomal details bearing disease therapeutics have been endorsed or are at present experiencing clinical preliminaries. Liposomal detailing of doxorubicin, an anthracycline-class tranquilize and topo-

Nanohybrids   Materials Research Forum LLC
Materials Research Foundations **87** (2021) 230-243   https://doi.org/10.21741/9781644901076-9

isomerase inhibitor with revealed irreversible cardiotoxicity has been effectively created to successfully treat malignant growths with a lot lesser-related symptoms.

4. Propelled application of liposomes

Liposomes are utilized to convey certain immunizations, catalysts, or medications (e.g., insulin and some malignancy drugs) to the body. At the point when utilized in the conveyance of certain disease drugs, liposomes help to shield solid cells from the medications' poisonous quality and forestall their focus in powerless tissues (e.g., the kidneys and liver), decreasing or wiping out the regular symptoms of sickness, weakness and balding. Liposomes are particularly successful in treating maladies that influence the phagocytes of the resistant framework since they will in general collect in the phagocytes, which remember them as remote trespassers. They have likewise been utilized tentatively to convey typical qualities into a cell so as to supplant faulty, ailment causing qualities. Liposomes are in some cases utilized in makeup on account of their saturating characteristics [19-21].

4.1 Liposome in immunization

Fusogenic liposomes are additionally powerful as a mucosal antibody bearer Liposomes are eagerly phagocytosed by macrophages and different cells of the reticuloendothelial framework. Thus, they make incredible adjuvants for some decontaminated antigens. A model that can be utilized to clarify this guideline is spoken to by a bacterial exopolysaccharide or a recombinant protein. These components are costly to deliver and refine, however when vaccinated into liposomes in modest quantities, a satisfactory insusceptible reaction can be accomplished.

4.2 Liposome in gene transfer

Direct quality exchange for the treatment of human illnesses requires a vector which can be regulated productively, securely and more than once. Cationic liposomes speak to one of only a handful hardly any models that can meet these prerequisites. As of now, in excess of twelve cationic liposome definitions have been accounted for. These liposomes tie and gather DNA precipitously to frame buildings with high partiality to cell films. Endocytosis of the edifices followed by disturbance of the endosomal film gives off an impression of being the significant system of quality conveyance. The adequacy and security of this DNA conveyance technique has been set up in numerous investigations. In light of these outcomes, two human quality treatment clinical preliminaries utilizing cationic liposomes have been led and more preliminaries will be begun sooner rather than later. The straightforwardness, effectiveness and security highlights have rendered the cationic liposome an alluring vehicle for human quality treatment.

Conclusion

The advancement of new-age pharmaceutical liposomes has denoted another period in tranquilize conveyance frameworks in disease therapeutics. Liposomes are adaptable medication conveyance frameworks that can be structured and altered so as to improve the effectively of the remedial medication. The wide cluster of liposomal tranquilize plans endorsed and experiencing clinical preliminaries focuses to the interpretation of liposomes from an object of research to favoured pharmaceutical transporters for clinical applications. A superior comprehension of liposomal sedate collaboration with the natural framework will encourage the rise of a novel class of hostile to malignancy therapeutics with improved efficacy and security. The immense range of liposome-based therapeutics in preclinical/clinical preliminaries and advertised details furnish another worldview in nanotherapeutics with center toward determination, treatment, and counteraction.

References

[1]   T.M. Allen, C. Hansen, J. Rutledge, Liposomes with prolonged circulation times—Factors affecting uptake by reticuloendothelial and other tissues, Biochimica et Biophysica Acta, 981 (1989) 27–35. https://doi.org/10.1016/0005-2736(89)90078-3

[2]   M. Alavi, N. Karimi, M. Safaei, Application of various types of liposomes in drug delivery systems, Advanced Pharmaceutical Bulletin, 7 (2017) 3-9. https://doi.org/10.15171/apb.2017.002

[3]   T.M. Allen, P.R. Cullis, Liposomal drug delivery systems: From concept to clinical applications, Advanced Drug Delivery Reviews, 65 (2013) 36-48. https://doi.org/10.1016/j.addr.2012.09.037

[4]   C. Chen, D. Han, C. Cai, X. Tang, An overview of liposome lyophilization and its future potential, Journal of Controlled Release, 142 (2010) 299–311. https://doi.org/10.1016/j.jconrel.2009.10.024

[5]   N. Duzgunes, G. Gregoriadis, Introduction: The origins of liposomes: Alec Bangham at Babraham, Liposomes, Pt E, 391 (2005) 1–3. https://doi.org/10.1016/S0076-6879(05)91029-X

[6]   J. Kunisawa, S. Nakagawa, T. Mayumi, Pharmacotherapy by intracellular delivery of drugs using fusogenic liposomes: application to vaccine development, Advanced

drug delivery reviews, 52 (2001) 177-186. https://doi.org/10.1016/S0169-409X(01)00214-9

[7] S. Hua, S.Y. Wu, The use of lipid-based nanocarriers for targeted pain therapies, Frontiers in Pharmacology, 4 (2013) 143. https://doi.org/10.3389/fphar.2013.00143

[8] H. Daraee, A. Etemadi, M. Kouhi, S. Alimirzalu, A. Akbarzadeh, Application of liposomes in medicine and drug delivery, Artificial Cells, Nanomedicine, and Biotechnology, 44 (2016) 381-391. https://doi.org/10.3109/21691401.2014.953633

[9] L. Sercombe, T. Veerati, F. Moheimani, S.Y. Wu, A.K. Sood, S. Hua, Advances and challenges of liposome assisted drug delivery, Frontiers in Pharmacology, 6 (2015) 286. https://doi.org/10.3389/fphar.2015.00286

[10] G. Bozzuto, A. Molinari, Liposomes as nanomedical devices, International Journal of Nanomedicine, 10 (2015) 975-999. https://doi.org/10.2147/IJN.S68861

[11] Y. Lee, D.H. Thompson, Stimuli-responsive liposomes for drug delivery, Wiley Interdisciplinary Reviews. Nanomedicine and Nanobiotechnology, 9 (2017) 1-40. https://doi.org/10.1002/wnan.1450

[12] L. Zhu, V.P. Torchilin, Stimulus-responsive nanopreparations for tumor targeting, Integrative Biology, 5 (2013) 96-107. https://doi.org/10.1039/c2ib20135f

[13] V. Torchilin, Multifunctional and stimuli-sensitive pharmaceutical nanocarriers, European Journal of Pharmaceutics and Biopharmaceutics, 71 (2009) 431-444. https://doi.org/10.1016/j.ejpb.2008.09.026

[14] H. Elsana, T.O.B. Olusanya, J. Carr-wilkinson, S. Darby, A. Farheem, A.A. Elkordy, Evaluation of novel cationic gene based liposomes with cyclodextrin prepared by thin film hydration and microfluidic systems, Scientific Reports, 9 (2019) 15120. https://doi.org/10.1038/s41598-019-51065-4

[15] M. Hara, J. Miyake, Calcium alginate gel-entrapped liposomes, Materials Science and Engineering C, 17 (2001) 101–105. https://doi.org/10.1016/S0928-4931(01)00316-2

[16] X. Li, L. Ding, Y. Xu, Y. Wang, Q. Ping, Targeted delivery of doxorubicin using stealth liposomes modified with transferrin, International Journal of Pharmaceutics, 373 (2009) 116–123. https://doi.org/10.1016/j.ijpharm.2009.01.023

[17] P. Pradhan, J. Giri, F. Rieken, Targeted temperature sensitive magnetic liposomes for thermo-chemotherapy, Journal of Controlled Release, 142 (2010) 108–121. https://doi.org/10.1016/j.jconrel.2009.10.002

[18]   S. Jain, P. Jain, R.B. Umamaheshwari, N.K. Jain, Transfersomes—A novel vesicular carrier for enhanced transdermal delivery: Development, characterization, and performance evaluation, Drug development and industrial Pharmacy, 29 (2003) 1013–1026. https://doi.org/10.1081/DDC-120025458

[19]   P.S. Zangabad, S. Mirkiani, S. Shahsavari, B. Masoudi, M. Masroor, H. Hamed, Z. Jafari, Y.D. Taghipour, H. Hashemi, M. Karimi, M.R. Hamblin, Stimulus-responsive liposomes as smart nanoplatforms for drug delivery applications, Nanotechnology Review, 7 (2018) 95-122. https://doi.org/10.1515/ntrev-2017-0154

[20]   B.M. Dicheva G.A. Koning, Targeted thermosensitive liposomes: an attractive novel approach for increased drug delivery to solid tumors, Expert opinion on drug delivery, 11 (2014) 83–100. https://doi.org/10.1517/17425247.2014.866650

[21]   A. Schnyder, J. Huwyler, Drug transport to brain with targeted liposomes, NeuroRx, 2 (2005) 99–107. https://doi.org/10.1602/neurorx.2.1.99

Materials Research Forum LLC
https://doi.org/10.21741/9781644901076-10

Chapter 10

Gelatin Based Hydrogels for Tissue Engineering and Drug Delivery Applications

Jhaleh Amirian[1,2,3]*, Gisya Abdi[4], Mehdihasan I. Shekh [1,2], Ehsan Amel Zendehdel[5], Bing Du[1,2], Florian J. Stadler[1,2]

[1] College of Materials Science and Engineering, Shenzhen Key Laboratory of Polymer Science and Technology, Guangdong Research Center for Interfacial Engineering of Functional Materials, Nanshan District Key Laboratory for Biopolymers and Safety Evaluation, Shenzhen University, Shenzhen, 518055, PR China

[2] Key Laboratory of Optoelectronic Devices and Systems of Ministry of Education and Guangdong Province, College of Optoelectronic Engineering, Shenzhen University, Shenzhen 518060, PR China

[3] Institute of Physical Chemistry, Polish Academy of Sciences, Kasprzaka 44/52, 01-224 Warsaw, Poland.

[4] Center of new technologies, University of Warsaw, ul. Banacha 2c, 02-097 Warsaw, Poland

[5] The Faculty of Art and Architecture, Eshragh Institute of Higher Education, Bojnord, Iran.

*jalehamirian@gmail.com

Abstract

Gelatin is one of the most popular natural polymers which is widely used for food, pharmaceutical, biomedical and cosmetic industries. Gelatin has been prepared from different sources such as porcine skin, cattle bone, and fish. Depending on the type of acidic and alkali extraction processes, type of gelatin A and B are obtained, respectively. This chapter provides a comprehensive overview of the preparation of gelatin-based hydrogels. Furthermore, we evaluate their method of crosslinking through Schiff-base, Michael addition reaction, light, UV, chemical, and physical crosslinking. Moreover, due to the unique properties of gelatin, they have the ability to immobilize cells and can be applied for stem cell and drug delivery for biomedical applications.

Keywords

Gelatin, Hydrogels, Bio-ink, Crosslinking, Tissue Engineering, Drug Delivery

Contents

1. Introduction

Gelatin is water-soluble protein fragments, which is obtained by partial hydrolysis of collagen. In this book chapter, we are dealing with the structural aspects and properties of gelatin [1-5]. As gelatin has been known as denatured collagen product, which can produce with various molecular weight (MWs) and isoionic points (IEPs) [1]. Gelatin due to unique active functional groups, chemical, rheological, and gel properties were chosen for different biomedical and drug delivery applications [6]. The gelatin types are dependent on the source of collagen and the methods of manufacturing. According to gelatin production, gelatin could be divided into two types A and B [1, 7], of which gelatin A were obtained through acid extraction, while gelatin-B is extracted under alkaline conditions [1].

These two types of gelatin show different physico-chemical and mechanical properties and could be used in different applications [6, 7]. Furthermore, gelatin was extracted from a different source of collagen. Among them, gelatin derived from bovine and porcine is widely used for biomaterials. However, recently some research groups extracted the gelatin from fish sources [6]. Growing applications of gelatin in food, beverage, pharmaceuticals, medical, tissue engineering, and nutraceuticals industries are critical factors for developing of gelatin extraction from different sources [6, 8]. However, fish gelatin has a higher price, variable quality, and different gelation and rheological behaviors properties compared to mammalian gelatin [6]. In addition, the rheological properties of gelatin are greatly affected by extraction process conditions such as temperature, pH, and species or tissue from which it is extracted [6]. The three most important functional properties of gelatin are gel strength, viscosity, and melting temperature. The typical gel strength rage of mammalian gelatin is 100-300 bloom. However, for fish gelatins, its much lower and is in range of 70 to 270 bloom [8]. In contrast, melting temperature for mammalian is in range of 20-25°C or 28-31°C, and for fish, gelatin is in the range of 8-25 °C or 11-28 °C [6, 9].

Moreover, gelatin is a biodegradable natural polymer with a wide range of properties such as low cost, excellent biocompatibility, and extracellular matrix (ECM) mimicking for cell growth [2, 10]. The gelatin hydrogels contain a large amount of water, which makes them ideal biomaterials for transporting oxygen, nutrient, and waste exchanges [11-13]. Having these properties make gelatin as superior biomaterials for tissue engineering and drug delivery application [11, 12]. However, the usage of gelatin was limited to applications at physiological conditions, including pH and 37°C temperature [8, 14]. Moreover, the gelatin like other hydrogels have some drawbacks, such as low stability, poor mechanical properties, and low elasticity [7, 14].

Furthermore, gelatin is water-soluble protein fragments, which is obtained by partial hydrolysis of collagen [7]. In this book chapter, we are dealing with the structural and properties of gelatin.

2.   Gelatin Source and Manufacturing

Generally, gelatin was manufactured from pigs, cows, and fish, as well as waste sources such as bovine skin [6, 8, 15]. Gelatin manufacturing aims to facilitate the conversion of collagen to gelatin and with considering to remove the impurities [1]. These impurities, including organic and inorganic components. Inorganic components, including ions such as sodium, potassium, calcium, magnesium, iron, chlorides, and phosphates, which can contribute to degradation [1]. On the other hand, organic components include proteins from blood, keratins, glycoproteins, as well as mucopolysaccharides such as hyaluronic acid, chondroitin sulfate, lipids, nucleic acids, and other cell components, which contribute to soluble degradation of gelatin [1]. The majority of collagen is located in the corium layer, which is a preferred source of pure skin collagen. Many factors such as species, age of animal as well as extraction techniques can affect the quantities of soluble collagen during collagen manufacturing. When we talk about insoluble collagen as in the case of bone, it means that this type of collagen associated with the presence of covalent bonds.

There are two approaches for the manufacturing and extraction of gelatin from collagen-based on pH[1]. The acid process was applied for preparation of gelatin A; alkaline process for gelatin B. Gelatin type A (cationic gelatin) and gelatin type B have an isoelectric point of 7-9 and 4.8-5, respectively (As shown in Figure 1) [1, 16]. Generally, gelatin can be identified by the characteristic amino acid sequence Arg-Gly-Asp (RGD) in its structure [16]. RGD plays an important role in cell adhesion. Subsequently, this resulted in the enhancement of the biological recital of gelatin compared with synthetic polymers [16, 17]. Gelatin is approved by the United States Food and Drug Administration (FDA), as a result, it has been applied in numerous pharmaceutical, foodstuff, drug, and tissue engineering applications [16].

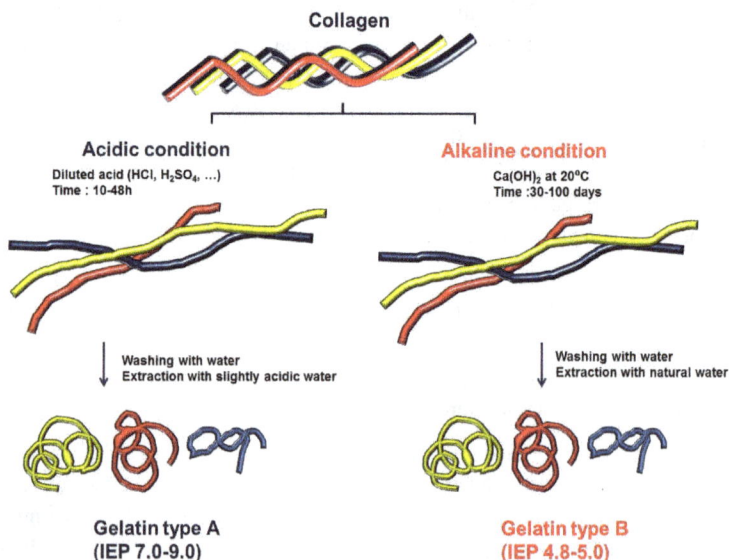

Figure 1. Gelatin type A and B extraction from collagen under acidic and alkaline conditions, respectively.

Nanohybrids   Materials Research Forum LLC
Materials Research Foundations 87 (2021) 244-270   https://doi.org/10.21741/9781644901076-10

Moreover, there are some conditions such as chemicals for pretreatment, temperature, and time of treatments, which can affect on the extracted gelatin polypeptide chain [1].

3. Crosslinking of Gelatin

Crosslinking defined as the creation of the chemical and physical bonds between the polymer chains, which regulates the mechanical strength, degradation rate, and biological properties of the hydrogel polymers [18]. Gelatin, as a natural polymer, does not possess the desired mechanical strength and degrades without appropriate crosslinking [18, 19]. Therefore, the applied crosslinking approach and crosslinker agents for gelatin-based scaffolds are very vital elements [18]. As the crosslinking approach and agents can tune the physio-chemical, mechanical, degradation, as well as cell-matric interactions [18, 19]. Therefore, it should consider that an ideal crosslinker agent should not have a negative effect on the biocompatibility of the scaffolds [18].

Table 1. Advantages and disadvantages of the chemical, photo, enzymatic, and physical crosslinking approaches.

Crosslinking approach	Advantages	Disadvantages
Chemical crosslinking	Causes strong bond formation. Control of degradation and mechanical properties.	Toxicity for cells Washing steps required to remove the remaining crosslinker Expensive than physical crosslinker
Photo-crosslinking	Crosslinking with UV and light by addition of photo-initiator and exposing time. Working in mild aqueous conditions	
Enzyme crosslinking	Working in mild aqueous conditions. Crosslinking by changing of pH, temperature, and ionic strength.	An expensive way for crosslinking Specific reaction and linkage
Physical crosslinking	Safe approach. Low toxicity compared with chemical crosslinking. Inexpensive	Weaker bonds formed compared with chemical crosslinking Lower crosslinking degree Increasing of the crosslinking time.

Different crosslinking approaches were used for the crosslinking of the gelatin-based scaffolds [18]. This may depend on the type, functional groups, and nature of the other biopolymers in composite with gelatin [6, 18]. During the crosslinking procedure, active groups react through chemical and physical bonds between the gelatin and other biopolymers chains. There are several techniques, which are successfully applied for crosslinking such as chemical crosslinking [19], photo-crosslinking [20], enzyme crosslinking [2, 21], and physical crosslinking [6]. Table 1. shows the positive and negative points of each crosslinking approach applied for crosslinking of a gelatin-based scaffold in biomedical applications.

3.1 Chemical Crosslinking

Gelatin consists of amine and the carboxylic acid group, which can chemically crosslink by different materials [4]. Many types of the crosslinking agents were used for crosslinking of the gelatin-based materials, namely 1-Ethyl-3-(3-dimethyl aminopropyl)-carbodiimide/N-hydroxy succinimide (EDC/NHS) [4] or carboiimide, genipin, glutaraldehyde, ethylene glycol diglycidyl ether, and citric acid, which are known as small molecule crosslinkers, reacting with different moieties in the gelatin structure. 1-Ethyl-3-(3-dimethyl aminopropyl)-carbodiimide (EDC) and N-hydroxysuccinimide (NHS) are water-soluble crosslinkers, which can react with many different types of the functional moieties such as hydroxyl, carboxyl, and sulfhydryl groups [4]. Furthermore, EDC/NHS chemistry is known as non-toxic, water-soluble, FDA-approved method for crosslinking [4]. On the other hand, the optimum pH for EDC for crosslinking is around 4.5; however, it can be used as a crosslinker in natural pH, too. In this crosslinking approach, NHS increases the reaction efficacy by preventing hydrolysis and rearrangement of the intermediate [4].

Glutaraldehyde (GA) is another crosslinking agent, consisting of two aldehyde groups [1]. It reacts with free amine groups of the gelatin, protein, or composite biopolymers with gelatin. This was occurred through Schiff-base reaction and made the strong crosslinked network. For long times, GA has been known as the gold standard approach for crosslinking of scaffold or thin film. However, it has was found to be cytotoxic and can cause severe inflammation.

In contrast, Genipin has been known as a natural crosslinking agent that is used in many different biomedical aspects. Genipin was extracted from the *Gardenia jasminoides Ellis* fruit. It consists of multiple active functional groups such as hydroxyl and carboxylic acid groups. When the carboxylic acid of the genipin reacts with amino groups of the gelatin or biopolymers, the material color changes from colorless to blue. According to previous studies, Genipin shows lower cytotoxicity compared to GA. Furthermore, there are some other natural, non-toxic, and biodegradable components such as D, L-glyceraldehyde, oxidized alginate, and Genipin, which can apply as protein crosslinker. Table 2. shows a summary of the advantages and disadvantages of the different chemical crosslinkers used for gelatin.

Table 2. Advantages and drawbacks of the chemical crosslinking agents used for gelatin in the biomedical application.

	Crosslinker	Chemical structure	Advantages	Disadvantages
Chemical crosslinking Agents	Carbodiimide		Less toxic than GA Extra remain crosslinker can be washed with DW after crosslinking	Expensive Lower crosslinking degree
	Genipin		Biocompatible, good crosslinking degree and less toxic	Very expensive After crosslinking scaffold color changes to blue
	Glutaraldehyde		Inexpensive, crosslinking polymer containing amine type I functional group Reduces the immunogenicity of scaffolds	High cell toxicity Hazard material for health Change of crosslinked scaffold to yellow Need more step to wash the remained extra crosslinker
	Ethylene glycol diglycidyl ether		Inexpensive	Very toxic and low crosslinking degree
	Citric acid		Inexpensive, good crosslinking degree and biocompatible	Need catalyst and temperature due to which polymer nature changes

3.2 Schiff-base Crosslinking

Schiff-base reaction has been exploited to create in situ forming adhesive hydrogel by creating imine bond formation between aldehyde groups and amines in the gelatin chain [22]. It is usually occurred by the condensation reaction of an aldehyde with a primary amine, according to Figure 2. This reaction can take place under both acidic or base catalysis or even with heating [23]. The amine acts as a nucleophile and reacts with an aldehyde to form the unstable compound called "carbinolamine" [22, 23]. The imine bonds were formed between the gelatin and biopolymers by acid- or base-catalyzed conditions [23].

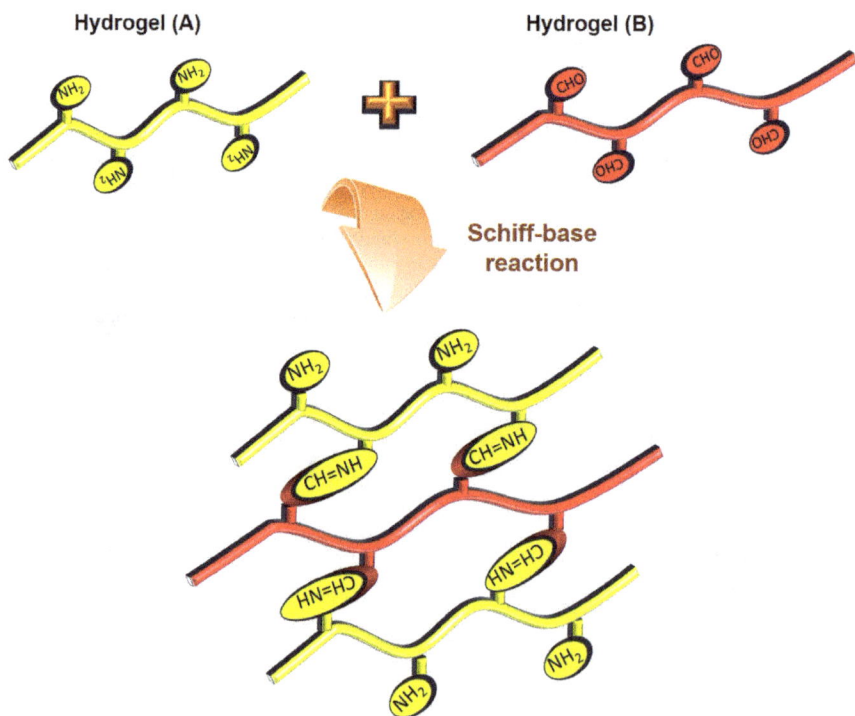

Figure 2. The formation of imine bond between amine of gelatin and aldehyde of the biopolymer via Schiff-base reaction.

One of the methods for making the aldehyde group is the oxidization of hydrogels, containing alcohol groups [24, 25]. Many researchers have focused on the oxidization of the polysaccharide-based materials. There are many different types of oxidization agents for oxidizing alcohol to aldehyde or acid. Periodates ($NaIO_4$ and KIO_4) selectively oxidize the polysaccharides' alcohol to aldehyde groups [24, 25]. The oxidization percentage can be controlled by temperature, time, and amount of periodate [25]. Oxidized hydrogels, such as sodium alginate, chondroitin sulfate, chitosan, dextran, and hyaluronic acid, were studied in great number. Indeed, many compositions have been developed based on this Schiff-base reaction such as gelatin/oxidized alginate [24], gelatin/oxidized hyaluronic acid [22, 26], gelatin/oxidized chondroitin [27], gelatin/oxidized chitin [28], gelatin/oxidized chitosan, and gelatin-chitosan oxidized sucrose [29]. Table 3 summarizes the type of injectable hydrogel, method of oxidization, oxidization agent, and application of them in tissue engineering.

Table 3. Gelatin based crosslinked hydrogel through Schiff-based addition (Method of preparation, oxidization agent types, and their application in biomedical purposes.

Composite	Method	Oxidization agent	Application	Ref
Oxidized chitosan (OCS)-Gelatin	5 g of CS dissolved in 100 mL of distilled water, reacted with different amounts of $NaIO_4$ for 6h at 20 °C in the dark. Solution neutralized with $NaHCO_3$ and KI. Solution dialyzed against distilled water for 2 days, frozen, and lyophilized. 1 mL OCS (10%) + 1 mL gelatin (20%)	$NaIO_4$		[27]
Oxidized sodium alginate (OSA)-Gelatin	Sodium alginate (1, 2, and 5%) mixed with different amounts of $NaIO_4$ (40, 60, and 80 mol %) at room temperature in the dark. Oxidization reaction stopped by ethylene glycol and dialyzed against distilled using dialysis bag (MWCO: 3500Da) for a day, frozen, and lyophilized. 1mL OSA (15%) + 1 mL gelatin (15%)	$NaIO_4$		[24]
Oxidized hyaluronic acid (OHA)-Gelatin microsphere (GMC)	HA (1g) dissolved in 10 mL deionized water, then 4mL of $NaIO_4$ (4M) added gradually. Reaction was continued in the dark for 2 h (T= 25 °C). Ethylene glycol added to solution for inactivation of unreacted $NaIO_4$ and stirred for 1h. Solution dialyzed (MWCO= 3500Da) against distilled water for 3 days, freeze and freeze-dried at -50 °C.	$NaIO_4$	Corneal stromal regenerative medicine.	[22, 26]

	GMC (300 mg) + OHA (160 mg) mixed in (ethanol: water with (v/v) ratio of 80:20) for 24 h at room temperature.			
Oxidized Sucrose (OS) /Chitosan-gelatin	Sucrose (1.026 g, 3 mol) and NaIO$_4$ (1.95 g, 15 mol) were dissolved in water: THF (v/v) (5:25), stirred at room temperature overnight in the dark. BaCl$_2$ (1.12 g, 4.5 mol) was added to solution and stirred at 0 °C for 2 h in order to stop the oxidization reaction. Stirring of solution stopped to precipitate, separated and dried.	NaIO$_4$	Ocular hypertension	[29]
	Chitosan (0.5, 0.375, 0.25 g)-Gelatin (0.25, 0.375, 0.5 g) in 50 mL water containing of 1% acetic acid/ water mixed with different amount of oxidized sucrose (0, 0.075, 0.15 g) at 40 °C.			
Oxidized urethane dextran (OUD)- /Gelatin	Urethane dextran (250mg) dissolved in 10 distilled water and NaIO$_4$ (60mg) added to the solution as oxidization agents and kept for 4h in the dark on ice.	NaIO$_4$	Promising tissue adhesive	[30]
	Gelatin (5 wt%) and oxidized urethane dextran (15, 25, 35, and 45 wt%) dissolved in PBS and stirred at 40 °C for 10 min. subsequently, the Irgacure 2959 was added to solution and hydrogel crosslinked by UV light source o 320-480 nm)			

3.3 Photo-Crosslinking

Light-induced photo-crosslinking has been widely applied for the fabrication of cell-laden injectable hydrogels with short gelation time and tunable physical properties [31] [20]. Among natural polymers, gelatin due to unique and favorable properties such as enzymatically degradable and water solubility, as well as its low cost of producing, can be used as biomedical hydrogel [32]. As previously mentioned, gelatin due to its abundance of inherent amine, hydroxyl, and carboxyl functional group can easily modify with different groups such as methacrylol and thiol groups (Figure 3) [32-34]. One of the ways for crosslinking of gelatin is modifying by grafting of methacrylol (methacrylamides) to create the photo-crosslinkable hydrogels [32, 33]. Among the method of crosslinking, photopolymerization showed the great degrees of crosslinking control light[35]. Photopolymerization of hydrogels was created by both UV(250nm < λ < 400nm) and visible light sources in order to create the cell-laden hydrogels and initiated by photo-initiator (e.g., Irgacure 2959) [20, 35]. In this crosslinking method, free-radical polymerization of photo-initiator can act as photo crosslinker agents between the GelMA chain to form a heterogeneous network [36]. Moreover, metal-free click reactions such as Diels Alder, strain-induced, and thiol-based reactions are important for biomaterials development [37]. Thiol-ene click chemistry presents an interesting alternative crosslinking mechanism, which is based on dimerization of thiols with reactive carbon-carbon double bonds called "-enes" [36]. Moreover, this reaction happens through a step-growth radical mechanism along with the high conversion of the functional group and a lower concentration of radical initiator as well as low polymerization shrinkage [36]. This, thiol-ene couples multifunctional molecules, resulted in preparing the more homogenous network [36]. Thiol-ene coupling systems follow a step-growth mechanism compared with methacrylate and acrylate systems, which followed chain growth [37].

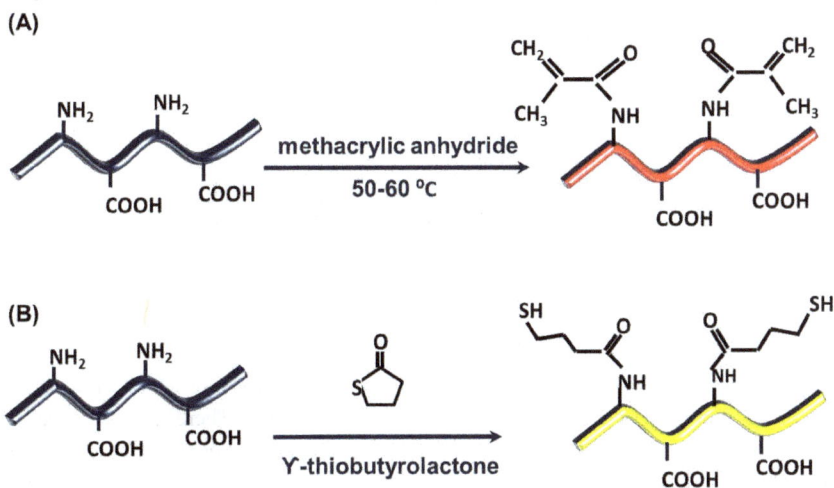

Figure 3. Synthesis of (A) gelatin methacrylol (GelMA) and (B) thiolated gelatin via functionalization of the primary amine group in gelatin.

Anseth *et al.* synthesized methacrylate modified gelatin (GelMA), which can be chained polymerized into covalent gelatin hydrogels for *in situ* cell encapsulation and long term 3D cell culture [38, 39]. They have shown the GelMA hydrogel with valvular interstitial cells laden. Moreover, Khademhosseini and his colleagues also utilized the GelMA hydrogel system as a cell-laden hydrogel for microscale tissue engineering [39]. They have applied different cell types [40] to encapsulate them inside GelMA hydrogels for different applications, as summarized in Table 4. Furthermore, GelMA in composite with other materials also shows the great properties which can be used in different aspect of tissue engineering and regenerative medicine. For instance, poly(ethylene glycol) (PEG) hydrogels known as non-toxic, non-immunogenic, with a strong tunable structure for nutrient and oxygen-transporting; however, PEG showed viod of bioactivity as drawback [41]. Therefore, Khademhosseini *et al.* fabricated the PEG-GelMA composite hydrogels, which shows the great tunable and biocompatibility for 3D cell culture[41]. There is another way to prepare hydrogel, which can be crosslinked by light.

Table 4. Summary of prepared GelMA and thiolated gelatin hydrogel and their composite applied for tissue engineering, cell, and drug deliveries.

Composition	Year	Cell type for laden in hydrogel (number of cells/mL)	Applications
GelMA	2009	Aortic valvular interstitial cells (VIC, 10×10^6)	Not mentioned
PEG-GelMA	2011	NIH3T3 (3.5×10^5)	Promising application in 3D cell culture and regenerative medicine applications [41]
GelMA	2012	human blood-derived endothelial colony-forming cells (ECFCs, 1×10^6) and bone marrow-derived mesenchymal stem cells (MSCs, 1×10^6)	microvascular networks and complex engineered tissue [40]

3.4 Michael Addition Reaction

The Michael addition occurred through the addition of carbanion or another nucleophile to α, β- unsaturated carbonyl compound (as shown in Figure 4). There are many pieces of research, which applied this reaction for the preparation of hydrogel, polymer film, and gold nanoparticles. Lee *et al.* have surface-modified polycaprolactone (PCL) fibrous membrane using gelatin through layer by layer assembly. They have modified PCL by hydrolysis and, subsequently, the aminolysis procedure. Afterward, surface modified PCL was modified by GelMA through Michael-type addition [42]. Michael addition reaction is a popular reaction for the preparation of hydrogel due to their controlling reaction time, different types of bonding with biomolecules, and mild reactivity [12]. Greene *et al.* synthesized the norbornene-functionalized gelatin(GelNB) and heparin functionalized-GelNB (GelNB-Hep) as crosslinked gelatin-based hydrogels for providing 3D cell culture platform for cancer cell *in vitro* study [43]. They have encapsulated the hepatocellular carcinoma cells inside GelNB hydrogel and evaluate them by in vitro study[43]. Mehrali *et al.* have employed a new approach for the preparation of crosslinked pectin hydrogel by methacrylation of pectin [44]. They have synthesized the

UV-crosslinked pectin methacrylate (PEMA) polymer and mix it with thiolated gelatin in order to endow this hydrogel matrix with cell differentiation purposes[44]. They have encapsulated the three wide ranges of cells, such as muscle progenitor (C2C12), neural progenitor (PC12), and human mesenchymal stem cells (hMSCs); however, this hydrogel indicates that MSCs encapsulation can facilitate the formation of bone-like apatite after 5 weeks of culturing.

Figure 4. Schematic reaction of the Michael-type addition

3.5 Enzymatic Crosslinking of Gelatin

An enzyme can make specific bonds. Therefore, they can induce crosslinking between protein chains. There are many types of enzymes, such as horseradish peroxidase (HRP) [21], tyrosinase, and transglutaminase, which have been extensively applied for *in situ* gel formation. HRP was known as enzymes that can crosslink phenol residues [2, 20]. HRP was used along with hydrogen peroxide (H_2O_2) as a source for oxygen to change the oxidization state OH group. The HRP reaction and mechanism were briefly summarized in Figure 5.

It has been reported that hydrogel, including living cells, can form a gel in the presence of HRP and hydrogen peroxide. Arai *et al.* synthesized the alginate and gelatin containing a phenolic hydroxyl (Ph), alginate-Ph, poly (vinyl alcohol) (PVA)-Ph, and gelatin-Ph, as bio-ink materials for 3D-bioprinter [2, 10, 21, 45]. These hydrogels can act as carriers for cell delivery to tissue and organs. Many research groups have been focused on the preparation of enzymatic injectable hydrogels for cell and drug deliveries [20]. Saki *et al.* prepared the L929 fibroblast cells encapsulated Gel-Ph gels with a gelation time of 10s [2]. The gels show 95% of cell viability as well as they have performed the *in vivo* study by injection in subcutaneous of mice [2]. The results showed that the hydrogel remained for 7 days without showing necrosis in the injected area [2]. In 2009, Hu *et al.* synthesized gelatin grafted with hydroxyphenyl propionic acid (Gel–HPA) hollow hydrogel fiber with immobilized cells inside which can enzymatically be crosslinked by HRP and H_2O_2. Their hydrogel system showed high cell viability and can be applied as a cell carrier for tissue engineering [10].

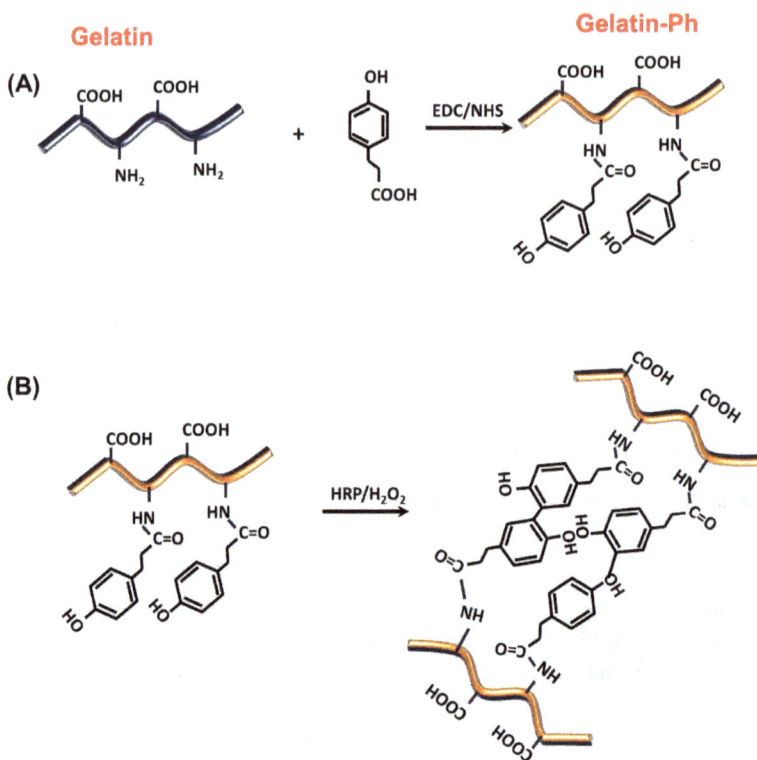

Figure 5. (A) Conjugation of Tyramine on gelatin chain using EDC/NHS and (B) crosslinking of the gelatin-Ph using HRP and H_2O_2.

Furthermore, they have reported that this hydrogel showed the cell growth and neurogenesis of human MSCs in the 3D-environment[46]. After the development of the GEL-HPA hydrogels, researchers put their efforts into getting better properties from this hydrogel with making a composition with other biopolymers. In this regard, Park and colleagues designed *in situ* cross-linkable gelatin-PEG-Tyr hydrogels and evaluated the biocompatibility of injectable hydrogel by *in vitro* (3D cell culture) and *in vivo* testing (subcutaneous injection in the back of mice) [47]. There are many kinds of research and studies for the preparation of the injectable enzymatic based hydrogel using gelatin.

Table 5 shows the type of the enzyme name, enzyme origin, type of the reaction, and their advantages. Furthermore, some of the research based on the injectable hydrogel through enzymatic crosslinking approach were classified in Table 6.

Table 5. Enzyme type, origin, advantages for crosslinking of the gelatin-based scaffold.

	Enzyme name	Origin	Reaction	advantages	Ref
Enzymatic agents	**Microbial TGase (mTGase)**	Streptoverticillium	Amide bond between a carboxylic acid group of glutamic acid and amine group of Lysine	Smaller molecular size and higher reaction rate Ca^{2+} independence Suitable for medical & industrial applications Improving the mechanical strength	[18]
	Horseradish peroxidase (HRP)/H$_2$O$_2$	Horseradish	Adding the phenolic hydroxyl groups of Tyramine to carboxyl side of gelatin	High crosslinking efficiency Mild reaction condition Good biocompatibility Applicable enzyme for tissue engineering	[18, 20]

Table 6. Injectable hydrogel by enzymatic crosslinking approach and their applications in tissue engineering and regenerative medicine.

Injectable hydrogel formulation (Bio ink Formulation	Crosslinking method	Application	Reference
Gelatin-Ph	Gelatin 2% (w/v) was dissolved in MES buffer and heated to 60 °C, then cool down to 25 °C. Tyramine hydrochloride, EDC, and NHS were added to solution and mixed for 12 h. subsequently, Na$_3$PO$_4$ (50 mM) added and stirred for 30 min. After that solution dialyzed against distilled water (with MWCO: 10,000) and then solution filtered, frozen, and freeze-dried.	Potential use in drug delivery and tissue engineering applications.	[2]
Alginate-Ph/ Gelatine-Ph Alginate and Gelatin-Tyramine containing phenolic hydroxyl (Ph) (Gelatin-Ph)	Gelatin and alginate dissolved in MES buffer and solution activated by EDC/NHS for 1h. Then tyramines added to solution and stirred for 24h. Purification of solution done by dialysis of solution against distilled water for 24 h.	Bio-ink material for the fabrication of cell-adhesive gel structure.	[45]

	A 1.5 wt% Alg-Ph and 1.5 wt% Alg-Ph/0.5 wt% Gelatin-Ph solution, including 50 units/mL of HRP for 30 min		
Gelatin/chitosan IPN	Chitosan-PA synthesized by phloretic acid (5.3 mmol), EDC (8 mmol), NHS (8 mmol) in aqueous solution.	Promising hydrogel for tissue engineering scaffolds and wound dressing	[19]
	Gelatin/Chitosan-PA hydrogel crosslinked by combination of 10 U/mL mTG , 5 U/mL HRP, and 0.8 mmol H_2O_2		
Gelatin/ Hyaluronic acid	15% gelatin, 1% HA, 2% EDC, and 2% NHS in the hydrogel	Hemorrhage control	[48]
Gelatin-HPA(Gtn-HPA)	HPA (20 mmol), NHS (27.8 mmol) and EDC (20 mmol) dissolved in 250 mL mixture of water and DMF (v/v) (3:2) at pH of 4.7 for 5 h. After that, 10 g gelatin dissolved in water and added to (EDC/NHS) solution and stirred overnight at pH of 4.7. Then the solution dialyzed against 100 mM NaCl (EtOH: water with ratio of 25:75) for 1 day and lyophilized.	Cell and drug immobilized hydrogel carrier, 3D culture, and co-culture, and cell transplantation for soft tissue engineering purposes.	[10, 46]
	Gtn-HPA, HRP, and H_2O_2 with Concentrations of 25 mg/ml, 6.25 unit/ml, and 0.5 wt% applied in order to make *in situ* hydrogels, respectively.		

3.6 Physical Crosslinking

The addition of electrolytes and non-electrolytes solutes are the simplest ways of improving the gel properties of gelatin-based materials. Salts are the most common electrolytes, which can modify the gelatin by the formation of salts bridges [49]. There are different salts such as $CaCl_2$, $MgCl_2$, and NaH_2PO_4, which can effect on gel properties [6, 49, 50]. Sarabia *et al.* claimed that positive and negative charges of salts could affect the gel properties of fish gelatin depend on the type of salt and amino acid content of gelatin [6, 51]. Gelatin and melting temperatures were decreased by adding NaCl and $MgCl_2$ to fish gelatin at a pH of 5 and 8 [6].

Moreover, adding $(NH_4)_2SO_4$ also decreased the gelation and melting temperature of fish gelatin with increasing pH [6, 52]. This happens because the H^+-content of the medium was decreased by adding ammonium ions. Positive charge and large size of ammonium lead to favor of electrostatic interactions between peptide chains. Subsequently, this leads to a higher degree of protein unfolding and highly subsequent gelation [6]. However, storage modulus (G'), gelation, and melting temperatures were increased by adding $MgSO_4$ to fish gelatin at high concentrations [6]. On the other hand, non-electrolytes solutes such as sugar, polysaccharides, glycerol, and proteins can be applied for modifying dish gelatin in order to obtain desired properties [6]. Many research groups discussed the effect of protein- polysaccharides system for food, biomedical, and pharmaceutical industries applications [6].

Furthermore, polysaccharides and gelatin molecules can make hydrogen bond interactions, and this can contribute to the improvement of gelation and rheological properties of composite systems [6]. The stability of these complex systems in aqueous solution highly depends on some parameters such as size, type, and reactive groups at molecular levels. Furthermore, mixing conditions such as temperature, pH, time, total biopolymer concentration, and mixing ratio also can effect on gelation and rheological properties of gelatin-polysaccharides systems [6].

4.   Applications

4.1   Gelatin-based hydrogel for drug release

The common method of drug usage requires high dosage or repeated plans in order to stimulate a therapeutic effect; however, this caused to lower efficiency as well as severe side effects and toxicity by drugs. There are different ways or methods, such as taking oral administrations as tablets, capsules, and syrup, as well as drug inoculation by intravenous (IV) and intramuscular (IM) injections [53]. For instance, IV injection of such interleukin-12 (IL-12) drug caused systematic toxicity, which results in deaths in a clinical trial [53, 54]. On the other hand, oral administration is known as a common approach for delivering drugs and vitamins. Nevertheless, the application is limited due to their poor targeting and short circulation time (less than 12 hours) [53]. Half-lives of peptides and protein drugs are only minutes to hours. Researchers have tried to design a controlled release system to address these issues [53]. Hydrogels have been mostly applied for drug delivery systems for cardiovascular treatment, cartilage-bone regeneration, wound healing, and hemostasis agents. Positive outcomes in therapeutic purposes were shown by improving their efficiency, decreasing toxicity, and essential dosage of drugs [53].

Furthermore, hydrogels, according to the size, porosity, architecture, and existing functional groups, determine how hydrogels can be applicable for drug delivery [53] [50].

Indeed, hydrogels consist mostly of water (70-99%); therefore, they shows the physical and structural similarity with body tissues. Moreover, this similarity of hydrogels offers excellent biocompatibility and great potential for encapsulating cells and hydrophilic drugs [34, 53, 55]. In addition to, risk of drug denaturation and aggregation is low as hydrogels forms in aqueous solution. The stiffness of hydrogel was tuned from 0.5kPa to 5MPa, which is matched with different soft tissues [53]. Many factors can affect the drug release, namely mesh size, network degradation, swelling behavior, and mechanical deformation. In general, hydrogels have mesh size in the range of 5 to 100 nm, which allows the small solute and liquid diffusion. Furthermore, the mesh size influences the drug diffusion in hydrogel through steric interactions between the drug-polymer networks. If the mesh is larger than the drug ($r_{Mesh}/r_{Drug} > 1$), the drug release is controlled by diffusion. In this case drug will freely migrate through hydrogel network; therefore, diffusion is extensively independent of mesh size (Figure 6).

Whenever the mesh size and drug size are approximately the same ($r_{Mesh}/r_{Drug} \sim 1$), the steric hindrance will dominate the drug release. In contrast, in the case of hydrogel with small mesh size along with large drug size ($1 > r_{Mesh}/r_{Drug}$), drugs were immobilized on the surface of hydrogels. Another parameter for controlling drug release is regulating of hydrogel degradation. During hydrogel degradation, the mesh size increases, thus facilitating drug diffusion out of hydrogel. The hydrogel degradation was known as erosion and can be used for controlled drug release. Many hydrogels bear erosion due to their permeability to water and enzymes, and this erosion can be applied for controlling drug release. Furthermore, oxidization of polysaccharides such as alginate and chitosan causes to balk erosion. Thus their degradation rate can control by regulating the oxidization degree of polysaccharides. Also, controlling the degradation rate of hydrogels, the drug release can regulate by controlling the swelling behavior of hydrogel. As hydrogels swell the mesh, overall size increases. As a result, it can affect the drug release behavior. Swelling behavior of hydrogels are changed by different external parameters such as temperature, pH [56], ionic strength, light [57] and glucose [58, 59] level etc. For instance, hydrogel with pH responsive swelling behavior are used in oral and cancer delivery systems. Some alginate-based hydrogels under acidic conditions change to condensed structure. However, under natural pH carboxylic acid of alginate were deprotonated, which led to large osmosis. As a result, the hydrogel swells, leading to drug release. Moreover, mechanical deformation of hydrogels relates to the releasing of drug through changing of hydrogel network structure, caused due to increased mesh size. This network deformation was created by different approaches such as mechanical, ultrasound- induced and magnetic field-induced deformation. The mechanical deformation of hydrogel networks is appropriate for enhancement of vascularization in

Nanohybrids   Materials Research Forum LLC
Materials Research Foundations **87** (2021) 244-270   https://doi.org/10.21741/9781644901076-10

tissue regeneration. There are some concerns about mechanical failure in implanted area, which can be overcome by designing of self-healing hydrogels. Crosslinking of hydrogel using different crosslinker materials can cause to inhibition in water entry and consequences leads to high surface-eroding hydrogel. One of the ways for controlling of the drug release, short- and long-term, in hydrogel is the controlling of hydrogel degradation rate or erosion mechanism at molecular level.

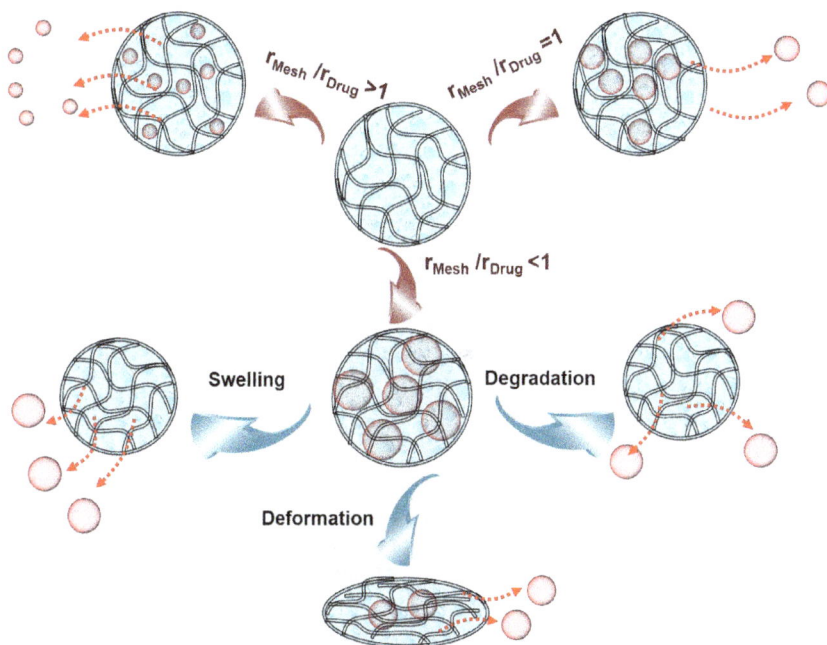

Figure 6. Relationship between mesh size, drug size, and drug diffusion (radius of the drug molecule (r_{Drug}) and radius of the polymer network (r_{Mesh})).

4.2   Polymer-drug interaction

Another approach for controlling the drug release is to the conjugation of drugs to hydrogel polymer chains. This way of drug conjugation is important in the case of small drugs, which can easily release from hydrogel within a short time without conjugation. However, the drug release can be tuned after conjugation of drugs with hydrogel polymer chains according to conjugation and interaction types such as electrostatic interactions and hydrophobic associations[59].

4.2.1 Covalent conjugation of drug-hydrogel

This method of drug conjugation is highly stable until hydrogel degraded. Different types of linkage were explored to immobilize the drug on polymer chains such as amide bond through carbodiimide, thiol-ene bonds, and metal-free click chemistry. Most of the growth factors such as transforming growth factor-β1 (TGFβ1), Bone morphogenetic protein (BMP-2) [3-5], Bone morphogenetic protein-7 (BMP-7) [60, 61], and vascular endothelial growth factor (VEGF) [4, 5, 62], have been conjugated through amine-carboxylic acid reaction to form amide bonds [59].

4.2.2 Electrostatic interactions of drug-hydrogel

This system was applied for hydrogels with positive and negative charges with a strong affinity to drugs with negative and positive or both charges, respectively. For example, alginate hydrogels which have negative charges were applied for delivering of cationic drugs. For instance, hydrogels with incorporated heparin have been applied to control the delivery of growth factors such as VEGF and BMP-2 through heparin-binding growth factors, which are applicable growth factors for vascularization and bone regeneration purposes. Moreover, the incorporation of sulfonate groups on hydrogels increases the electrostatic interactions between hydrogels and protein drugs in order to extend the drug release duration [59].

4.2.3 Hydrophobic association of drug-hydrogel

Hydrogels possess a large amount of water; thus, their hydrophilic nature becomes problematic for encapsulating hydrophobic drugs. Therefore, the phase separation between encapsulated hydrophobic drugs and hydrophilic matrix may lead to deterioration of hydrogel strength and stability. Therefore, researchers have attempted to design hydrogels containing hydrophobic components, monomers, or polymers to provide sites for hydrophobic drugs [59].

Conclusions

This chapter focuses on the gelatin-based materials and their characteristic properties, including their modification with different methods for biomedical aspects. Furthermore, different crosslinking methods such as chemical, physical, enzymatic, photo-crosslinking, Schiff-base reaction, and Michael addition were investigated for crosslinking of the gelatin-based materials. Moreover, the advantages and disadvantages of each method applied for the crosslinking of gelatin base materials were discussed separately. Furthermore, expanding applications for applied gelatin-based biomaterials for different tissue engineering, 3D-printing, cell delivery, and drug delivery purposes were

Nanohybrids   Materials Research Forum LLC
Materials Research Foundations **87** (2021) 244-270   https://doi.org/10.21741/9781644901076-10

investigated. In the end, the studies done on gelatin-based materials show that gelatin is known as promising materials in tissue engineering, drug, and cell delivery applications. To date, both *in vitro* and *in vivo* test results suggest that gelatin is not immunogenic or toxic and, with some modification, can be used as a bio-ink material for 3D printing of the tissue and organs.

References

[1]   Selestina Gorgieva, V.K., *Collagen-vs . Gelatine-Based Biomaterials and Their Biocompatibility : Review and Perspectives.* Biomaterials Applications for Nanomedicine, 2012: p. 17-52. https://doi.org/10.5772/24118

[2]   Sakai, S., et al., An injectable, in situ enzymatically gellable, gelatin derivative for drug delivery and tissue engineering. Biomaterials, 2009. 30(20): p. 3371-3377. https://doi.org/10.1016/j.biomaterials.2009.03.030

[3]   Song, M.-J., et al., Bone morphogenetic protein-2 immobilization on porous PCL-BCP-Col composite scaffolds for bone tissue engineering. Journal of Applied Polymer Science, 2017. 134(33): p. 45186. https://doi.org/10.1002/app.45186

[4]   Amirian, J., et al., Bone formation of a porous Gelatin-Pectin-biphasic calcium phosphate composite in presence of BMP-2 and VEGF. International Journal of Biological Macromolecules, 2015. 76: p. 10-24. https://doi.org/10.1016/j.ijbiomac.2015.02.021

[5]   Amirian, J., et al., The effect of BMP-2 and VEGF loading of gelatin-pectin-BCP scaffolds to enhance osteoblast proliferation. Journal of Applied Polymer Science, 2015. 132(2). https://doi.org/10.1002/app.41241

[6]   Huang, T., et al., *Fish gelatin modifications: A comprehensive review.* Trends in Food Science & Technology, 2019. 86: p. 260-269. https://doi.org/10.1016/j.tifs.2019.02.048

[7]   Sharma, G., et al., Applications of nanocomposite hydrogels for biomedical engineering and environmental protection. Environmental Chemistry Letters, 2018. 16(1): p. 113-146. https://doi.org/10.1007/s10311-017-0671-x

[8]   Mariod, A.A. and H.F. Adam, *Review: gelatin, source, extraction and industrial applications.* Acta Scientiarum Polonorum - Technologia Alimentaria, 2013. 12(2): p. 135-147.

Materials Research Forum LLC
https://doi.org/10.21741/9781644901076-10

[9] See, S.F., et al., Effect of different pretreatments on functional properties of African catfish (Clarias gariepinus) skin gelatin. J Food Sci Technol, 2015. 52(2): p. 753-62. https://doi.org/10.1007/s13197-013-1043-6

[10] Hu, M., et al., Cell immobilization in gelatin–hydroxyphenylpropionic acid hydrogel fibers. Biomaterials, 2009. 30(21): p. 3523-3531. https://doi.org/10.1016/j.biomaterials.2009.03.004

[11] Pathania, D., et al., Preparation of a novel chitosan-g-poly(acrylamide)/Zn nanocomposite hydrogel and its applications for controlled drug delivery of ofloxacin. International Journal of Biological Macromolecules, 2016. 84: p. 340-348. https://doi.org/10.1016/j.ijbiomac.2015.12.041

[12] Li, Y., J. Rodrigues, and H. Tomás, *Injectable and biodegradable hydrogels: gelation, biodegradation and biomedical applications.* Chemical Society Reviews, 2012. 41(6): p. 2193-2221. https://doi.org/10.1039/C1CS15203C

[13] Amirian, J., et al., Examination of In vitro and In vivo biocompatibility of alginate-hyaluronic acid microbeads As a promising method in cell delivery for kidney regeneration. Int J Biol Macromol, 2017. 105(Pt 1): p. 143-153. https://doi.org/10.1016/j.ijbiomac.2017.07.019

[14] Xing, Q., et al. Increasing mechanical strength of gelatin hydrogels by divalent metal ion removal. Scientific reports, 2014. 4, 4706. https://doi.org/10.1038/srep04706

[15] Jongjareonrak, A., et al., Skin gelatin from bigeye snapper and brownstripe red snapper: Chemical compositions and effect of microbial transglutaminase on gel properties. Food Hydrocolloids, 2006. 20(8): p. 1216-1222. https://doi.org/10.1016/j.foodhyd.2006.01.006

[16] Madkhali, O., G. Mekhail, and S.D. Wettig, *Modified gelatin nanoparticles for gene delivery.* International Journal of Pharmaceutics, 2019. 554: p. 224-234. https://doi.org/10.1016/j.ijpharm.2018.11.001

[17] Linh, N.T.B., et al., Augmenting in vitro osteogenesis of a glycine–arginine–glycine–aspartic-conjugated oxidized alginate–gelatin–biphasic calcium phosphate hydrogel composite and in vivo bone biogenesis through stem cell delivery. Journal of Biomaterials Applications, 2016. 31(5): p. 661-673. https://doi.org/10.1177/0885328216667633

[18] Oryan, A., et al., Chemical crosslinking of biopolymeric scaffolds: Current knowledge and future directions of crosslinked engineered bone scaffolds.

International Journal of Biological Macromolecules, 2018. 107: p. 678-688.
https://doi.org/10.1016/j.ijbiomac.2017.08.184

[19] Zhang, Y., et al., Tough biohydrogels with interpenetrating network structure by
bienzymatic crosslinking approach. European Polymer Journal, 2015. 72: p. 717-725.
https://doi.org/10.1016/j.eurpolymj.2014.12.038

[20] Chuang, C.-H., et al., Enzymatic regulation of functional vascular networks using
gelatin hydrogels. Acta Biomaterialia, 2015. 19: p. 85-99.
https://doi.org/10.1016/j.actbio.2015.02.024

[21] Roberts, J.J., et al., A comparative study of enzyme initiators for crosslinking
phenol-functionalized hydrogels for cell encapsulation. Biomaterials research, 2016.
20: p. 30-30. https://doi.org/10.1186/s40824-016-0077-z

[22] Lai, J.-Y., Biofunctionalization of gelatin microcarrier with oxidized hyaluronic
acid for corneal keratocyte cultivation. Colloids and Surfaces B: Biointerfaces, 2014.
122: p. 277-286. https://doi.org/10.1016/j.colsurfb.2014.07.009

[23] da Silva, C.M., et al., *Schiff bases: A short review of their antimicrobial activities.*
Journal of Advanced Research, 2011. 2(1): p. 1-8.
https://doi.org/10.1016/j.jare.2010.05.004

[24] Emami, Z., et al., Controlling alginate oxidation conditions for making alginate-
gelatin hydrogels. Carbohydrate Polymers, 2018. 198: p. 509-517.
https://doi.org/10.1016/j.carbpol.2018.06.080

[25] Amirian, J., et al., In-situ crosslinked hydrogel based on amidated pectin/oxidized
chitosan as potential wound dressing for skin repairing. Carbohydrate Polymers, 2021.
251: p. 117005. https://doi.org/10.1016/j.carbpol.2020.117005

[26] Lai, J.-Y. and D.H.-K. Ma, Ocular biocompatibility of gelatin microcarriers
functionalized with oxidized hyaluronic acid. Materials Science and Engineering: C,
2017. 72: p. 150-159. https://doi.org/10.1016/j.msec.2016.11.067

[27] Dawlee, S., et al., Oxidized Chondroitin Sulfate-Cross-Linked Gelatin Matrixes:
A New Class of Hydrogels. Biomacromolecules, 2005. 6(4): p. 2040-2048.
https://doi.org/10.1021/bm050013a

[28] Ge, Y., et al., Intelligent gelatin/oxidized chitin nanocrystals nanocomposite films
containing black rice bran anthocyanins for fish freshness monitorings. International
Journal of Biological Macromolecules, 2019.
https://doi.org/10.1016/j.ijbiomac.2019.11.101

[29]   El-Feky, G.S., et al., Chitosan-Gelatin Hydrogel Crosslinked With Oxidized Sucrose for the Ocular Delivery of Timolol Maleate. Journal of Pharmaceutical Sciences, 2018. 107(12): p. 3098-3104. https://doi.org/10.1016/j.xphs.2018.08.015

[30]   Wang, T., J. Nie, and D. Yang, *Dextran and gelatin based photocrosslinkable tissue adhesive.* Carbohydrate Polymers, 2012. 90(4): p. 1428-1436. https://doi.org/10.1016/j.carbpol.2012.07.011

[31]   Noshadi, I., et al., In vitro and in vivo analysis of visible light crosslinkable gelatin methacryloyl (GelMA) hydrogels. Biomaterials Science, 2017. 5(10): p. 2093-2105. https://doi.org/10.1039/C7BM00110J

[32]   Brown, G.C.J., et al., Covalent Incorporation of Heparin Improves Chondrogenesis in Photocurable Gelatin-Methacryloyl Hydrogels. Macromolecular Bioscience, 2017. 17(12): p. 1700158. https://doi.org/10.1002/mabi.201700158

[33]   Russo, L., et al., *Gelatin hydrogels via thiol-ene chemistry.* Monatshefte für Chemie - Chemical Monthly, 2016. 147(3): p. 587-592. https://doi.org/10.1007/s00706-015-1614-5

[34]   Sharma, G., et al., *Sodium Dodecyl Sulphate-Supported Nanocomposite as Drug Carrier System for Controlled Delivery of Ondansetron.* International journal of environmental research and public health, 2018. 15(3): p. 414. https://doi.org/10.3390/ijerph15030414

[35]   Hao, Y., et al., Visible light cured thiol-vinyl hydrogels with tunable degradation for 3D cell culture. Acta Biomaterialia, 2014. 10(1): p. 104-114. https://doi.org/10.1016/j.actbio.2013.08.044

[36]   Bertlein, S., et al., Thiol–Ene Clickable Gelatin: A Platform Bioink for Multiple 3D Biofabrication Technologies. Advanced Materials, 2017. 29(44): p. 1703404. https://doi.org/10.1002/adma.201703404

[37]   Daniele, M.A., et al., Interpenetrating networks based on gelatin methacrylamide and PEG formed using concurrent thiol click chemistries for hydrogel tissue engineering scaffolds. Biomaterials, 2014. 35(6): p. 1845-1856. https://doi.org/10.1016/j.biomaterials.2013.11.009

[38]   Benton, J.A., et al., Photocrosslinking of Gelatin Macromers to Synthesize Porous Hydrogels That Promote Valvular Interstitial Cell Function. Tissue Engineering Part A, 2009. 15(11): p. 3221-3230. https://doi.org/10.1089/ten.tea.2008.0545

[39]   Mũnoz, Z., H. Shih, and C.-C. Lin, Gelatin hydrogels formed by orthogonal thiol–norbornene photochemistry for cell encapsulation. Biomaterials Science, 2014. 2(8): p. 1063-1072. https://doi.org/10.1039/C4BM00070F

[40]   Chen, Y.-C., et al., Functional Human Vascular Network Generated in Photocrosslinkable Gelatin Methacrylate Hydrogels. Advanced Functional Materials, 2012. 22(10): p. 2027-2039. https://doi.org/10.1002/adfm.201101662

[41]   Hutson, C.B., et al., Synthesis and Characterization of Tunable Poly(Ethylene Glycol): Gelatin Methacrylate Composite Hydrogels. Tissue Engineering Part A, 2011. 17(13-14): p. 1713-1723. https://doi.org/10.1089/ten.tea.2010.0666

[42]   Lee, J.W. and H.S. Yoo, Michael-Type Addition of Gelatin on Electrospun Nanofibrils for Self-Assembly of Cell Sheets Composed of Human Dermal Fibroblasts. ACS Omega, 2019. 4(20): p. 18677-18684. https://doi.org/10.1021/acsomega.9b02602

[43]   Greene, T. and C.-C. Lin, Modular Cross-Linking of Gelatin-Based Thiol–Norbornene Hydrogels for in Vitro 3D Culture of Hepatocellular Carcinoma Cells. ACS Biomaterials Science & Engineering, 2015. 1(12): p. 1314-1323. https://doi.org/10.1021/acsbiomaterials.5b00436

[44]   Mehrali, M., et al., Pectin Methacrylate (PEMA) and Gelatin-Based Hydrogels for Cell Delivery: Converting Waste Materials into Biomaterials. ACS Applied Materials & Interfaces, 2019. 11(13): p. 12283-12297. https://doi.org/10.1021/acsami.9b00154

[45]   Arai, K., et al., The development of cell-adhesive hydrogel for 3D printing. 2016, 2016. 2(2): p. 10. https://doi.org/10.18063/IJB.2016.02.002.

[46]   Wang, L.-S., et al., Injectable biodegradable hydrogels with tunable mechanical properties for the stimulation of neurogenesic differentiation of human mesenchymal stem cells in 3D culture. Biomaterials, 2010. 31(6): p. 1148-1157. https://doi.org/10.1016/j.biomaterials.2009.10.042

[47]   Park, K.M., et al., In situ cross-linkable gelatin–poly(ethylene glycol)–tyramine hydrogel viaenzyme-mediated reaction for tissue regenerative medicine. Journal of Materials Chemistry, 2011. 21(35): p. 13180-13187. https://doi.org/10.1039/c1jm12527c

[48]   Luo, J.-W., et al., *In situ injectable hyaluronic acid/gelatin hydrogel for hemorrhage control.* Materials Science and Engineering: C, 2019. 98: p. 628-634. https://doi.org/10.1016/j.msec.2019.01.034

[49]	Amirian, J., et al., Examination of In vitro and In vivo biocompatibility of alginate-hyaluronic acid microbeads As a promising method in cell delivery for kidney regeneration. International Journal of Biological Macromolecules, 2017. 105: p. 143-153. https://doi.org/10.1016/j.ijbiomac.2017.07.019

[50]	Amirian, J., et al., Incorporation of alginate-hyaluronic acid microbeads in injectable calcium phosphate cement for improved bone regeneration. Materials Letters, 2020. 272: p. 127830. https://doi.org/10.1016/j.matlet.2020.127830

[51]	Sarabia, A.I., M.C. Gómez-Guillén, and P. Montero, *The effect of added salts on the viscoelastic properties of fish skin gelatin.* Food Chemistry, 2000. 70(1): p. 71-76. https://doi.org/10.1016/S0308-8146(00)00073-X

[52]	Karayannakidis, P.D. and A. Zotos, Physicochemical Properties of Yellowfin Tuna (Thunnus albacares) Skin Gelatin and its Modification by the Addition of Various Coenhancers. Journal of Food Processing and Preservation, 2015. 39(5): p. 530-538. https://doi.org/10.1111/jfpp.12258

[53]	Li, J. and D.J. Mooney, *Designing hydrogels for controlled drug delivery.* Nature Reviews Materials, 2016. 1: p. 16071. https://doi.org/10.1038/natrevmats.2016.71

[54]	Cohen, J., *IL-12 Deaths: Explanation and a Puzzle.* Science, 1995. 270(5238): p. 908-908. https://doi.org/10.1126/science.270.5238.908a

[55]	Gaurav, S., et al., Fabrication and Characterization of Polysorbate/Ironmolybdophosphate Nanocomposite: Ion Exchange Properties and pH-responsive Drug Carrier System for Methylcobalamin. Current Analytical Chemistry, 2020. 16(2): p. 138-148. https://doi.org/10.2174/1573411014666180727144746

[56]	Zhang, S., et al., A pH-responsive supramolecular polymer gel as an enteric elastomer for use in gastric devices. Nature Materials, 2015. 14: p. 1065. https://doi.org/10.1038/nmat4355

[57]	Yan, B., et al., Near Infrared Light Triggered Release of Biomacromolecules from Hydrogels Loaded with Upconversion Nanoparticles. Journal of the American Chemical Society, 2012. 134(40): p. 16558-16561. https://doi.org/10.1021/ja308876j

[58]	Obaidat, A.A. and K. Park, Characterization of protein release through glucose-sensitive hydrogel membranes. Biomaterials, 1997. 18(11): p. 801-806. https://doi.org/10.1016/S0142-9612(96)00198-6

[59]	Kokufata, E., Y.-Q. Zhang, and T. Tanaka, *Saccharide-sensitive phase transition of a lectin-loaded gel.* Nature, 1991. 351(6324): p. 302-304. https://doi.org/10.1038/351302a0

[60]   Jung, A., et al., A novel hybrid multichannel biphasic calcium phosphate granule-based composite scaffold for cartilage tissue regeneration. Journal of Biomaterials Applications, 2017. 32(6): p. 775-787. https://doi.org/10.1177/0885328217741757

[61]   Jung, A., et al., A novel hybrid multichannel biphasic calcium phosphate granule-based composite scaffold for cartilage tissue regeneration. J Biomater Appl, 2018. 32(6): p. 775-787. https://doi.org/10.1177/0885328217741757

[62]   Amirian, J., et al., In vitro endothelial differentiation evaluation on polycaprolactone-methoxy polyethylene glycol electrospun membrane and fabrication of multilayered small-diameter hybrid vascular graft. Journal of Biomaterials Applications, 2020. 34(10): p. 1395-1408. https://doi.org/10.1177/0885328220907775

Keyword Index

About the Editors

Dr. Gaurav Sharma research activity started in 2009 at Shoolini University (India) as master of philosophy student, and then, he continued his research work as PhD student. He is working on the preparation and characterization of diverse multifunctional nanomaterials, and their composites, specially focused for their potential applications in environmental remediation and biological field. At School of chemistry, Shoolini University (India) he carried out diverse research lines, interrelated to each other based on synthesis and characterization of nanocomposites, hydrogels, bi and trimetallic nanoparticles, ion exchangers, drug delivery nanomaterials, adsorbents and photocatalysts etc. Moreover, he performed and taught different courses as nanochemistry, polymer chemistry, spectroscopy and natural products, among others. On the other hand, he supervised 5 Master of Philosophy, and more than 25 Master and Bachelors students. He established collaborative research with various groups in countries such as Finland, Saudi Arabia, China, Spain, Vietnam and South Africa. In this context, he was invited as visiting research professor from University of KwaZuklu-Natal (South Africa) in 2017 and 2019. In 2017, he joined as postdoctoral fellow at college of materials science and engineering, Shenzhen University. He got a project from China postdoctoral science foundation in 2018. He is a reviewer of more than 36 journals.

The outcome of his research work includes more than 130 publications-SCI, in reputed journals such as Renewable and Sustainable Energy Reviews, Chemical Engineering Journal, Journal of Cleaner Production, Carbohydrate Polymers, ACS Applied Materials and Interfaces, Journal of Hazardous Materials and International Journal of Biological Macromolecules etc. He contributed to 9 book chapters and edited 5 books.

Furthermore, the impact of his research has been highlighted too by fact, that he has been recognized as Web of Science Highly Cited Researcher-2020. For instance, few of his research paper in International Journal of Biological Macromolecules, Process Safety and

Environmental Protection, Materials Science and Engineering: C; Journal of King Saud University-Science, Journal of Photochemistry and Photobiology A: Chemistry; are most cited articles published since 2017.

His h-index is 47, citations: 5050 (web of science). He is Associate Editor of International Journal of Environmental Science and Technology (Springer); Chemical Papers (Springer). Editorial Board member of Current Organic Chemistry, Innovations in Corrosion and Materials Science, Journal of Nanostructure in Chemistry (Springer), Nanotechnology for Environmental Engineering(Springer), Letters in Applied NanoBioScience etc, and Academic Editor of Journal of Nanomaterials and Advances in Polymer Technology.

Dr. Alberto García Peñas is Assistant Professor in the Department of Materials Science and Engineering at University Carlos III of Madrid (Spain), where he is working on smart polymers, composites, blends and biopolymers. Importance of this work is endorsed by numerous articles in high impact journals, and participation in different international and national conferences. Furthermore, quality of his global research is reflected in the numerous awards that this Thesis received: the "2015 Student Innovation Award" international award by multinational Borealis to the best Doctoral Thesis worldwide in the area of polyolefins; Awards to Best Thesis granted by Specialized Groups in Polymers and in Calorimetry and Thermal Analysis, both groups from the Royal Spanish Societies of Chemistry and Physics. Merits reached during the 2015-2016 academic year were also recognized by the Council for Scientific Research. Nowadays, he has 10 awards and honors which were collected along these years.

He is Executive Board Member of Network of Researchers China-Spain. He organized the First Meeting of Ibero-American Researchers and a EURAXESS Researchers' Night in Shenzhen (China). Dr. García-Peñas is leading research works and seminars associated with education and training for Chemical Engineers and Chemists.

www.ingramcontent.com/pod-product-compliance
Lightning Source LLC
Chambersburg PA
CBHW071335210326
41597CB00015B/1464